ADVANCES IN CHEMICAL PHYSICS

VOLUME 115

Advances in
CHEMICAL PHYSICS

Edited by

I. PRIGOGINE

Center for Studies in Statistical Mechanics and Complex Systems
The University of Texas
Austin, Texas
and
International Solvay Institutes
Université Libre de Bruxelles
Brussels, Belgium

and

STUART A. RICE

Department of Chemistry
and
The James Franck Institute
The University of Chicago
Chicago, Illinois

VOLUME 115

AN INTERSCIENCE® PUBLICATION
JOHN WILEY & SONS, INC.
NEW YORK • CHICHESTER • WEINHEIM • BRISBANE • SINGAPORE • TORONTO

For ordering and customer service, call 1-800-CALL WILEY

Library of Congress Catalog Number: 58-9935

ISBN 0-471-39331-2

Printed in the United States of America.

10 9 8 7 6 5 4 3 2 1

CONTRIBUTORS TO VOLUME 115

S. A. ADELMAN, Department of Chemistry, Purdue University, West Lafayette, IN

JONATHAN P. K. DOYE, University Chemical Laboratories, Cambridge, United Kingdom

D. A. KOFKE, Department of Chemical Engineering, State University of New York at Buffalo, Buffalo, NY

JOHN J. KOZAK, Department of Chemistry, Iowa State University, Ames, IA

MARK A. MILLER, University Chemical Laboratories, Cambridge, United Kingdom

P. A. MONSON, Department of Chemical Engineering, University of Massachusetts, Amherst, MA

PAUL N. MORTENSON, University Chemical Laboratories, Cambridge, United Kingdom

R. RAVI, Department of Chemical Engineering, Indian Institute of Technology, Kanpur, India

DAVID J. WALES, University Chemical Laboratories, Cambridge, United Kingdom

TIFFANY R. WALSH, University Chemical Laboratories, Cambridge, United Kingdom

v

INTRODUCTION

Few of us can any longer keep up with the flood of scientific literature, even in specialized subfields. Any attempt to do more and be broadly educated with respect to a large domain of science has the appearance of tilting at windmills. Yet the synthesis of ideas drawn from different subjects into new, powerful, general concepts is as valuable as ever, and the desire to remain educated persists in all scientists. This series, *Advances in Chemical Physics*, is devoted to helping the reader obtain general information about a wide variety of topics in chemical physics, a field that we interpret very broadly. Our intent is to have experts present comprehensive analyses of subjects of interest and to encourage the expression of individual points of view. We hope that this approach to the presentation of an overview of a subject will both stimulate new research and serve as a personalized learning text for beginners in a field.

I. Prigogine
Stuart A. Rice

CONTENTS

ADVANCES IN CHEMICAL PHYSICS

VOLUME 115

ENERGY LANDSCAPES: FROM CLUSTERS TO BIOMOLECULES

DAVID J. WALES, JONATHAN P. K. DOYE, MARK A. MILLER,
PAUL N. MORTENSON, and TIFFANY R. WALSH

*University Chemical Laboratories, Lensfield Road,
Cambridge CB2 1EW, UK*

I. INTRODUCTION

This review is about potential energy surfaces. In general, the potential energy, V, of a system of N interacting particles, is a function of $3N$ spatial coordinates: $V = V(\{\mathbf{r}_i\}), (1 \leq i \leq N)$, or $V = V(\mathbf{X})$, where $\mathbf{X}$ is the $3N$-dimensional configuration vector. The potential energy surface (PES) is therefore a $3N$-dimensional object embedded in a $(3N + 1)$-dimensional space, where the extra dimension corresponds to the value of the potential energy function.

In the present work we will be concerned only with the ground state PES in the Born–Oppenheimer approximation [1], although extensions to excited electronic states are certainly possible in principle. We will also employ classical mechanics and neglect quantum effects, which are expected to be small for the systems of interest here. The PES then determines, either directly or indirectly, the structure, dynamics, and thermodynamics of the system. Mechanically stable configurations correspond to local minima of V, while the gradient of V tells us the forces on the various atoms. Thermodynamic properties also depend upon V via ensemble averages. Even systems composed of hard spheres or discs have a PES, albeit a rather strange one, since the energy is either zero or infinity [341]. For the more realistic potentials considered in this review, it is possible to provide further insight into dynamics and thermodynamics by considering the densities of states associated with local minima and elementary events consisting of transitions between minima. The theory and validity of this approach is considered in Section I.C.

Advances in Chemical Physics, Volume 115, edited by I. Prigogine and Stuart A. Rice.
ISBN 0-471-39331-2 © 2000 John Wiley & Sons, Inc.

For many small molecules, relatively accurate quantum mechanical calculations provide a useful global picture of the PES [2,3]. However, here we are interested in treating systems with far more degrees of freedom, where it is often necessary to find different approaches (Section I.C). Two areas of current research have stimulated our interest, namely protein folding and the properties of glasses. Understanding the behavior of such systems at a fundamental level, in terms of the underlying PES, has motivated us to study larger systems. Of course, much thought has already been given to the question of how dynamics and thermodynamics emerge from the PES. However, it is only relatively recently that increased computer power and improved algorithms have combined to make detailed studies of realistic systems possible. These advances apply particularly to the characterization of pathways between minima and hence the *connectivity* of the PES, which is critical in determining relaxation dynamics. It is not enough simply to characterize a large sample of minima or investigate the effects of a particular barrier height distribution.

A. Clusters

Clusters, or finite systems in general, provide a useful way to approach the relation of properties to the PES. They exhibit a wide variety of behavior, depending upon size and composition, and can therefore provide insight into more complex systems that behave in the same way. Finite size effects upon thermodynamics are now well understood (Section III.C) [4], and a corresponding framework for dynamics is now developing. Finite systems are certainly more convenient to deal with than bulk matter, since there are fewer stationary points, and free energy barriers, which scale extensively with system size, are easier to overcome. Time scales for certain processes may therefore be reduced without changing the essential nature of the phenomenon. For example, many systems exhibit "magic number" sizes of pronounced intensity when free clusters are formed in a molecular beam. To explain these peaks we must understand why the magic number clusters exhibit special stability, and why they are kinetically accessible in the experiment. These are essentially the same questions that must be addressed to understand protein folding and glass formation.

A brief outline of the issues involved in protein folding and the properties of structural glasses is given in the following sections. Although all the examples we will consider in this review involve relatively small systems, the results are directly relevant to the behavior of larger biomolecules and bulk material. In each case a fundamental understanding of structure, dynamics, and thermodynamics, may be developed from the underlying PES. Perhaps the most important question is how some systems can locate their global energy minimum efficiently, while others cannot. Native biomolecules, crystalline materials, and magic number clusters presumably lie at the opposite end of the scale from glasses. We will provide examples of finite systems that span this range of behavior, and provide detailed explanations for them.

Secondary issues concern the properties of systems that have failed to locate the global free energy minimum on cooling and are trapped elsewhere on the PES for experimentally observable periods. Relaxation times and thermodynamic properties vary widely, depending upon the system. Classes of behavior can be identified empirically, but in our view real understanding can only come by explaining how these properties are determined by the PES. Although we have identified complex relaxation and non-Arrhenius behavior in finite systems, extrapolation of these results to bulk is a continuing research effort.

B. Protein Folding

The difficulty of finding a particular structure among an exponentially large number of local minima was first recognized in the context of protein folding by Levinthal in 1969 [5]. It was around this time that the first attempts were being made to predict the three-dimensional structure of globular proteins from their amino acid sequence. It was known that some denatured proteins could regain their native structure reversibly and reproducibly on a laboratory time scale [6]. By considering a grid in configuration space, Levinthal estimated that the number of possible conformations (not necessarily locally stable) for a typical protein is of the order of 10^{300}. However, the observed time taken for a typical protein to reach the native state would permit no more than approximately 10^8 of these configurations to be searched. This discrepancy has come to be known as Levinthal's "paradox", although it has now been resolved in principle from general considerations of the PES.

It was first proposed that perhaps not all the possible configurations have to be searched. Levinthal himself suggested that folding might occur via a specific pathway, initiated by the formation of a condensed nucleus and proceeding through a well-defined sequence of events [7]. A reduction in the search space may also result from rapid collapse to a compact state, followed by rearrangement. However, even the number of compact states is probably still too large for a random search to find the native state on the observed time scale. Further reduction of the search space has been postulated by several authors [8,9]. Although simulations of simple lattice models suggest that the initial collapse is indeed fast and insensitive to the amino acid sequence [10,11], experiments on cytochrome c do not seem to support this view [12].

A reduction in the search space probably does not provide a satisfactory resolution of Levinthal's paradox. However, the implicit notion that somehow the search is not random does appear to be important. Levinthal's analysis assumes that the PES is flat, like a golf course with a single hole corresponding to the native state [13]. Bryngelson et al. further suggested that a funnel structure arises for naturally occurring proteins because they are "minimally frustrated" [13]. In fact, a simple model that includes an energetic bias toward the lowest energy minimum can reduce search times to physically meaningful scales [14,15]. A more detailed analysis of model potential energy surfaces showed that the potential energy bias is indeed important and highlighted some other features that are significant [16]. Leopold et al. considered the organization of the PES more explicitly, and proposed that naturally occurring proteins exhibit a collection of convergent kinetic pathways that lead to a unique native state, which is thermodynamically the most stable [17]. This structure was termed a "folding funnel" because the misfolded configurations are focused toward the correct target.

The folding funnel approach has since been developed in terms of a *free* energy surface by Wolynes, Onuchic, and co-workers [13,18–20]. To obtain a free energy surface it is necessary to define order parameters and to average over the remaining degrees of freedom. For example, the number of amino acid contacts that are the same as in the native state is often a useful order parameter. Such free energy surfaces are now often referred to as free energy "landscapes". One could equally well call the PES a potential energy landscape, but without the necessity to perform any averaging. We will use the terms PES and potential energy landscape interchangeably, but it must be remembered that, in contrast, one cannot have a free energy surface or landscape without introducing order parameters of some sort. The free energy surface is also temperature-dependent, of course, whereas the PES is not.

Once a free energy surface has been defined, it may be possible to locate a free energy transition state between the folded and unfolded states. "Folding funnels" are identified with free energy minima. Alternatively, a "funnel" can be defined directly in terms of the underlying PES, as we will see in Section II.C. The funneling properties of the PES give rise to a deep free energy minimum and so these descriptions are probably equivalent. The above view of protein folding can be reconciled with a description based upon a folding pathway if we allow the "pathway" to be an ensemble of paths that all lead to the native state [21].

Simulations have shown that the folding ability can be quantified in terms of the ratio of the folding temperature, T_f, where the native state becomes thermodynamically the most stable, to the glass transition temperature, T_g, where the kinetics slow down dramatically because of large free energy barriers [11,22]. T_g is usually defined as the temperature for which the folding time passes through a certain threshold. Folding is fastest for large T_f/T_g, since the native state is then statistically populated at temperatures where it is kinetically accessible.

The field of protein folding is far from closed by this resolution of the basic issues at the heart of the Levinthal paradox. The predicted appearance of the PES for systems that can locate their global potential minimum efficiently [16] is just beginning to be tested for biomolecules. It has also been suggested that the necessary and sufficient condition for folding in a simple lattice model is a pronounced global minimum in the potential energy [10,23]. Such a feature certainly increases the value of T_f, but the strength of the correlation with the energy gap separating the global minimum from the next lowest state has been questioned, and so has the validity of assessing folding ability from such a small region of the PES [24]. After some discussion, it was suggested that one should consider the energy gap between the native state and structurally dissimilar states. The lowest energy minima that are structurally dissimilar from the native state could act as effective kinetic traps if they lie at the bottom of their own deep funnels. The higher their energy, the less effective they can be as traps, and so this criterion is in better accord with the conclusions based upon analysis of general PES's [16]. However, it may still be insufficient to ensure rapid folding [25].

The major challenge remaining is the same one that Anfinsen addressed in 1964 [6], namely the prediction of the native state of a protein given only its amino acid sequence. The importance of this global optimization problem is reflected in the number of groups actively working on it, but we will not pursue it at length here. However, it is important to note that the study of PES's is directly relevant to this effort (Section III.E). If the free energy minimum of interest does not correspond to the global potential energy minimum we are forced to explore wider regions of the PES.

C. Structural Glasses

In the solid phase, some materials, such as TiO_2 and Al_2O_3, are always found in a crystalline form, and it is difficult or impossible to produce them in an amorphous (glassy) state [26]. Given that the disordered minima vastly outnumber the ordered ones, this phenomenon is analogous to Levinthal's paradox in protein folding. However, many materials, such as SiO_2 and B_2O_3, readily form glasses, becoming trapped in an amorphous liquid-like state. If cooled sufficiently, they take on many of the properties of a solid. The understanding of glass formation and processes in viscous liquids in terms of the PES is an ambitious goal. Previous work on energy landscapes in glassy materials has already been applied to the study of protein (Section I.B) and cluster dynamics [27].

Goldstein was the first to propose a connection between the properties of glasses and the PES [28]. He asserted that a proper understanding can only be gained by considering the detailed nature of the interactions between the constituent particles as manifested in the PES. In his description, a glass undergoes intrawell oscillations about a potential energy minimum and, at sufficiently high thermal energies, can surmount potential barriers. These barriers separate configurations in which only a few particles have moved significantly. The corresponding transitions are therefore local, but such processes may occur simultaneously in different parts of the material. Two time scales are distinguishable: the rapid intrawell vibrational period and the time between interwell hops in a given part of the material. As the temperature is raised, the potential barrier treatment remains applicable until these time scales are no longer distinct, and the thermal energy is significantly larger than the barriers. By this point, the liquid is so fluid that the system does not have time to equilibrate within individual minima. At sufficiently low temperatures, particle rearrangements give rise to viscous flow. The application of stress biases the rearrangements toward minima that dissipate the stored energy. Through a continuous series of dissipative rearrangements, the system loses memory of its original state, so that when the stress is released this state is not recovered.

Johari and Glodstein subsequently measured the dielectric relaxation time (which probes the response of a material's polarizability to an applied electric field) of a wide variety of glassy systems [29]. They detected a bifurcation of relaxation times at low temperature. The "primary," or α, relaxation becomes slower and non-Arrhenius in behavior—i.e., different from $\tau \propto \exp(A/T)$—as the glass transition is approached from above, leaving the faster "secondary," or β, relaxation to dominate. At temperatures sufficiently above the glass transition, the α and β relaxation times are indistinguishable. Although these authors did not suggest a mechanism for the α and β processes, they attributed the "freezing out" of the α processes at low temperatures to insurmountable potential barriers.

The crucial concept of inherent structure, introduced by Stillinger and Weber, has allowed direct verification of some aspects of Goldstein's picture. The inherent structure of an instantaneous configuration is the local minimum reached by following the steepest-descent path [30]. Monitoring the transitions between inherent structures in simulations revealed the onset of slow barrier crossings and localized rearrangements at low temperature [31]. The fact that these transitions occur essentially independently in different parts of the sample was also demonstrated by the extensive behavior of the transition rate [32].

A widely used basis for the classification of glass-forming liquids is the strong/fragile scheme of Angell [33]. Strong liquids are characterized by a viscosity that has an Arrhenius temperature dependence: $\eta \propto \exp(A/T)$, where A is a constant. At the glass transition, defined as the temperature below which the viscosity exceeds some large threshold, a strong liquid undergoes smooth changes of thermodynamic quantities, and the jump in the heat capacity is small. Conversely, fragile liquids deviate from Arrhenius behavior in a way that is usually well described by the Vogel–Tammann–Fulcher (VTF) expression [34], $\eta \propto \exp[A/(T - T_0)]$, where T_0 is the VTF temperature of the material and A is another constant. Fragile liquids experience sudden changes in thermodynamic properties and a large jump in the heat capacity at a well-defined glass transition. Certain other properties also appear to be related to the fragility, notably the response to a mechanical or dielectric perturbation. The resulting relaxation can usually be modelled by a stretched exponential, or Kohlrausch–Williams–Watts [35], function: $g \propto \exp[-(t/\tau)^\theta], 0 \leq \theta \leq 1$, where τ is a constant that determines the relaxation time, and pure exponential (Debye) behavior corresponds to $\theta = 1$. Although there is considerable scatter, data for a wide variety of liquids show a correlation of increasing fragility with departure from Debye relaxation, i.e., small θ [36].

Strong liquids tend to have open network structures, like SiO_2, which resist structural degradation with temperature changes, whereas fragile liquids have less directional interactions, such as Coulomb or van der Waals. Beyond these observations, most inferences as to the nature of the energy landscape at the fragile and strong extremes are somewhat speculative. Stillinger [37] associated the α and β processes with a "cratering" of the PES, similar to the notion of multiple funnels. The β processes correspond to hops between neighboring potential wells, while α processes involve a concerted series of such rearrangements, and take the configuration point from one crater to another. The larger energy and entropy of activation for the intercrater motion cause these processes to be frozen out at low temperature. Stillinger envisaged that, at the strong liquid extreme, the energy landscape is uniformly rough, with no organization into craters, so that only β processes are significant. In contrast, fragile liquids might have significant landscape cratering and a pronounced α/β bifurcation at low temperature.

To explain the large change in heat capacity at the glass transition in fragile liquids, Angell associated fragility with a high density of minima per unit energy increase, and low barriers between them [38]. Angell also pointed out that a liquid may be converted between fragile and strong by changing its density or the conditions under which it is prepared, in which case different fragilities can be associated with different well-separated regions of the energy landscape. For example, increasing the density of silica (the archetypal strong liquid) by 30% causes the diffusivities of the silicon and oxygen components to vary in a strongly non-Arrhenius manner characteristic of a fragile liquid [38].

Direct evidence for some of these interpretations has recently come from simulations. Sastry et al. [39] have performed molecular dynamics (MD) simulations of a periodic glass-forming binary Lennard-Jones mixture, monitoring relaxation in terms of the autocorrelation function of the atomic positions. On cooling, the liquid passes into a "landscape influenced" regime, in which the relaxation becomes nonexponential (placing it in the fragile category), and then into a "landscape dominated" regime, in which the configuration point explores regions where the minima are separated by higher barriers, and activated dynamics are observed. Doliwa and Heuer [40] have interpreted α and β processes in terms of the cage effect, where a tagged particle is confined to a small region of space by its neighbors. Inside this cage, motion is local and β processes dominate, but at increasing temperatures, escape from the cage becomes possible and larger-scale α processes come into play, until the dynamics eventually become diffusive.

Analysis of stationary points on the energy landscape has also provided considerable insight. Lacks [41] and Malandro [42] have examined mechanical instabilities in monatomic glasses and silica as the pressure is increased, finding that the pressure generally lowers barriers between minima, and eventually destabilizes them, leading to an irreversible structural transition. In the case of silica, this transition involves an increase in coordination of the silicon. Heuer [43] has attempted a full enumeration of minima for a periodic system of 32 Lennard-Jones atoms. He presented an ordering scheme and analysis that permits some insight into the length scales and transitions associated with the glass at different densities. We note, however, that a preliminary investigation of the same system using the techniques to be described in Section III revealed that there are at least ten times as many minima as found by Heuer. Angelani et al. [44] have also used an exhaustive approach for periodic Lennard-Jones systems of very few atoms (11 to 29), combined with the master equation. These authors monitored the relaxation of the autocorrelation function of components of the stress tensor. Their model exhibited the stretched exponential relaxation dynamics, temperature dependence of the stretching parameter (θ), and viscosity expected for a fragile liquid, as well as (perhaps surprisingly) good agreement with molecular dynamics simulations.

Daldoss et al. have investigated the possibility of tunnelling between pairs of minima connected by very low barriers in Lennard-Jones clusters [45]. Such pairs are generally referred to as "two-level systems" in the glasses literature, and are invoked to explain the low temperature specific heat and thermal conductivity.

II. DESCRIBING A MULTIDIMENSIONAL POTENTIAL ENERGY SURFACE

A. Features of the Landscape

There are no strict rules for the growth in the number of stationary points corresponding to different structures with system size, although some bounds are provided by the Morse rules [46–48]. The evolution is system dependent, but empirical observations and theoretical arguments suggest an exponential growth in the number of structurally distinct minima [31,32,49].

For a system composed of N_A atoms of type A, N_B atoms of type B, etc., the Hamiltonian is invariant to all permutations of equivalent nuclei and to inversion of all coordinates through a space-fixed origin. The number of permutation–inversion isomers of any given configuration could therefore be as large as

$$2 \times N_A! \times N_B! \times N_C! \times \dots, \tag{1.1}$$

but is reduced by a factor of h^{PG}, the order of the prevailing point group [50].

Putting together these two results we find that the number of minima on the PES for a homonuclear system of N atoms is likely to increase as

$$2N! \exp(aN^b), \tag{1.2}$$

where $a > 0$ and b is of order unity.

For our purposes the most important points on a PES are the minima and the transition states that connect them. Following Murrell and Laidler we define a transition state as a stationary point where the Hessian matrix, $H_{ij} = \partial^2 V / \partial X_i \partial X_j$, has exactly one negative eigenvalue whose eigenvector corresponds to the reaction coordinate [51]. (A stationary point is defined by the vanishing of the gradient vector: $\nabla V = \mathbf{0}$.) Minima linked by higher-index saddles (the index being the number of negative Hessian eigenvalues) must also be linked by one or more true transition states of lower energy [51]. The connectivity of these stationary points defines what is often loosely referred to as the topology of PES. We can then reserve the term *topography* to refer to the relative energies of the stationary points. In fact, Mezey has provided a rigorous topological analysis of the PES in terms of *catchment regions* of the local minima defined by steepest-descent paths, and readers are referred to his book for details [3].

Rearrangements that connect permutational isomers of the same structure are termed *degenerate* [52], and the corresponding pathways may exhibit certain symmetry properties [53–55]. In an asymmetric degenerate rearrangement the two sides of the path are inequivalent, while in a symmetric degenerate rearrangement they are related by a symmetry operation [56]. The transition state corresponding to a symmetric degenerate rearrangement exhibits additional point group symmetry elements that are not conserved along the path [53–55].

B. Monotonic Sequences

A monotonic sequence is an ordered set of connected minima (and the intervening transition states) where the energy of the minima always decreases [57,58]. Kunz and Berry describe the collection of monotonic sequences that lead to a particular minimum as a "basin." By this definition a basin is analogous to the folding funnel described in terms of a collection of convergent kinetic pathways. Unfortunately other authors have used the term basin in different ways. "Basins of attraction" are an appealing alternative to Mezey's notion of catchment regions for local minima, defined in terms of steepest-descent paths, or some other local minimization procedure. Becker and Karplus employ "superbasins" in their disconnectivity graph approach to the PES described in the next section (Section I.C) [59]. Saito et al. define basins in terms of sets of minima connected by low barriers [60]. To avoid confusion we will refer to the Berry–Kunz collection of monotonic sequences as a monotonic sequence basin (MSB). The MSB leading to the global minimum is termed "primary," and is separated from neighboring "secondary" MSB's by transition states lying on a "primary divide," and so on. Above such a divide, it is possible for a minimum to belong to more than one MSB through different monotonic sequences.

Monotonic sequences provided some of the first insight into how different systems relax to their global minima. A comparison between Ar_{19} and $(KCl)_{32}$ clusters revealed differences that were used to explain the relaxation of these two systems from high energy states [27]. In simulations a cooling rate of $10^9 \, Ks^{-1}$ usually leads to trapping in local minima for Ar_{19}, while $(KCl)_{32}$ generally finds its way to a rocksalt morphology even for a cooling rate of $10^{12} \, Ks^{-1}$.

Some monotonic sequences are illustrated in Figure 1.1 and Figure 1.2. We see that the downhill barrier height is usually large compared with the energy difference between successive minima for Ar_{19}, but small for $(KCl)_{32}$. The energies of the successive minima of the Ar_{19} cluster do not increase rapidly, except at the step up from the global minimum. The corresponding energies of successive minima on the $(KCl)_{32}$ cluster differ significantly, except at high energy. The number of ions typically involved in any given rearrangement of $(KCl)_{32}$ is also greater than the number of atoms involved in an Ar_{19} mechanism.

Ball et al. associated the "staircase" profile of the $(KCl)_{32}$ monotonic sequences with an efficient "structure-seeker" and contrasted this picture

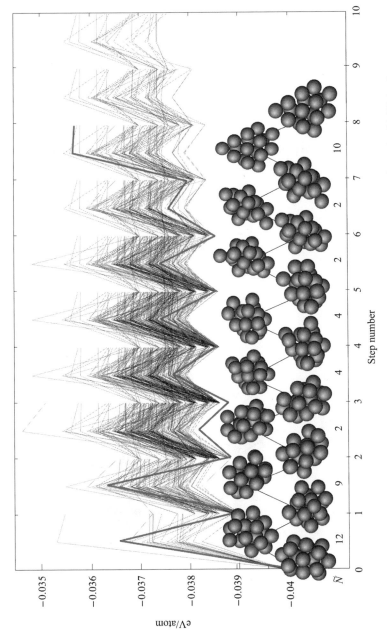

Figure 1.1. Monotonic sequences for Ar_{19} [27]. The structures correspond to the highlighted sequences and $\tilde{N}$ is the number of atoms participating in each rearrangement (Section III.A).

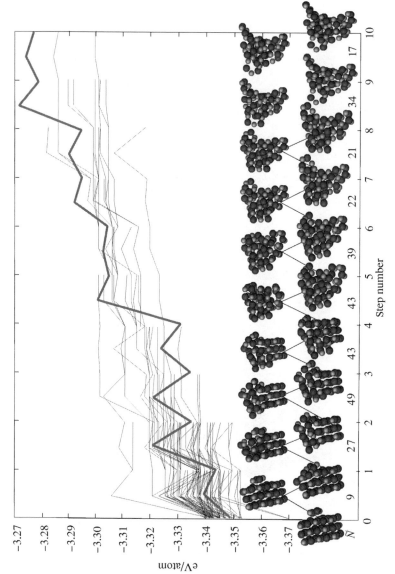

Figure 1.2. Monotonic sequences for $(KCl)_{32}$ [27]. The structures correspond to the highlighted sequences and $\tilde{N}$ is the number of atoms participating in each rearrangement (Section III.A).

12

with the "sawtooth" profile observed for Ar_{19}. However, the monotonic sequences do not give a good picture of the global connectivity of the PES, and we have recently revised our interpretation of these results. In fact, we will see in Section IV.B that Ar_{19}, as modelled by the Lennard-Jones potential, can locate its global minimum double icosahedron structure quite easily. In contrast, alkali halide clusters relax efficiently to a rocksalt morphology of some type, but there is then a separation of time scales to find the true global potential energy minimum (Section VII). The alternative rocksalt morphologies actually act as efficient kinetic traps.

The "staircase" profiles for $(KCl)_{32}$ therefore provide for efficient relaxation to a rocksalt morphology, but not necessarily the global minimum, while the overall connectivity of the Ar_{19} PES means that the global minimum is easily located for reasonable cooling rates (Section IV.B).

C. Disconnectivity Graphs

An alternative visual representation of a PES is provided by the disconnectivity graph approach of Becker and Karplus [59]. This technique was first introduced to interpret a structural database for the tetrapeptide isobutyryl-(ala)$_3$-NH-methyl, produced by Czerminski and Elber [61], and was subsequently applied to study the effects of conformational constraints in hexapeptides [62]. Essentially complete disconnectivity graphs have been constructed for a variety of small molecules by Westerberg and Floudas [63]. Garstecki et al. have shown that the disconnectivity graphs of disordered ferromagnets exhibit funneling properties, while those of spin glasses do not [64], showing how the energy landscape approach can serve to unify apparently disparate fields.

The method is formally expressed in the language of graph theory [59], but can easily be summarized as follows. The analysis begins by mapping every point in configuration space onto its inherent structure (Section I.C) by following the steepest-descent path [30]. Thus, configuration space is represented by the discrete set of minima, each of which has an associated catchment region or basin of attraction. Although this approach does not indicate the volume of phase space associated with each minimum, the density of minima can provide a qualitative impression of the volumes associated with the various regions of the energy landscape.

At a given total energy, E, the minima can be grouped into disjoint sets, called "superbasins" by Becker and Karplus, whose members are mutually accessible at that energy. In other words, each pair of minima in a superbasin is connected directly or through other minima by a path whose energy never exceeds E, but would require more energy to reach a minimum in another superbasin. If two minima are connected by more than one transition state, only the lowest affects the appearance of the graph, because it defines the energy at which the minima become mutually accessible. At low energy there is just one superbasin—that containing the global minimum. At successively higher energies, more superbasins come into play as new minima are reached. At still higher energies, the superbasins coalesce as higher barriers are overcome, until finally there is just one containing all the minima (provided there are no infinite barriers). Note that a superbasin is defined at a specified energy, in contrast to a monotonic sequence basin (Section II.B), which is a fixed feature of the PES.

The disconnectivity graph is constructed by performing the superbasin analysis at a series of energies, plotted on a vertical scale. At each energy, a

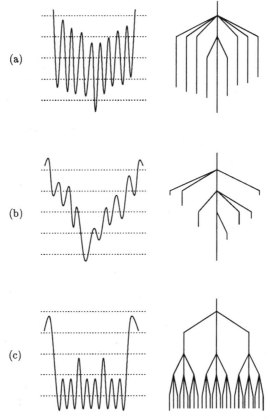

Figure 1.3. Schematic examples of potential energy surfaces (potential energy as a function of some generalized coordinate) and the corresponding disconnectivity graphs. In each case, the dotted lines indicate the energy levels at which the superbasin analysis has been made. (a) A gently sloping funnel with high barriers ("willow tree"); (b) a steeper funnel with lower barriers ("palm tree"); and (c) a "rough" landscape ("banyan tree").

superbasin is represented by a node, with lines joining nodes in one level to their daughter nodes in the level below. The choice of the energy levels is important; too wide a spacing and no topological information is left, while too close a spacing produces a vertex for every transition state and may hide the overall structure of the landscape. The horizontal position of the nodes is arbitrary, and can be chosen for clarity. In the resulting graph, all branches terminate at local minima, and all minima connected directly or indirectly to a node are mutually accessible at the energy of that node. The strength of the disconnectivity graph in representing the topology of the PES is that it does not depend upon the dimensionality of the system, whereas schematic plots of the potential energy itself are restricted to one or two dimensions. This topological mapping shares some common features with the "energy lid" and "energy threshold" approaches of Sibani, Schön, and co-workers [65–67], where regions of phase space connected below a fixed energy are considered.

The resulting graphs, which are connected but contain no cycles, are known as "trees" for reasons that should be clear from Figure 1.3. In Figure 1.3a we see a gently sloping funnel with high barriers resembling a weeping willow, in Figure 1.3b a steeper funnel with lower barriers resembles a palm tree, and in Figure 1.3c the "rough" landscape produces a banyan tree[1] [68,69]. The palm and willow trees both have a single stem leading to the global minimum; minima are cut off a few at a time as the energy decreases. Higher barriers produce the long dangling branches of the willow tree in Figure 1.3a.

The banyan tree is qualitatively different. Here the PES possesses a hierarchical arrangement of barriers, giving rise to multiple sub-branching in the graph. Previous work on the dynamics of hierarchical systems suggests that such a landscape will exhibit a multiplicity of time scales and power law decay of correlation functions [70,71].

III. METHODS

A. Geometry Optimization and Pathways

Finding a local minimum starting from an arbitrary configuration is relatively straightforward; one must merely proceed "downhill' somehow. Locating a true transition state is considerably harder, since the stationary point is a local maximum in precisely one direction. A Newton–Raphson algorithm, for example, will tend to converge to a stationary point with the same number of negative Hessian eigenvalues as the starting point, as

---

[1]Banyan trees have branches that drop shoots to the ground.

explained below [72]. The eigenvector-following technique [73–76] overcomes this difficulty by searching uphill along one specified eigenvector and downhill in all other directions. Refinements of this method [72,77–79] have made it robust and efficient enough for routine application to systems containing hundreds of atoms [80–82].

The minima connected to a given transition state are defined by the end points of the two steepest-descent paths commencing parallel and antiparallel to the transition vector (i.e., the Hessian eigenvector whose eigenvalue is negative) at the transition state. In practice, a small displacement of order 0.01 is taken parallel or antiparallel to the transition vector followed by energy minimization. Although eigenvector-following can be used for these minimizations, the pathways are not strictly identical to the SD paths [72], and occasionally lead to different minima. Since both the details of the pathways and the connectivity are of interest here, we have used a steepest-descent technique for most of the path calculations, employing analytic second derivatives, as detailed by Page and McIver [83]. For systems containing more than 200 atoms, Polak–Ribiere conjugate-gradient minimization [84] was used since it is faster than the second-order method and usually provides a good approximation to the true steepest-descent path. To ensure that true minima had indeed been located, each conjugate-gradient minimization was followed by a few steps of eigenvector-following minimization. We emphasize that, within the Born–Oppenheimer approximation, steepest-descent paths (and other properties of the PES) are independent of mass, temperature, and coordinate system when defined properly in terms of covariant derivatives [85,86].

The standard minimization techniques employed in the present work will not be reviewed here. Instead we will concentrate upon the location of transition states. Many algorithms have been suggested to perform this task [77,87–113], but we will only outline the approach used in the present work, which is based upon eigenvector-following. The simplicity of the force fields used for most of the present systems has enabled us to build up databases containing hundreds of thousands of minima and transition states, all converged to high accuracy using analytic second derivatives. The present methods have therefore been optimized for such systems, but have also been employed successfully in *ab initio* calculations [114–119]. We note that other workers are now beginning to address the problems posed by larger systems too [120–123].

1. Eigenvector-Following

Consider the Taylor expansion of the potential energy around a general point in nuclear configuration space, $\mathbf{X}$, truncated at second order:

$$V(\mathbf{X} + \mathbf{h}) = V(\mathbf{X}) + \mathbf{g}(\mathbf{X})^T \mathbf{h} + \tfrac{1}{2}\mathbf{h}^T \mathbf{H}(\mathbf{X})\mathbf{h}, \qquad (1.3)$$

where $\mathbf{g}(\mathbf{X})$ is the gradient vector at $\mathbf{X}$, $g_\alpha(\mathbf{X}) = \partial V(\mathbf{X})/\partial X_\alpha$, and $\mathbf{H}(\mathbf{X})$ is the second derivative matrix without mass-weighting.

Applying the condition $dV(\mathbf{X}+\mathbf{h})/d\mathbf{h} = 0$ leads us to the standard Newton–Raphson step:

$$\mathbf{h}_{NR} = -\mathbf{H}^{-1}\mathbf{g}. \tag{1.4}$$

However, $\mathbf{h}_{NR}$ is not useful in this form because the inverse Hessian of a free molecule is undefined due to zero eigenvalues [124]. Internal coordinates or projection operators [95,125] may be used to circumvent this problem. However, we have recently moved to a shifting procedure that is more efficient for large systems. Since analytic forms are known for the Hessian eigenvectors corresponding to overall translations and infinitesimal rotations, the corresponding eigenvalues can be shifted arbitrarily. For each such eigenvector, $\mathbf{e}$, we simply add a large multiple of $e_i e_j$ to Hessian element H_{ij}. Steps along the eigenvectors corresponding to these easily recognized eigenvalues are then ignored.

Solving the eigenvalue problem

$$\sum_{\beta=1}^{3N} H_{\alpha\beta} B_{\beta\gamma} = \varepsilon_\gamma^2 B_{\alpha\gamma}, \tag{1.5}$$

for the matrix $\mathbf{B}$ enables us to transform to new orthogonal coordinates $W_\alpha = \sum_{\beta=1}^{3N} B_{\beta\alpha} X_\beta$. The Newton–Raphson step and the corresponding energy change are now:

$$h_{NR,\alpha} = -g_\alpha(\mathbf{W})/\varepsilon_\alpha^2 \quad \text{and} \quad \Delta V_{NR} = -\sum_{\alpha=1}^{3N} g_\alpha(\mathbf{W})^2/2\varepsilon_\alpha^2 \tag{1.6}$$

where $g_\alpha(\mathbf{W}) = \partial V(\mathbf{W})/\partial W_\alpha$. Hence, contributions from terms with $\varepsilon_\alpha^2 > 0$ and $\varepsilon_\alpha^2 < 0$ lower and raise the energy, respectively. To find a minimum we may need to start at a point where all the non-zero Hessian eigenvalues are positive, and to find a transition state we may need to start at a point where precisely one of the ε_α^2 is negative [125]. However, introducing an additional Lagrange multiplier gives increased flexibility that can be exploited to find transition states systematically [73–76], and this is the basis of the eigenvector-following approach.

In the present work, a separate Lagrange multiplier is used for each eigendirection [115]. For the Lagrangian

$$L = -V(\mathbf{W}) - \sum_{\alpha=1}^{3N}\left[g_\alpha(\mathbf{W})h_\alpha + \frac{1}{2}\varepsilon_\alpha^2 h_\alpha^2 - \frac{1}{2}\mu_\alpha(h_\alpha^2 - c_\alpha^2)\right], \tag{1.7}$$

the step that is optimal in all directions is

$$h_\alpha = g_\alpha(\mathbf{W})/(\mu_\alpha - \varepsilon_\alpha^2), \tag{1.8}$$

and the energy change corresponding to this step is

$$\Delta V = \sum_{\alpha=1}^{3N}(\mu_\alpha - \varepsilon_\alpha^2/2)g_\alpha(\mathbf{W})^2/(\mu_\alpha - \varepsilon_\alpha^2)^2. \tag{1.9}$$

Now we must make a choice for μ_α. We clearly need $\mu_\alpha - \varepsilon_\alpha^2/2 < 0$ for minimization and $\mu_\alpha - \varepsilon_\alpha^2/2 > 0$ for maximization of the energy with respect to any eigendirection. We also require $\mu_\alpha \to 0$ as $g_\alpha(\mathbf{W}) \to 0$, so that the Newton–Raphson step is recovered in the vicinity of a stationary point.

The present work employs

$$\mu_\alpha = \varepsilon_\alpha^2 \pm \tfrac{1}{2}|\varepsilon_\alpha^2|\left(1 + \sqrt{1 + 4g_\alpha(\mathbf{W})^2/\varepsilon_\alpha^4}\right), \tag{1.10}$$

plus for maximization, minus for minimization, which gives steps

$$h_\alpha = \frac{\pm 2g_\alpha(\mathbf{W})}{|\varepsilon_\alpha^2|\left(1 + \sqrt{1 + 4g_\alpha(\mathbf{W})^2/\varepsilon_\alpha^4}\right).} \tag{1.11}$$

For systems described by simple empirical potentials containing fewer than 100 atoms we simply calculate analytic first and second derivatives at every step in a transition state search. Because we routinely start these searches from minima, Hessian update techniques do not seem to be competitive. We use the gradients at the present point, n, and the previous point $n - 1$, to estimate the corresponding eigenvalue:

$$\varepsilon_\alpha^2(\text{est}) = [g_\alpha(n) - g_\alpha(n - 1)]/h_\alpha(n - 1). \tag{1.12}$$

Comparing $\varepsilon_\alpha^2(\text{est})$ with the true value of ε_α^2 gives us some idea of how reliable our steps are for this eigendirection. A trust ratio for each direction r_α, is defined by

$$r_\alpha = \left|\frac{\varepsilon_\alpha^2(\text{est}) - \varepsilon_\alpha^2(n)}{\varepsilon^2(n)}\right|. \tag{1.13}$$

The maximum allowed step for each eigendirection is then updated according to whether r_α exceeds a specified trust radius. Most of our results employ a common trust radius of two, with the maximum allowed step increased or decreased by 10% according to the value of r_α.

2. Methods for Larger Systems

For simple empirical potentials, analytical second derivatives of the potential are often computationally inexpensive and the rate-determining step in calculations using the eigenvector-following technique is diagonalizing the Hessian matrix. However, it is possible to use an iterative method [126] to find the largest eigenvalue of a symmetric matrix, such as the Hessian. By a standard shifting technique we can then find the smallest eigenvalue and associated eigenvector, thus avoiding diagonalization.

An arbitrary vector, $\mathbf{y}$, can be written as a linear combination of the Hessian eigenvectors, $\mathbf{e}_i$:

$$\mathbf{y} = \sum_i a_i \mathbf{e}_i. \qquad (1.14)$$

Hence, repeated application of the Hessian matrix on $\mathbf{y}$ gives

$$\mathbf{H}^n \mathbf{y} = \lambda_1^n \left[a_1 \mathbf{e}_1 + \left(\frac{\lambda_2}{\lambda_1} \right)^n a_2 \mathbf{e}_2 + \left(\frac{\lambda_3}{\lambda_1} \right)^n a_3 \mathbf{e}_3 + \cdots \right]. \qquad (1.15)$$

In the limit of large n, the term that dominates is

$$\mathbf{y}_n = \lambda_1^n a_1 \mathbf{e}_1, \qquad (1.16)$$

where λ_1 is the eigenvalue with the largest magnitude. Renormalizing $\mathbf{y}_n$ to unity at each interation prevents it from growing too large. The convergence of this algorithm may be speeded up by shifting the Hessian eigenvalues to reduce the fractions λ_k / λ_1 toward zero [126]. To find the smallest eigenvalue, $\lambda_{\min}$, we shift all the eigenvalues down by λ_1 so that $\lambda_{\min}$ has the largest magnitude, and iterate again.

In the eigenvector-following/conjugate-gradient (EF/CG) hybrid approach [82], we use iteration and shifting to find the value of the smallest eigenvalue and the corresponding eigenvector that we intend to follow uphill. We then utilize the eigenvector-following formulation described above to calculate the size of the uphill step. The Polak–Ribiere conjugate-gradient (CG) method [84] was used to minimize the energy in the tangent space. Here we simply apply the projector $\hat{\mathcal{P}} \mathbf{x} = \mathbf{x} - (\mathbf{x} \cdot \hat{\mathbf{e}}_{\min}) \hat{\mathbf{e}}_{\min}$ to all gradients throughout the CG optimization. The unit vector $\hat{\mathbf{e}}_{\min}$ is then updated and the sequence of steps is repeated until various convergence criteria are satisfied. It is only important to converge fully the CG minimization near to the desired transition state, and so a maximum of ten CG steps or fewer can be used. After the first cycle the eigenvectors found in the previous step are employed as input to the iterative procedure, which speeds up the process considerably. Unit vectors with randomly chosen components are used to initiate the algorithm. For each transition state the converged eigenvector corresponding to the unique negative eigenvalue is saved for use in subsequent pathway calculations.

Interference from Hessian eigenvectors corresponding to overall translation or rotation is prevented by orthogonalization to these degrees of freedom.

We have also developed a second hybrid EF/CG scheme for use when the Hessian is unavailable [82]. A variational approach is used to find the smallest eigenvalue and corresponding eigenvector [127]; Voter has recently employed a similar method to accelerate molecular dynamics simulations of rare events in solids [128].

Consider taking a step $\mathbf{y}$ from the current position and define

$$\lambda(\mathbf{y}) = \frac{\mathbf{y'Hy}}{\mathbf{y}^2}. \tag{1.17}$$

Expanding $\mathbf{y}$ in terms of the unknown Hessian eigenvectors as above reveals a lower bound for $\lambda(\mathbf{y})$:

$$\lambda(\mathbf{y}) = \frac{\sum_i a_i^2 (\lambda_i - \lambda_{\min})}{\sum_j a_j^2} + \lambda_{\min} \geq \lambda_{\min}, \tag{1.18}$$

By differentiating this expression with respect to a_α, we obtain

$$\frac{\partial \lambda}{\partial a_\alpha} = \frac{2a_\alpha}{\sum_j a_j^2} \left(\lambda_\alpha - \frac{\sum_i a_i^2 \lambda_i}{\sum_j a_j^2} \right). \tag{1.19}$$

Setting this derivative equal to zero, we find nontrivial solutions whenever all the a_α vanish except for one. In practice we have found that minimizing $\lambda(\mathbf{y})$ with respect to the a_α generally leads to the global minimum.

The numerical second derivative of the energy may be used as an approximation to $\lambda(\mathbf{y})$, with $E(\mathbf{X}_0)$ the energy at point $\mathbf{X}_0$ in nuclear configuration space and $\xi \ll 1$:

$$\lambda(\mathbf{y}) \approx \frac{E(\mathbf{X}_0 + \xi\mathbf{y}) + E(\mathbf{X}_0 - \xi\mathbf{y}) - 2E(\mathbf{X}_0)}{(\xi\mathbf{y})^2}. \tag{1.20}$$

Differentiating (1.20) gives

$$\frac{\partial \lambda}{\partial \mathbf{y}} = \frac{\nabla E(\mathbf{x}_0 + \xi \mathbf{y}) - \nabla E(\mathbf{x}_0 - \xi \mathbf{y})}{\xi \mathbf{y}^2} - 2\lambda \hat{\mathbf{y}}, \qquad (1.21)$$

where $\hat{\mathbf{y}} = \mathbf{y}/|\mathbf{y}|$. We used Polak–Ribiere CG minimization [84] and the above approximation to the first derivative to minimize $\lambda(\mathbf{y})$. Once the smallest eigenvalue and the corresponding eigenvector are known we can proceed to find transition states using eigenvector-following for the uphill step and CG minimization in the tangent space, as for the EF/CG method above. Henkelman and Jónsson have described an alternative implementation of the above algorithm and applied it to Al adatom diffusion on an Al (100) surface [129].

To prevent contamination of the desired eigenvector by modes corresponding to overall translation or rotation, a simple orthogonalization scheme was applied both to the vector $\mathbf{y}$ and the derivative $\partial \lambda / \partial \mathbf{y}$. After the calculations described in this review were completed we switched all minimizations from conjugate-gradient to a new approach that is significantly faster [342].

3. Properties of Rearrangement Pathways

Several parameters are useful in describing rearrangement mechanisms. The first is the integrated path length, S, approximated as a sum over steps, m:

$$S \approx \sum_m |\mathbf{X}(m+1) - \mathbf{X}(m)|, \qquad (1.22)$$

where $\mathbf{X}(m)$ is the $3N$-dimensional position vector for N nuclei in Cartesian coordinates at step m. The second parameter is the distance between the two minima in nuclear configuration space, D:

$$D = |\mathbf{X}(s) - \mathbf{X}(f)|, \qquad (1.23)$$

where $\mathbf{X}(s)$ and $\mathbf{X}(f)$ are the $3N$-dimensional position vectors of the minima at the start and finish of the path. The third is the moment ratio of displacement [31], γ, which gives a measure of the cooperativity of the rearrangement:

$$\gamma = \frac{N \sum_i [\mathbf{r}_i(s) - \mathbf{r}_i(f)]^4}{(\sum_i [\mathbf{r}_i(s) - \mathbf{r}_i(f)]^2)^2}, \qquad (1.24)$$

where $\mathbf{r}_i(s)$ is the position vector of atom i in starting minimum s, etc. If every atom moves through the same distance then $\gamma = 1$, whereas if only one atom moves then $\gamma = N$. Alternatively, we may use $\bar{N} = N/\gamma$, which gives a more direct measure of the cooperativity.

B. Sampling the Surface

A number of similar approaches have been developed for systematically exploring a PES by hopping between potential wells [98,122,130,131], and these are easily adapted to produce a flexible algorithm. In our scheme, we commence at a known minimum and proceed as follows.

1. Search for a transition state along the Hessian eigenvector with the smallest nonzero eigenvalue.

2. Follow steepest-descent paths from the transition state to find the two connected minima, as described in Section III.A.

3. There are various possible outcomes from Step 2:

   (a) In most cases, one of the connected minima is the minimum from which the transition state search was initiated. If this is so, we must decide whether to stay at the original minimum or to move to the connected one.

   (b) Sometimes the transition state is not connected to the minimum from which the search started. If neither minimum has been found before, the pathway is then isolated from the rest of the databse. Since we want to explore patterns of connectivity on the PES, we discard the transition state and both minima under these circumstances. Such searches can be repeated later, when the database has grown and a connection may be found.

   (c) If the original minimum is not connected, but one or both minima have already been visited, the pathway is recorded, but we remain at the original minimum.

4. The procedure continues from Step 1, searching in both direction (parallel and antiparallel) along eigenvectors with successively larger eigenvalues from the modified position.

5. When a specified number, n_{ev}, of eigenvectors of a given minimum have been searched for transition states, we move to the lowest-energy minimum for which fewer than n_{ev} eigenvectors have been searched.

A finishing criterion is required. For some systems it is possible to let the process continue until all n_{ev} eigenvectors of every minimum have been searched. We call this approach an "n_{ev}-complete" search. However, even searching all $3N - 6$ eigenvectors with nonzero eigenvalues from each minimum does not guarantee that all transition states will be found. In fact, it is possible to find all the stationary points using a branch-and-bound technique [63], but this approach is only feasible for fairly small systems. In practice, searches along eigenvectors with small eigenvalues are most likely to converge to a transition state in a reasonable number of iterations. For larger systems, there are too many minima and transition states for an n_{ev}-complete search, and the process must be stopped when the database reaches a certain size. It is then important to bear in mind the incompleteness of the sample when interpreting the results. Even for an n_{ev}-complete search, the value of n_{ev} affects how thorough the survey is, because it determines the number of transition state searches that are performed.

For an n_{ev}-complete search, the basis for the decision in Step 3a is irrelevant, since all n_{ev} eigenvectors of all minima are eventually searched. Otherwise, the decision affects whether a localized area is probed thoroughly (if moves are often rejected), or a wider area is explored more superficially (if moves are frequently accepted). The present sampling schemes are primarily designed to collect samples with which to visualize the connectivity of the potential energy surface using disconnectivity graphs. Although we have used the resulting samples to calculate relaxation rates and thermodynamic properties, the bias introduced by the sampling schemes is unknown. Hence we will focus mainly on qualitative aspects of the dynamics and thermodynamics, and employ harmonic densities of states. The issue of sampling is further discussed in the following section with reference to thermodynamic properties.

C.   Thermodynamics from the Superposition Approach

The superposition method extends Stillinger and Weber's division of the PES into catchment basins for each minimum to calculate thermodynamic properties. In this approach the configurational part of the phase space integral in the definition of the density of states, $\Omega(E)$, or partition function, $Z(T)$, is divided into separate integrals for each minimum, giving

$$\Omega(E) = \sum_i \Omega_i(E) \quad \text{and} \quad Z(T) = \sum_i Z_i(T), \tag{1.25}$$

where the sum is over all the *geometrically distinct* minima on the PES, and Ω_i and Z_i are the density of states and partition function, respectively, for the catchment basin of the set of permutational isomers of minimum i. Ω_i and Z_i must include a factor, n_i, to account for the equivalent integrals corresponding to each permutational isomer of a geometrically distinct minimum i, where $n_i = 2N!/h_i^{PG}$ and h_i^{PG} is the order of the point group of i.[2]

---

[2] In their formulation of the superposition method Hoare and McInnes omitted the effect of symmetry [132]. This omission explains why they predicted that for LJ_{13} the cluster would reside in the icosahedral global minimum up to a reduced temperature of 0.6. In fact, LJ_{13} melts at $T \sim 0.29$. Without the inclusion of symmetry, the density of states of the icosahedron is overestimated by a factor of 120.

These expressions are in principle exact, and furthermore provide a natural way to divide up the calculation of the thermodynamics, because the harmonic approximation provides a good first approximation to Ω_i and Z_i. Hence, it is not surprising that this approach has been independently proposed by a number of authors [132–136]. However, it is only since increased computational power has made large-scale characterizations of the PES feasible that the method has been applied successfully to larger systems.

As there are well-established simulation techniques to calculate accurate thermodynamic properties of clusters, biomolecules and liquids, we must explain the utility of the more qualitative superposition approach. Firstly, it provides a clear connection between the PES and the thermodynamics, and so enables us to obtain physical insight into how the energy landscape determines the temperature-dependent properties of the system. In particular, we can easily examine the contribution of a particular region of configuration space to the thermodynamics by restricting the sum in Eq. (1.25) [137]. For example, the microcanonical and canonical probabilities of finding the system in a region "A" are

$$P_A = \frac{\sum_{i\in A} \Omega_i(E)}{\Omega(E)} \quad \text{and} \quad P_A = \frac{\sum_{i\in A} Z_i(T)}{Z_i(T)}. \tag{1.26}$$

Secondly, for systems with large free energy barriers between low-energy states it can be difficult to calculate the low-temperature thermodynamic properties with standard simulation techniques, because the barriers hinder ergodicity. Although there are methods that address these problems, such as umbrella sampling [138] and jump-walking [139], they can be computationally intensive and often require an order parameter to provide a natural pathway between the two states. The superposition approach does not suffer from such problems, because it enables absolute densities of states and partition functions to be calculated for the different regions, albeit approximately. Finally, the superposition method is needed to implement the master equation approach for calculating dynamical properties in terms of well-to-well rate constants, as described in Section III.D.

To obtain more accurate thermodynamic properties from the superposition method there are a number of more technical issues that must be addressed. Firstly, the sum in Eq. (1.25) is over all the minima. For small enough systems this range is not a problem because (virtually) all the minima can be catalogued. The size limit at which it becomes unfeasible to catalogue all the minima depends on the system of interest; for example, the current limit for Lennard-Jones clusters lies not far beyond $N = 13$. However, Eq. (1.25) can also be applied to larger systems if the number of thermodynamically *relevant* minima is small enough to be catalogued. This condition might be satisfied, for example, when the equilibrium between a number of low-energy states is of interest. However, in general we are faced with the problem of calculating thermodynamic properties from an incomplete, but statistically representative, sample of minima [140].

Of course, no simulation technique actually samples all of phase space: only a representative sample that has the distribution of the relevant ensemble is required. Similarly, we can obtain a representative sample of minima by performing minimizations from a set of points generated by a Monte Carlo (MC) or MD simulation. To compensate for the incompleteness, we weight the density of states or partition function for each *known* minimum by g_i, the number of minima of energy $\sim E_i$ for which the minimum i is representative [136]. Hence,

$$\Omega(E) \approx \sum_i{}' g_i \Omega_i(E) \quad \text{and} \quad Z(T) \approx \sum_i{}' g_i Z_i(T) \qquad (1.27)$$

where the prime indicates that the sum is now over an *incomplete* but *representative* sample of minima. The effect of g_i can be incorporated using the probability that minimization from a configuration generated by the simulation at energy E led to minimum i. If the system is ergodic on the time scale of the simulation, then for the microcanonical ensemble we obtain

$$P_i(E^*) = g_i \Omega_i(E^*)/\Omega(E^*). \qquad (1.28)$$

Hence,

$$\Omega(E) = \Omega(E^*) \sum_i{}' P_i(E^*) \frac{\Omega_i(E)}{\Omega_i(E^*)} \qquad (1.29)$$

$$= \frac{\Omega_1(E^*)}{P_1(E^*)} \sum_i{}' P_i(E^*) \frac{\Omega_i(E)}{\Omega_i(E^*)} \qquad (1.30)$$

where $i = 1$ corresponds to the global minimum. The constant of proportionality in the above equation was obtained by noting that the limit of $\Omega(E)$ as $E \to 0$ is $\Omega_1(E)$. By comparing Eq. (1.30) with the first of Eqs. (1.27): we find

$$g_i = \frac{P_i(E^*) \Omega_1(E^*)}{P_1(E^*) \Omega_i(E^*)}. \qquad (1.31)$$

If the minima are generated from configurations saved in a canonical simulation at temperature T^* then the analogous expression is

$$g_i = \frac{P_i(T^*) Z_1(T^*)}{P_1(T^*) Z_i(T^*)}. \qquad (1.32)$$

The above reweighting technique [136] is analogous to the histogram Monte Carlo approach [141–143], but instead of determining the configurational density of states from the canonical potential energy distribution, g, effectively a density of minima, is obtained from the occupation probability of the different basins of attraction. A similar approach has been used to calculate the density of minima as a function of the potential energy for a bulk liquid [144,145].

The accuracy of this reweighting method depends on the statistical accuracy of $P_i(E^*)$ or $P_i(T^*)$. In particular, one needs to choose E^* or T^* so that all the relevant minima are significantly sampled. For most of our applications to clusters this condition is easily met: at the upper end of the melting region, the cluster frequently passes back and forth between the solid-like and liquid-like states. Unfortunately, it is not always possible to find such a temperature or energy. For example, in larger clusters we do not expect to find a point where all the thermodynamic states are in "dynamic coexistence" in the manner of small clusters. In such cases, one could develop an extension to the superposition method, analogous to multi-histogram Monte Carlo, in which the information from samples of minima generated at different energies or temperatures is combined. Alternatively, one could attempt to find a non-Boltzmann ensemble, for example, a multicanonical ensemble, in which all the states are sampled. In this case, g_i would have to be obtained by the inversion of the equivalent expression to Eq. (1.28) for the new ensemble.

The second difficulty with the superposition method is obtaining expressions for Ω_i and Z_i. As we noted earlier, the natural starting point is to truncate a Taylor expansion for the potential energy around each minimum at second order giving the harmonic approximation. This truncation gives

$$\Omega_i = \frac{n_i(E - E_i)^{\kappa-1}}{\Gamma(\kappa)(h\bar{\nu}_i)^\kappa} \quad \text{and} \quad Z_i = \frac{n_i \exp(-\beta E_i)}{(\beta h\bar{\nu}_i)^\kappa,} \tag{1.33}$$

where E_i is the potential energy of minimum i, $\bar{\nu}_i$ is the geometric mean vibrational frequency, κ is the number of vibrational degrees of freedom, and $\beta = 1/kT$. For a cluster, $\kappa = 3N - 6$ but for a bulk system $\kappa = 3N - 3$. We have ignored the rotational and translational degrees of freedom since we generally consider systems with zero linear and angular momentum.

Although the harmonic approximation usually gives a qualitatively correct picture, to obtain accurate thermodynamic properties one needs to have a good anharmonic expression for Ω_i or Z_i. Of course, the superposition of harmonic densities of states is itself anharmonic, but further corrections to the harmonic densities of states of the individual minima may also be necessary. There have been many attempts to model anharmonicity [146] and various different functional forms have been proposed [137,147–149]. Here we consider one form that we have found to give accurate results for clusters [137]. Following the method of Haarhoff [150], the classical density of states of a one-dimensional Morse oscillator can be approximated by differentiating the expression for the quantum number n with respect to energy:

$$\Omega(E) = \frac{1}{h\nu \sqrt{1 - \dfrac{E}{D_e}}} = \frac{1}{h\nu}\left(1 + \frac{E}{2D_e} + \frac{3}{8}\left(\frac{E}{D_e}\right)^2 + \cdots\right), \qquad (1.34)$$

where D_e is the dissociation energy of the Morse oscillator and we have used the binomial expansion for the square root. The first term in the series is harmonic and the second represents a first-order anharmonic correction, whose magnitude is given by the anharmonicity parameter $a = 1/2D_e$. If we ignore higher order terms and assume that a is the same for all normal modes, we obtain anharmonic expressions for Ω_i and Z_i:

$$\Omega_i(E) = \frac{n_i}{(h\bar{\nu}_i)^\kappa} \sum_{l=0}^{\kappa} \frac{C_l^\kappa a_i^l (E - E_i)^{\kappa + l - 1}}{\Gamma(\kappa + l)}. \qquad (1.35)$$

and

$$Z_i(T) = \frac{n_i \exp(-\beta E_i)}{(h\bar{\nu}_i)^\kappa} \sum_{l=0}^{\kappa} \frac{C_l^\kappa a_i^l}{\beta^{\kappa + l}}, \qquad (1.36)$$

where C_l^κ is a binomial coefficient. The first terms in these series correspond to the harmonic approximation. Although various methods for obtaining the a_i from the PES have been explored [137,149,151], no effective method has been found. Instead, much better results are obtained by assuming an average value of a_i for particular regions of the PES, and then fitting these values to reproduce some thermodynamic property obtained from simulation such as the curvature of the caloric curve [137].

Although we have only discussed the quantities $\Omega(E)$ and $Z(T)$, many other thermodynamic properties can be obtained by application of standard thermodynamic relations. For example, the microcanonical and canonical caloric curves follow from $1/kT(E) = \partial \ln \Omega(E)/\partial E$ and $U(T) = -\partial \ln Z(T)/\partial \beta$, respectively. The resulting harmonic expressions can be found in Ref. 136 and Ref. 152 and the anharmonic expressions in Ref. 137.

D. Master Equation Dynamics

The interwell dynamics of a complex PES represents a set of coupled differential equations that we solve using the master equation [153,154]. The first applications to clusters were made by Kunz and co-workers [57,58,155,156]. Since then the method has been pursued by our group [16,149,157–160] and a number of others [17,59,63,71,140,148,151, 161,162].

Let $\mathbf{P}(t)$ be a vector whose components, $P_i(t)(1 \leq i \leq n_{\min})$, are the probabilities of the cluster residing in a potential well corresponding to minimum i at time t. This description immediately raises the question of how we determine which well a given configuration should be associated with. Some sort of order parameter or energy minimization is clearly required.

The time evolution of the probabilities is governed by

$$\frac{dP_i(t)}{dt} = \sum_{j \neq i}^{n_{\min}} [k_{ij}P_j(t) - k_{ji}P_i(t)], \qquad (1.37)$$

where k_{ij} is the first order rate constant for transitions from well j to well i. The sum is restricted to wells other than i, since transitions between permutational isomers do not affect P_i. We can set up a transition matrix, $\mathbf{W}$, with components

$$W_{ij} = k_{ij} - \delta_{ij} \sum_{m=1}^{n_{\min}} k_{mi}, \qquad (1.38)$$

so that the diagonal components W_{ii} contain minus the total rate constant out of minimum i. This definition allows us to write the set of coupled linear differential equations (1.37)—the "master equation"—in matrix form:

$$\frac{d\mathbf{P}(t)}{dt} = \mathbf{W}\mathbf{P}(t). \qquad (1.39)$$

If $\mathbf{W}$ cannot be decomposed into block form then the system has a uniquely defined equilibrium state, $\mathbf{P}^{\mathrm{eq}}$, for which $(d\mathbf{P}/dt)|_{\mathbf{P}=\mathbf{P}^{\mathrm{eq}}} = 0$, that is, $\mathbf{W}$ has a single zero eigenvalue whose eigenvector is the equilibrium probability distribution. $\mathbf{W}$ is asymmetric, but can be symmetrized using the condition of detailed balance: at equilibrium,

$$W_{ij}P_j^{\mathrm{eq}} = W_{ji}P_i^{\mathrm{eq}}, \qquad (1.40)$$

so that $\bar{W}_{ij} = (P_j^{eq}/P_i^{eq})^{1/2} W_{ij}$ is symmetric. $\mathbf{W}$ and $\tilde{\mathbf{W}}$ have the same eigenvalues, λ_i, and their normalized eigenvectors, $\mathbf{u}^{(i)}$ and $\tilde{\mathbf{u}}^{(i)}$, respectively, are related by $\mathbf{u}^{(i)} = \mathbf{S}\tilde{\mathbf{u}}^{(i)}$, where $\mathbf{S}$ is the diagonal matrix $S_{ii} = \sqrt{P_i^{eq}}$. Hence, individual components of the eigenvectors are related by $u_j^{(i)} = \tilde{u}_j^{(i)} \sqrt{P_j^{eq}}$. The solution of Eq. (1.39) is then [153,154]

$$P_i(t) = \sqrt{P_i^{eq}} \sum_{j=1}^{n_{min}} \tilde{u}_i^{(j)} e^{\lambda_j t} \left[\sum_{m=1}^{n_{min}} \tilde{u}_m^{(j)} \frac{P_m(0)}{\sqrt{P_m^{eq}}} \right], \qquad (1.41)$$

where $\tilde{u}_m^{(j)}$ is component m of $\tilde{\mathbf{u}}^{(j)}$.

Apart from the zero eigenvalue, all the λ_j are negative [153]. We adopt the convention of labeling the eigenvalues and eigenvectors in decreasing order according to the value of λ_i, so that $\lambda_1 = 0$, and $\lambda_j < 0$ for $2 \leq j \leq n_{min}$. As $t \to \infty$, only the $j = 1$ term in Eq. (1.41) survives, and $\mathbf{P}(t) \to \mathbf{P}^{eq}$. This limit defines the baseline to which the remaining modes decay exponentially. The size of the contribution of mode j to the evolution of the probability of minimum i depends on component i of eigenvector j, and on the weighted overlap between the initial probability vector and eigenvector j, that is, the term in square brackets in Eq. (1.41). The sign of the product of these two quantities determines whether the mode makes an increasing or decreasing contribution with time. Combinations of modes with different signs give rise to the possibility of accumulation and subsequent decay of transient populations as probability flows from the initial state to equilibrium via intermediates.

Eq. (1.41) requires the diagonalization of the matrix $\tilde{\mathbf{W}}$, whose dimension is the number of isomers in the database. The time required to compute the eigenvectors scales as the cube of the dimension, and the storage requirements scale as its square. (Although $\tilde{\mathbf{W}}$ may be sparse, its eigenvectors are not.) However, once diagonalization has been achieved, $P_i(t)$ can be calculated independently for any isomer i at any instant t. The only restriction on t comes from the accuracy to which the eigenvalues, λ_i, can be obtained; if the error is of the order $\delta\lambda$, Eq. (1.41) may diverge as t approaches $1/\delta\lambda$.

An alternative way of solving the master equation is to integrate Eq. (1.39) numerically. This approach has the advantage of not requiring diagonalization of $\tilde{\mathbf{W}}$, and is therefore the only way to proceed for large databases of minima. However, it has a number of disadvantages. Firstly, knowledge of the eigenvalues and eigenvectors of $\tilde{\mathbf{W}}$ is useful in interpreting the time evolution of $\mathbf{P}(t)$. Secondly, accurate integration over long periods can be very slow, since the accumulation of numerical error can cause the probabilities to diverge rapidly. Thirdly, the full probability vector $\mathbf{P}(t)$

(rather than selected components) must be propagated, and the integration must start from the time at which the initial probabilities are specified.

If the initial probability vector, $\mathbf{P}(0)$, corresponds to a strongly nonequilibrium distribution, many components of $\mathbf{P}(t)$ change rapidly as soon as the integration starts, and then relax more slowly toward equilibrium. Therefore, when numerically integrating the master equation, the step size required for a given accuracy is usually smaller when t is closer to zero, and can be enlarged as t grows. To take advantage of this, the numerical integration in the present work was performed using a Bulirsch–Stoer algorithm with an adaptive step size [84]. Results from this method coincided precisely with those of the analytic solution, where the latter could be determined.

The linearity of the master equation rests on the assumption that the underlying dynamics are Markovian. The probability of the transition $i \rightarrow j$ must not depend on the history of reaching minimum i, so that the elements of the transition matrix are indeed constants for a given temperature or total energy. For this restriction to apply, states within a potential well must equilibrate on a time scale faster than transitions to different minima, so that the transitions are truly stochastic. The Markovian requirement imposes an upper limit to the temperatures at which the master equation can be applied to transitions between minima, since at high temperatures the phase point does not reside in any one well long enough to establish equilibrium within it. This limit indicates that the master equation approach should be complementary to the standard mode-coupling theory of supercooled liquids [163–165].

1. Rate Constants

To use the master equation, one needs a general formula for the rate constant, $k_j^\dagger$, out of minimum j through transition state $\dagger$. In the microcanonical ensemble this relation is provided by Rice–Ramsperger–Kassel–Marcus (RRKM) theory [166]:

$$k_j^\dagger(E) = \frac{W^\dagger(E)}{h\Omega_j(E)}. \tag{1.42}$$

Here,

$$W^\dagger(E) = \int_{V^\dagger}^{E} \Omega^\dagger(E')dE' \tag{1.43}$$

is the sum of states at the transition state with the reactive mode removed; $\Omega^\dagger(E)$ is the density of states at the transition state excluding this mode, and $V^\dagger$ is the potential energy of the transition state. The total rate constant k_{ij}, for the process $j \rightarrow i$, is then obtained by summing Eq. (1.42) over all transition states linking j and i.

Rates in the canonical ensemble, that is, as a function of temperature rather than energy, can be obtained from Eq. (1.42) by Boltzmann weighting:

$$k_j^\dagger(T) = \frac{\int_{V^\dagger}^\infty k_j^\dagger(E)\Omega_j(E)e^{-E/k_BT}dE}{\int_{V_j}^\infty \Omega_j(E)e^{-E/k_bT}dE}, \tag{1.44}$$

where V_j is the potential energy of minimum j. Since the Laplace transforms

$$Z_j(T) = \int_{V_j}^\infty \Omega_j(E)e^{-E/k_BT}dE \quad \text{and} \quad Z^\dagger(T) = \int_{V^\dagger}^\infty \Omega^\dagger(E)e^{-E/k_BT}dE \tag{1.45}$$

are the vibrational partition functions of the minimum and transition state respectively, we obtain, after some manipulation of the integrals in the numerator,

$$k_j^\dagger(T) = \frac{k_BT}{h}\frac{Z^\dagger(T)}{Z_j(T)}. \tag{1.46}$$

The harmonic approximations to the density of states and partition function given in Eq. (1.33) lead to microcanonical and canonical rate constants

$$k_j^\dagger(E) = \frac{h_j^{PG}}{h^{PG\dagger}}\frac{\bar{v}_j^\kappa}{\bar{v}^{\dagger(\kappa-1)}}\left(\frac{E - V^\dagger}{E - V_j}\right)^{\kappa-1}, \tag{1.47}$$

and

$$k_j^\dagger(T) = \frac{h_j^{PG}}{h^{PG\dagger}}\frac{\bar{v}_j^\kappa}{\bar{v}^{\dagger(\kappa-1)}}e^{-(V^\dagger-V_j)/k_BT}. \tag{1.48}$$

Hence, in conjunction with the results of Section III.C, the choice of model for the density of states or partition function dictates both the dynamic and the equilibrium thermodynamic behavior of the system, the latter emerging from the master equation as $t \rightarrow \infty$. Attempts to incorporate anharmonic effects do not necessarily succeed in improving the description of both these properties. In the context of the master equation, Ball and Berry have developed and tested anharmonic partition functions by first concentrating on the equilibrium probabilities [151] and then on the dynamics [148]. No single model performed best for all the properties and systems tried, but the most consistent improvements were obtained using a model based on the expansion of the Morse oscillator density of states described in Section III.C, introducing an empirical parameter to limit excessively large anharmonic effects. Despite the increased complexity of this model, some thermodynamic properties show little improvement over the harmonic expressions [149]. The difficulties probably lie in trying to model the intricately shaped catchment basins surrounding individual minima using analytic formulae [149]. In the work reviewed in this article, we have chosen to use the harmonic approximation for all master equation results for simplicity.

2. *Relaxation Function*

Much attention has been given to the relaxation of various properties toward their equilibrium values, especially in the glasses literature. Kohlrausch first proposed a stretched exponential form as a description of viscoelasticity, while Williams and Watts suggested the same form for dielectric relaxation: $\exp[-(t/\tau)^\theta]$, where $0 \leq \theta \leq 1$ and $\theta = 1$ corresponds to the Debye limit. The master equation solution, Eq. (1.41), has a decaying multiexponential form that could lead to a wide variety of behavior depending upon the system.

Palmer et al. [70] showed how a hierarchically constrained model could produce relaxation corresponding to single exponential, power-law, or the stretched exponential form above. Huberman and Kerszberg report a universal power-law decay of autocorrelation functions for their hierarchical model [71]. León and Ngai have interpreted dynamic light scattering and dielectric relaxation data for "fragile" glass-formers in terms of a cross-over between two stretched exponential forms with different θ parameters [167].

The master equation result shows that single exponential Debye-type relaxation is only expected at times where all but one of the contributions to Eq. (1.41) have already decayed. Simple kinetics such as this are the expectation for biomolecular relaxation; Sabelko et al. refer to the non-Debye behavior that they have observed in protein folding as "strange" [168]. Given the form of Eq. (1.41) we would expect different fitting schemes to succeed for different systems, and, furthermore, we expect the results to depend upon the time scale of observation, as well as the property in question. However, there is at least one case where the form of the relaxation suggests an interpretation in terms of the underlying PES. Single exponential decay is expected for a surface with two distinct regions separated by a significant barrier. Equilibration may then occur more rapidly within each region than between them, and so an observer may find simple kinetics once initial transients have decayed. Under experimental conditions peptides in aqueous solution are only marginally stable and hence they may satisfy this two-state model. However, we might expect experiments on short time scales to observe more complex kinetics [168]. Given that much of the relaxation behavior observed is system specific, there seems no reason to expect the fitting functions that are employed to provide much insight. We will discuss several systems that exhibit time scale separation in the following sections.

3.   Simplified Master Equations, Barriers, and Connectivity

The master equation approach has been applied before to protein folding and glass dynamics along with several simplifying assumptions. A "global" master equation is one where any two states can be interconverted in a single transition, and analytical solutions have been achieved using Laplace transformation [169]. An analytical solution is also available in the "effective medium approximation" where a problem with randomly missing connections is replaced by one in which all states are connected by the same reduced effective rate constant [170]. We expect such global models to exhibit "small world" behavior, which is more appropriate for the transmission of epidemics and other processes described by random graphs in which the system can leap between widely separated regions of phase space in a single bound [171]. In a large biomolecule or a solid we do not expect such processes to exist, and hence the dynamics exhibited by global master equations may not really be typical of such systems. Unfortunately, analytical solutions to the master equation disappear as soon as any further complexity is introduced.

Diezemann and co-workers have recently presented numerical solutions of a master equation for the orientational relaxation of rotational correlation functions in glass-formers such as salol [172,173]. They have considered both globally and locally connected models, where in the latter case only states similar in energy to the starting point can be reached in one hop. They report that the main features of their results are qualitatively the same for these two extremes of connectivity.

Energy master equations have also been considered by several workers [174–177]. The solutions generally involve a number of additional simplifications, such as global connectivity, in order to simplify the initial integro-differential equation. However, none of the above studies really account for the *connectivity* of the PES properly. To include connectivity in a model with continuous distributions of energy levels requires the introduction of at least one more parameter to provide a measure of *location*. It is difficult to see how this could be achieved in a general way.

Another common feature of much of the above work, and of previous theoretical efforts focusing upon the role of the PES [28,178], is that they have been handicapped by lack of knowledge about the distribution and connectivity of the local minima. Even the association of peaks in dielectric loss spectra with α and β "processes" [29] contains an implicit assumption that there is some unknown mechanistic basis for this result. The Adam–Gibbs model of relaxation, which makes a connection to the entropy [178,179], has achieved some success, but failures have also been reported [180]. The model assumes that relaxation is due to "cooperatively rearranging regions" whose size increases with decreasing temperature, but we currently have no results that support this picture. However, on a hierarchical PES longer (sequential) pathways with higher barriers would generally be needed to achieve relaxation, and such surfaces may also give rise to non-Arrhenius behavior as discussed in Section VIII.

Although the routine characterization of transition states and rearrangement pathways in solids by eigenvector-following was demonstrated by Wales and Uppenbrink in 1994 [80], the new insight offered by such methods has yet to be fully exploited. We expect to see such developments in the next few years. In particular, we wish to emphasize that the connectivity is especially important in determining relaxation properties. Here we define connectivity in terms of single rearrangement processes mediated by a true transition state in the Murrell–Laidler sense [51]. It is not enough to know the distribution of barrier heights because surfaces with the same distribution could have entirely different kinetics. Here, by barrier heights we mean the energy differences between a true transition state and the minima that it connects. We must distinguish this use from barrier distributions corresponding to multiple step paths between minima that are not connected directly. For example, Sastry et al. [39] have interpreted their quench results for a binary Lennard-Jones system in terms of increasing "barrier heights" with decreasing temperature. Their use of barrier height includes multistep rearrangements (and free energy, rather than potential energy barriers). This terminology involves a convolution of what we prefer to call the connectivity and the energy differences between transition states and connected minima.

E. Global Optimization

The global optimization problem is intimately linked to the nature of the underlying PES [16,181,182]. This is an area of intense activity, due to the economic importance of solving traveling salesman-type problems in the design of microprocessor circuitry, and the prediction of folded protein structures from amino acid sequence alone. Anfinsen's "thermodynamic principle"[13,183] for proteins has gained widespread acceptance, and holds that the native state lies at the global free energy minimum at the appropriate temperature.[3] Global optimization is also an issue in crystal engineering, where the design of materials with particular physical properties would be assisted if it were possible to predict crystal structures from the intermolecular potential alone.

A physically intuitive approach to global optimization was provided by simulated annealing [185]. The system is equilibrated at a high energy and cooled slowly enough to maintain equilibrium until the low-energy state is reached. Problems may arise when the identity of the global free energy minimum changes as the temperature is decreased. The system may then be trapped in what becomes only a local minimum. The multiple-funnel landscapes discussed in some of the following sections correspond to such a scenario.

Many different algorithms have been suggested to solve the global optimization problem, and here we will mention only methods that fall into the category of "hypersurface deformation." In the diffusion equation approach the true PES is used as the initial boundary condition for a multidimensional diffusion equation [186]. As time progresses the surface becomes smoother until it contains just one minimum. Having located this minimum, the deformation is reversed, tracking the minimum backward in the hope that it maps back onto the global minimum of the original function. This technique fails when the global minimum of the deformed surface does not originate from the global minimum of the actual PES. In an attempt to avoid this problem, the distance scaling method deforms the surface in such a way that the equilibrium separation for pair potentials is unaffected, but the range of the potential is increased [187]. This scaling also reduces the number of minima on the PES, and the method was successful in locating the global minimum of some difficult Lennard-Jones clusters [188]. Since the range of the potential can completely alter the identity of the global minimum [189], the problem of mapping back structures from the deformed surface onto the true global minimum must be addressed in such procedures.

In the present work we have generally located the global minimum of the PES using a "basin-hopping" approach [190–192]. The following transformation is applied to the energy landscape:

$$\tilde{E}(\mathbf{X}) = \min\{E(\mathbf{X})\}, \tag{1.49}$$

where $\mathbf{X}$ represents the $3N$-dimensional vector of nuclear coordinates and min signifies that an energy minimization is carried out starting from $\mathbf{X}$. The

---

[3] Some interesting exceptions have been found, for example, the protein plasminogen activator inhibitor 1 readily folds to the active state, but converts to an inactive form (presumably of lower free energy) on the time scale of hours [184].

transformed energy, $\tilde{E}(\mathbf{X})$, at any point, $\mathbf{X}$, becomes the energy of the structure obtained by minimization. Each local minimum is therefore surrounded by a catchment basin of constant energy consisting of all the neighboring geometries from which that particular minimum is obtained. The overall energy landscape becomes a set of plateaus, one for each catchment basin, but the energies of the local minima are unaffected.

Monte Carlo sampling at a fixed temperature provides a simple but quite effective means to search the transformed surface, $\tilde{E}(\mathbf{X})$ [193]. Steps are proposed by perturbing the current coordinates and carrying out a minimization from the resulting geometry. A step is accepted if the energy of the new minimum, E_{new}, is lower than the starting point, E_{old}. If $E_{new} > E_{old}$ then the step is accepted if $\exp[(E_{old} - E_{new})/k_B T]$ is greater than a random number drawn from the interval [0,1]. Better results have been obtained if the structure is reset to that of the current local minimum at each step [192,194], and this corresponds to the "Monte Carlo plus energy minimization" (MCM) procedure of Li and Scheraga [190,191].

The success of the basin-hopping approach has been explained in terms of the thermodynamics of the transformed landscape [182,195]. The plateau transformation broadens the temperature range of thermodynamic transitions, so that the global minimum has a significant occupation probability at temperatures where the free energy barriers between different funnels are still surmountable.

IV.   LENNARD-JONES CLUSTERS

In this section we will focus on clusters bound by the Lennard-Jones (LJ) potential [196]:

$$E_c = 4\epsilon \sum_{i<j} \left[\left(\frac{\sigma}{r_{ij}}\right)^{12} - \left(\frac{\sigma}{r_{ij}}\right)^{6} \right], \qquad (1.50)$$

where ϵ is the pair well depth and $2^{1/6}\sigma$ is the equilibrium pair separation. Their structural, thermodynamic, and dynamic properties have been much studied to explore the size dependence of various properties. Furthermore, LJ clusters provide a reasonable model of rare gas clusters, a system that is readily accessible by experiment. This background makes LJ clusters an ideal model system on which to test the energy landscape approach.

A.   Structure of LJ Clusters

The global minimum of the PES provides the starting point for under-standing structure, because at zero Kelvin it must have the lowest free

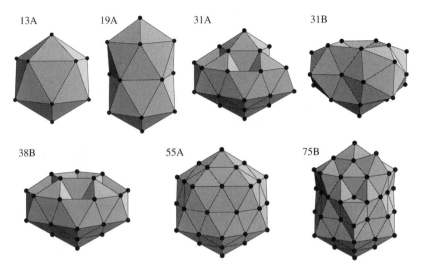

13A   19A   31A   31B

38B   55A   75B

Figure 1.4. A selection of low-energy icosahedral LJ minima. The number denotes the number of atoms, and the letter denotes the energetic rank of the minimum (the global minimum has the label "A," etc.).

energy. At higher temperature other minima become increasingly populated, and fundamental changes of structure may occur, even before melting.

As LJ clusters have become a test system for global optimization methods [188,197], putative global minima have been found up to $N = 309$ [192,198,199]. In this size range most clusters are based on Mackay icosahedra [200]. Complete Mackay icosahedra are possible at $N = 13, 55, 147, 309, \ldots$ and two examples are shown in Figure 1.4. The Mackay icosahedra consist of 20 face-centered-cubic (fcc) tetrahedra sharing a common point and have six five-fold axes of symmetry. Icosahedra are lowest in energy because they have a larger number of nearest-neighbor contacts than other ordered structures (due to their spherical shape and the predominance of {111} faces) and because the strain can be accommodated without too large an energetic penalty in this size range. This preference for icosahedral structure is in agreement with electron diffraction experiments on rare gas clusters [201], mass spectral abundances [202–205], and x-ray absorption spectroscopy [206].

Figure 1.5 provides confirmation that the complete Mackay icosahedra are particularly stable. Between these "magic numbers," the stability of the clusters first decreases as atoms are added to the surface of the icosahedron, and then increase as the overlayer nears completion. When especially stable

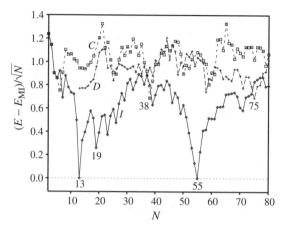

Figure 1.5. Comparison of the energies of icosahedral (I), decahedral (D), and close-packed (C) LJ_N clusters. The energy zero is E_{MI}, a function fitted to the energies of the first four Mackay icosahedra at $N = 13$, 55, 147, and 309; $E_{MI} = -2.3476 - 5.4633N^{1/3} + 14.8814N^{2/3} - 8.5699N$.

nonicosahedral clusters coincide with a size when the icosahedral structures have an incomplete overlayer, the global minimum may be nonicosahedral. There are eight such cases known for $N < 150$, and these are illustrated in Figure 1.6. At $N = 38$ the global minimum is an fcc truncated octahedron [187,189,207]. For $N = 75$–77 [189] and 102–104 [208] the global minima are Marks decahedra [209]. These structures are based on a decahedron consisting of five edge-sharing fcc tetrahedra but with the equatorial edges

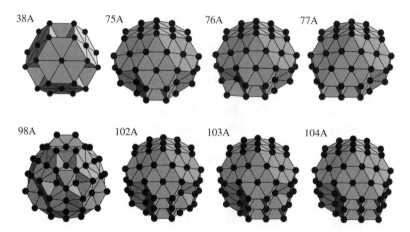

Figure 1.6. The nonicosahedral LJ global minima for $N < 150$. The number denotes the number of atoms and the letter denotes the energetic rank of the minimum.

truncated and reentrant faces added. The 98-atom global minimum has an unusual structure with tetrahedral symmetry, which is again formed from a combination of fcc tetrahedra, some incomplete [210].

B. Evolution of the PES with Size

In this section we present disconnectivity graphs for a selection of LJ clusters to illustrate the effects of size on the PES [211]. These effects are both general—the number of stationary points on the PES increases exponentially with size—and specific—the clusters we have chosen illustrate a number of interesting features.

The six clusters we analyse in detail are LJ_{13}, LJ_{19}, LJ_{31}, LJ_{38}, LJ_{55}, and LJ_{75}. The complete Mackay icosahedra at LJ_{13} and LJ_{55} are particularly stable compared to other sizes, and their PES's have a large energy gap between the two lowest-energy minima (2.85ϵ and 2.64ϵ for LJ_{13} and LJ_{55}, respectively). We therefore expect their behavior to be governed by a single deep funnel. At certain sizes between LJ_{13} and LJ_{55}, particularly stable overlayers with no low-coordinate atoms can be formed, giving rise to subsidiary minima in Figure 1.5. For example, for LJ_{19} the global minimum is a double icosahedron, and we expect LJ_{19} to exhibit a single-funnel PES but to relax less efficiently to the global minimum than LJ_{13} or LJ_{55}. On the other hand, LJ_{31} is an example where there are many low-energy icosahedral minima associated with different sorts of overlayer.

We also choose to study two of the clusters that do not have an icosahedral global minimum: LJ_{38} and LJ_{75}. For these clusters the second-lowest energy minimum is based upon icosahedral packing and we expect these surfaces to have two funnels: one that leads to the low-energy icosahedral minima and one that leads to the global minimum.

We used the methods described in Section III.B to obtain samples of stationary points for these clusters. Only LJ_{13} is small enough for an exhaustive search of all minima and transition states to be possible. For the other clusters, the searches were terminated once we were confident that we had obtained an accurate representation of the low-energy regions of the PES. The disconnectivity graphs constructed from these samples are shown in Figure 1.7.

For $N = 13$ it is possible to include all the minima that we have found in the disconnectivity graph, which therefore provides a practically complete global picture of the PES. The graph has the form expected for an almost ideal funnel: there is a single stem, representing the superbasin of the global minimum, with branches sprouting directly from it at each level indicating the progressive exclusion of minima as the energy decreases. The form of the LJ_{13} graph implies that all the minima are connected directly to the superbasin associated with the global minimum. In fact, 99% of the minima

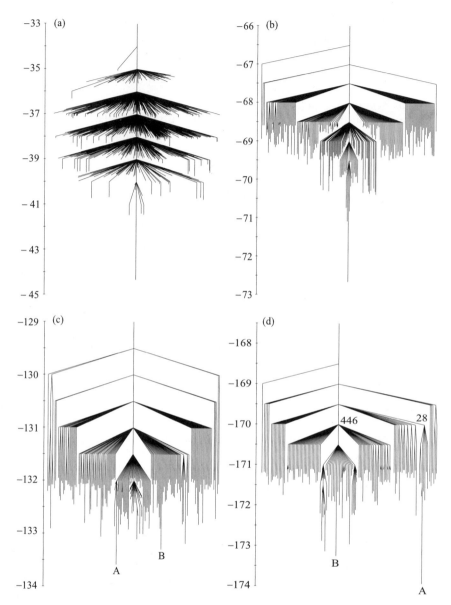

Figure 1.7. Disconnectivity graphs for (a) LJ$_{13}$, (b) LJ$_{19}$, (c) LJ$_{31}$, (d) LJ$_{38}$, (e) LJ$_{55}$, and (f) LJ$_{75}$. In (a) all the minima are represented. In the other parts only the branches leading to the (b) 250, (c) 200, (d) 150, (e) 900, and (f) 250 lowest-energy minima are shown. The numbers adjacent to the nodes indicate the number of minima that the nodes represent. The branches associated with the minima depicted in Figure 1.4 and Figure 1.6 are labeled by the letters that indicate their energetic rank. Energy is in units of ε.

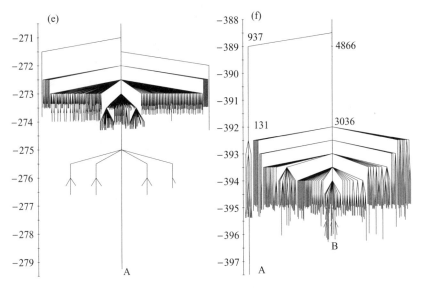

Figure 1.7. (*Continued*)

are within two rearrangements of the global minimum and none is further than three steps away. Relaxation to the global minimum is therefore relatively easy for this cluster, although there is some hindrance from the fairly large barriers $(1-2\epsilon)$ that exist for escaping from some of the minima into the superbasin of the global minimum.

For all the other clusters it is not computationally feasible to obtain a nearly complete set of minima. Moreover, if we attempt to represent all the minima of our samples on a disconnectivity graph, the density of lines simply becomes too great. Instead, we only represent those branches that lead to a specified number of the lowest-energy minima, both for clarity and because our samples of minima and transition states are likely to be most complete for the low-energy regions of the PES. Minima that are not represented can still contribute to the appearance of the graph if they mediate low barrier paths between minima that are included. However, this approach does have the consequence that, as the size increases, the graphs increasingly focus on a smaller and smaller proportion of the surface.

We can see this effect if we examine the graph for LJ_{55}. Although there is more fine structure than for LJ_{13}, the form is again indicative of a single funnel. However, all the minima represented in the disconnectivity graph have relatively ordered structures and so, unlike the graph for LJ_{13}, this representation does not tell us whether there is a funnel leading from the disordered liquid-like minima to the global minimum Mackay icosahedron.

Instead, it only shows that the low-energy region of the PES, associated with structures based on the Mackay icosahedron, is funnel-like. The graph probably only represents the bottom of a larger funnel leading down from the huge number of disordered minima.

The fine structure of the LJ_{55} disconnectivity graph reveals some interesting features. The minima separate into bands related to the number of defects present in the Mackay icosahedron [152,212]. For example, all the minima in the first band above the global minimum are Mackay icosahedra with a missing vertex and an atom on the surface. The 11 minima in this band correspond to the 11 possible sites for this atom that are unrelated by symmetry. In the disconnectivity graph these minima split into four groups corresponding to the four symmetry-unrelated faces on which the adatom can be located. The splitting occurs because the barriers for rearrangements in which the adatom passes between faces are larger than the barriers for changing the position of the adatom on a face. The second band of minima consists of Mackay icosahedra with two missing vertices and two surface atoms. The lower-lying minima in this band have the two adatoms in contact, either on the same face or bridging an edge.

The disconnectivity graph for LJ_{19} shows that the PES is again funnel-like (Fig. 1.7b), as expected. The branches for most of the minima connect directly to the superbasin containing the global minimum. Although our graph only shows branches leading to the lowest 250 minima, a plot of distance from the global minimum against potential energy reveals that the funnel continues up to higher energies [211]. Figure 1.2 in Section II.B showed that the downhill pathways to the global minimum often had a "sawtooth" profile, because the barrier heights are relatively large compared to the energy differences between the minima. In comparison to the "staircase" profiles of $(KCl)_{32}$ (Fig. 1.2) Ball et al. concluded that LJ_{19} has topographical features typical of a glass-former [27]. The LJ_{19} disconnectivity graph also shows that some of the downhill barriers are quite large. However, as the global minimum is at the bottom of a funnel the barriers only slow down the rate of relaxation toward the double icosahedron, rather than prevent it. The appearance of $N = 19$ as a magic number in mass spectra of rare gas clusters [202] confirms that the global minimum is kinetically accessible.

The disconnectivity graph for LJ_{31} (Fig. 1.7c) is fundamentally different from those considered above. The energetic bias toward the global minimum is smaller. In fact, there are a number of minima with energies close to that of the global minimum that are separated from it by fairly large energy barriers. This situation is partly the result of competition between two distinct types of overlayer. In the first type, the anti-Mackay overlayer, atoms are added to the faces and vertices of the underlying 13-atom icosahedron (giving rise, for example, to the double icosahedron, 19A). In the second type, the Mackay overlayer atoms are added to the edges and vertices. The completion of the Mackay overlayer leads to the next Mackay icosahedron. LJ_{31} is the first size for which a cluster with the Mackay overlayer is the global minimum [213]. It can be seen from Figure 1.4 that minimum 31A is a fragment of the 55-atom Mackay icosahedron. The second lowest-energy minimum, 31B, has an anti-Mackay overlayer.

We can deduce something of the relaxation behavior of LJ_{31} from its disconnectivity graph. Once the cluster has reached a low-energy configuration, presumably by rapid descent of a funnel from the liquid, subsequent relaxation toward the global minimum may be considerably slower. There is little energetic bias at the bottom of the PES to guide the system toward the global minimum and the barriers for interconversion of the low-energy minima can be relatively large. Therefore, it is not surprising that the time required to find the global minimum using the basin-hopping global optimization algorithm shows a maximum at LJ_{31} [188].

The effects of competing structures that we noted for LJ_{31} appear in more extreme form for LJ_{38} and LJ_{75}. The disconnectivity graphs clearly partition the low-energy minima into two main groups, namely those associated with the global minimum and those with icosahedral structures. These two groups of minima are separated by a large energy barrier, so the graph splits into two stems at high energy leading to two structurally distinct sets of low-energy minima. This splitting is characteristic of a multiple funnel PES. The separation is particularly dramatic for LJ_{75}, where the decahedral to icosahedral barrier is over 3ϵ larger than any of the other barriers between the 250 lowest-energy minima.

From the relative numbers of minima associated with each funnel it is clear that the icosahedral funnel is much wider. This is one of the reasons why relaxation is much more likely to lead to the icosahedral minima. In Section IV.E we will see some of the thermodynamic factors that also underlie this behavior. Once the cluster enters the icosahedral funnel it is likely to be trapped because of the large energy and free energy barriers between the two states. The interfunnel paths with the lowest barriers are shown in Figure 1.8. The corresponding energy barriers are 4.22ϵ and 3.54ϵ for LJ_{38} and 8.69ϵ and 7.48ϵ for LJ_{75}.

The two pathways differ in a number of ways. First, the LJ_{75} path is longer, more complicated and higher in energy. Second, at its highest points the LJ_{38} pathway passes through disordered structures [130,214], whereas all the minima along the LJ_{75} pathway are ordered. Indeed, a quasi-fivefold axis is retained throughout the pathway, as has also been observed in MD simulations of a decahedral to icosahedral transition in gold [215,216] and in LJ_{55} [218].

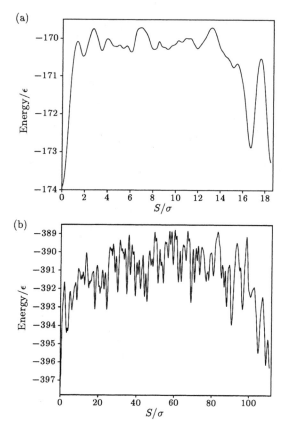

Figure 1.8. The lowest-barrier path between the global minimum and the second lowest-energy minimum for (a) LJ$_{38}$ and (b) LJ$_{75}$. The pathways pass over (a) 13 and (b) 65 transition states.

C. Coarse-Graining the PES

As the size of the cluster increases, our disconnectivity graphs focus on an increasingly small proportion of the whole PES to avoid being swamped by the rapidly increasing number of minima. However, it would be desirable to retain a more global picture of the PES. To do so, the disconnectivity graphs would need to be based not on the barriers between minima, but between larger topographical units.

Here, we use monotonic sequence basin (MSB's) to produce a more coarse-grained picture of the PES. Disconnectivity graphs that only include the minimum at the bottom of each MSB can be produced by excluding all the minima directly connected to a lower-energy minimum. In the resulting graphs, branches are joined by a node at the energy of the lowest transition state on the divide between the MSB's. For the clusters we consider here, the number of MSB's is small enough for all of them to be represented on the graph. Two examples are shown in Figure 1.9.

For LJ_{13} there is only one MSB, reflecting the ideal funnel character and the remarkable degree of connectivity. For LJ_{19} there are only a few MSB's, and these are all directly connected to the primary MSB, again reflecting the single funnel character of the PES that we noted earlier. The double-funnel character of the LJ_{38} PES is still apparent in the MSB disconnectivity graph, but now we get a better impression of the overall shape of the PES (Fig. 1.9a). There is a wide, gently sloping funnel down from the higher-energy minima toward the low-energy icosahedral funnel, whereas the funnel down to the global minimum is much narrower. A total of 2292 of the minima lie on monotonic sequences to the lowest-energy icosahedral minimum, whereas only 518 lie on sequences leading to the global minimum. Only 12 of the minima lie on sequences to both, showing that there is little overlap between the two MSB's.

For LJ_{55} the MSB analysis leads to a remarkable simplification of the disconnectivity graph (Fig. 1.9b). The single-defect minima produce just one MSB, and the fine structure of the two-defect minima collapses onto the band of MSB's that branches off at -273ϵ and -272ϵ. The remaining

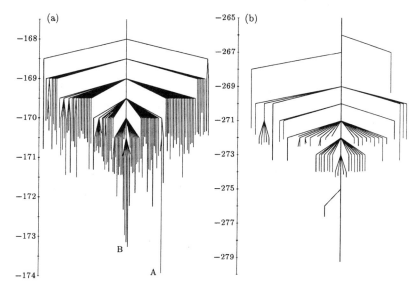

Figure 1.9. Disconnectivity graphs for (a) LJ_{38} and (b) LJ_{55}. Only the branches corresponding to monotonic sequence basin bottoms are shown. Energy is in units of ϵ.

branches are mostly three-defect minima with the three surface atoms close together. The graph clearly shows the single-funnel character of the PES.

D. LJ$_{55}$: A Single-Funnel Landscape

As we saw in the previous sections, the PES of LJ$_{55}$ has a single funnel. Here we examine the thermodynamics for this cluster to illustrate the behavior associated with this type of landscape and the utility of the superposition method (Section III.C). LJ$_{55}$ exhibits a finite-size analogue of the first-order melting transition [217]. This transition gives rise to the peak in the LJ$_{55}$ heat capacity (Fig. 1.10a), which is spread over a range of temperature because of the cluster's finite size. The Landau free energy is defined as $A_L = A - k_B T \ln P(Q)$, where A is the Helmholtz free energy and $P(Q)$ is the canonical probability distribution for the order parameter Q. A_L can be

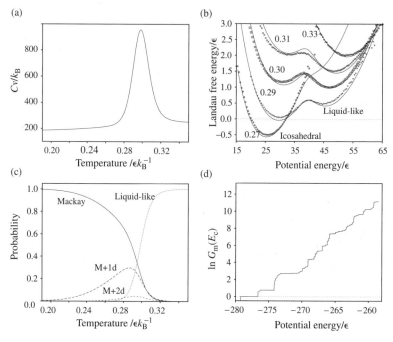

Figure 1.10. Thermodynamic properties of LJ$_{55}$ in the canonical ensemble. (a) The heat capacity, C_v. (b) The Landau free energy, $A_L(E_c)$. (c) The probability of the cluster being in the Mackay icosahedron, with one or two defects, and "liquid-like" regions of configuration space. (d) $G_m(E_c)$, the number of minima with energy less than E_c. All these properties were calculated using the anharmonic form of the superposition method from a sample of 1153 minima generated from a microcanonical MD simulation. In (b) the results are compared with simulation data obtained by Lynden-Bell and Wales (data points) [218] with the zero of free energy chosen for clarity.

used to detect the two free-energy minima (i.e., two thermodynamically stable states) in the melting region when the potential energy is used as an order parameter (Fig. 1.10b) [137,218]. These free-energy minima correspond to ordered structures based upon the Mackay icosahedron and the liquid-like state. Note the excellent agreement between results from the superposition method and the simulations. If the contributions to the ordered state are examined more closely, one finds that below the melting temperature the bands of minima identified in the LJ_{55} disconnectivity graph corresponding to Mackay icosahedra with one or two surface defects begin to be populated (Fig. 1.10c). Molecular dynamics simulations in this temperature range show that the cluster resides in each state for periods of the order of nanoseconds (using potential parameters appropriate to argon) and that during these periods the surface atoms and vacancies diffuse across the surface [219,220].

In energy landscape terms this melting transition is driven by the configurational entropy associated with the huge number of disordered liquid-like minima at the top of the funnel. The superposition method allows us to examine quantitatively this increase in minima as the potential energy is increased [137]. In Figure 1.10d we show $G_m(E_c)$, the number of minima with energy less than E_c: $G_m(E_c) = \sum_{E_i < E_c} g_i$. The number of minima increases approximately exponentially with energy in the range sampled by the simulation used to generate the sample of minima. At higher potential energies $G_m(E_c)$ must approach a limit.

As one expects for a single-funnel energy landscape, relaxation to the global minimum of LJ_{55} is relatively easy. To illustrate this, we performed a series of annealing runs, in which we cooled liquid clusters to zero temperature. For a linear cooling schedule of 10^6 MC cycles, 29% of the runs terminated at the global minimum. This figure increased to 94% for a 10^7 cycle cooling schedule.

E. LJ_{38}: A Double-Funnel Landscape

A double-funnel energy landscape gives rise to more complex behavior, and the LJ_{38} cluster provides a helpful example. As well as the large heat capacity peak associated with melting, we find a small peak at lower temperature (Fig. 1.11a) [182,195]. By examining the occupation probability of different regions of configuration space (Fig. 1.11b), this peak can be identified with a transition from the fcc global minimum to the low-energy icosahedral minima. This transition is entropically driven by the larger number of low-energy icosahedral minima and by their lower vibrational frequencies. Earlier we noted that the greater width of the icosahedral funnel is one reason why relaxation down the PES is more likely to take the system into this funnel. Here we see that it is also

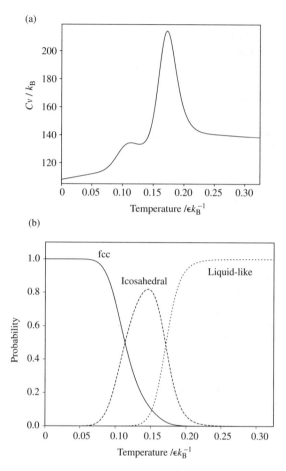

Figure 1.11. Equilibrium thermodynamic properties of LJ_{38} in the canonical ensemble. (a) The heat capacity, C_v. (b) The probability of the cluster being in the fcc, icosahedral and "liquid-like" regions of bound configuration space. These results were obtained using an anharmonic form of the superposition method (Section III.C).

thermodynamically more favorable for a liquid cluster to enter the icosahedral funnel on cooling below the melting point.

Once in the icosahedral funnel the cluster is likely to be trapped there even when the fcc funnel becomes more stable because of the large energy barrier between the two funnels (Section IV.B). This energy barrier gives rise to a free energy barrier between the two funnels. Using a bond-orientational order parameter, Q_4 [221,222], and umbrella sampling [138] we obtained free energy profiles at a number of temperatures (Fig. 1.12) [223]. As expected, the two funnels give rise to two free energy minima.

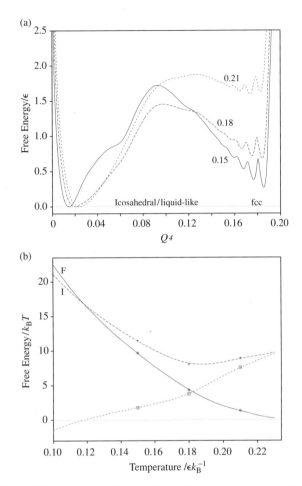

Figure 1.12. (a) Free energy profiles for LJ_{38} at three temperatures, as labelled. (b) The height of the free energy barriers relative to the fcc (F) and icosahedral (I) free energy minima and the free energy difference as a function of temperature. The data points are obtained from (a) and the interpolating lines from a histogram approach [141].

The size of the barrier separating the free energy minima is large with respect to the thermal energy and increases rapidly as the temperature decreases, thus making the interfunnel dynamics increasingly slow. For example, at the center of the fcc to icosahedral transition the free energy barriers are $1.90\epsilon = 16.9k_BT$.

As the order parameter Q_4 is unable to distinguish between the icosahedral and liquid structures, the low Q_4 free energy minimum evolves from representing the icosahedral funnel to representing the liquid-like states as the system passes through the melting transition. Interestingly, the free energy barrier for entering the fcc funnel has its minimum value at a temperature just above the melting point (Fig. 1.12b), and at this temperature the system does occasionally pass into the fcc funnel from the liquid [223].

We can also study the dynamics of the fcc to icosahedral transition at low temperature [149], where the number of thermodynamically and dynamically relevant minima is small enough that the master equation approach can be applied. The sample of stationary points that we obtained in order to construct the disconnectivity graph in Figure 1.7d provides a suitable representation of the low-energy regions of the PES for use in the master equation. To reduce numerical problems associated with the diagonalization of the transition matrix and to increase computational efficiency, we also removed some of the "dead-end" minima (i.e., minima only connected to one other minimum) that had a very low equilibrium occupation probability.

The solution of the master equation can provide the time evolution of the occupation probabilities of all the minima, but it is not immediately obvious how to extract interfunnel rate constants from this information when there are many different pathways between the funnels. To proceed we consider a general scheme for equilibrium between two funnels A and B, i.e., $A \rightleftharpoons B$ with "forward" and "reverse" rate constants k_+ and k_-. The rate of change of the occupation probability of funnel A is accordingly

$$\frac{dP_A(t)}{dt} = -k_+ P_A(t) + k_- P_B(t). \tag{1.51}$$

At sufficiently low temperature only the two funnels will be significantly occupied at equilibrium. Therefore, provided that the initial probability is itself confined to the funnels, $P_A(t) + P_B(t) = 1$. Using this equation and the detailed balance condition $(k_+ P_A^{eq} = k_- P_B^{eq})$, integration of Eq. (1.51) gives the basic result of first-order kinetics for a two-state model:

$$\ln\left[\frac{P_A(t) - P_A^{eq}}{P_A(0) - P_A^{eq}}\right] = -(k_+ + k_-)t. \tag{1.52}$$

There is an analogous equation for P_B.

To apply this equation to LJ_{38} we first use the disconnectivity graph in Figure 1.7d to assign 446 minima to the icosahedral funnel and 28 to the fcc funnel. When we calculate the quantity on the left-hand side of Eq. (1.52) at sufficiently low temperature we find that it is linear in time for both fcc and icosahedral funnels, enabling $k_+ + k_-$ to be extracted. However, at temperatures closer to the melting point this relationship begins to break down for the icosahedral funnel because there are contributions to the time evolution of P_{icos} arising from the liquid-like minima.

The rate constants can also be obtained by identifying the eigenvalue and eigenvector that correspond to interfunnel flow. The probability flow associated with a particular eigenvector occurs between those minima for which the corresponding components of the eigenvector have opposite signs [Eq. (1.41)]. This observation forms the basis for Kunz and Berry's net-flow index, which quantifies the contribution of an eigenvector i to flow into or out of a funnel A [57,58]. The index is defined by

$$f_i^A = \sum_{j \in A} \tilde{u}_j^{(i)} \sqrt{P_j^{\text{eq}}}. \tag{1.53}$$

The values of f^A and f^B for the interfunnel modes will be large and of opposite sign.

The rate constants from these two methods agree well, particularly at low-temperature or energy. For example, at an energy of -160ϵ in the micro-canonical ensemble Eq. (1.51) yields $k_+ + k_- = 4.99 \times 10^{-12} (\epsilon/m\sigma^2)^{1/2}$ and the eigenvalue of the mode with the largest net flow index ($f^{\text{fcc}} = -f^{\text{icos}} = 0.495$) is $-4.98 \times 10^{-12} (\epsilon/m\sigma^2)^{1/2}$. Using the detailed balance condition, the individual rate constants k_+ and k_- can then be obtained. For example, at the center of the fcc to icosahedral transition $k_+ = k_- = 43s^{-1}$ (using potential parameters appropriate to argon), so the interfunnel dynamics lie well beyond the time scales accessible to molecular dynamics. Figure 1.13a shows the temperature dependence of k_+ and k_-. From this plot it is clear that they obey an Arrhenius law fairly well. Some slight negative curvature is visible in the k_- results. Fitting to the form $k = A \exp(-E_a/k_B T)$ gives the pre-exponential factors and activation energies for the forward and reverse processes:

$$k_+ : \text{fcc} \rightarrow \text{icos} \qquad A = 11.1(\epsilon/m\sigma^2)^{1/2} \qquad E_a = 4.12\epsilon$$

$$k_- : \text{icos} \rightarrow \text{fcc} \qquad A = 3.18(\epsilon/m\sigma^2)^{1/2} \qquad E_a = 3.19\epsilon$$

Interestingly, the effective activation energy for fcc $\rightarrow$ icos is close to the overall barrier on the lowest-energy path between the funnels, starting from the global minimum, which is 4.22ϵ. This result suggests that the pathways passing over the highest-energy transition state on the lowest-energy pathway determine the interfunnel dynamics. E_a for icos $\rightarrow$ fcc, however, is significantly lower than the overall potential barrier for the reverse process, which is 3.54ϵ, starting from the lowest-energy icosahedral minimum.

We can analyze these results further from a simple model of the dynamics between two funnels, A and B. First, we assign the label 1 to the lowest-

DAVID J. WALES ET AL.

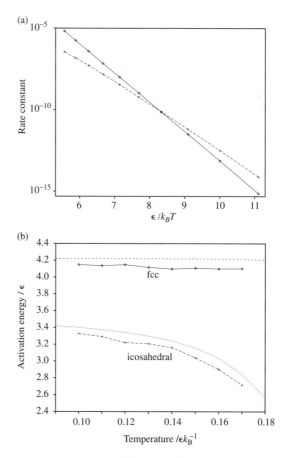

Figure 1.13. (a) Arrhenius plot of k_+ ($\diamond$) and k_- ($+$) for the interfunnel dynamics of LJ$_{38}$. The lines are fits to the form $k = A\exp(-E_a/k_BT)$. The units of the rate constant are $(\epsilon/m\sigma^2)^{1/2}$. (b) $E_a(T)$ for escape from the two funnels, as labeled. The lines with data points are estimates of the gradient of the lines in (a) and are compared to $E^\dagger - E_{fcc}$ and $E^\dagger - E_{icos}$ (lines without data points).

energy minimum in funnel A, the label $\dagger$ to the highest energy transition state on the lowest barrier path between the two funnels, and the labels a and b to the minima on either side of this transition state, where a is in funnel A and b is in funnel B. Then if we assume that, as the above results suggest, all the interfunnel probability flow passes through transition state $\dagger$ and that the minima in funnel A are in local equilibrium, it is easy to show [224] that the rate of interfunnel flow, k_+P_A, is given by

$$k_+P_A = k_{ba}^\dagger P_a \propto P_1\exp\left(-(E^\dagger - E_1)/k_BT\right). \tag{1.54}$$

Therefore, the slope of an Arrhenius plot is given by

$$\frac{d \ln k_+}{d(1/T)} = -\frac{E^\dagger - E_1}{k_B} + \frac{d \ln(P_1/P_A)}{d(1/T)} \tag{1.55}$$

where we have assumed that the other factors in Eq. (1.54) are temperature independent. This assumption is correct in the harmonic approximation.

Therefore, if the occupation probability for funnel A is dominated by the occupation probability for the lowest-energy minimum in the funnel, that is, $P_A \approx P_1$, then for this temperature range the dynamics should obey an Arrhenius law with $E_a = E^\dagger - E_1$. However, if P_1/P_A has a significant temperature dependence, then an Arrhenius plot will no longer produce a straight line. This nonlinearity can be determined if we assume the minima to be harmonic. It is easy to show that Eq. (1.55) then becomes

$$\frac{d \ln k_+}{d(1/T)} = -\frac{E^\dagger - E_A}{k_B} \qquad \text{where} \qquad E_A = \sum_{i \in A} P_i E_i / P_A. \tag{1.56}$$

In this case the slope of an Arrhenius plot is given by the difference in energy between the transition state † and the average energy of the minima in funnel A weighted by their occupation probabilities. Although the plot is nonlinear this quantity still corresponds physically to an activation energy, and so it is appropriate to think about a temperature-dependent activation energy given by $E_a(T) = -k_B d \ln k_+(T)/d(1/T)$. This quantity is plotted in Figure 1.13b. When we compare the values of $E_a(T)$ to the estimates from Eq. (1.56) we see that there is good agreement apart from a small offset. This comparison confirms that the origin of the temperature dependence of $E_a(T)$ is the temperature dependence of E_{fcc} and E_{icos}. There is little temperature dependence in E_a for escape from the fcc funnel because there is a large energy gap between the global minimum and the second lowest energy fcc minimum, which causes the global minimum to be the only fcc minimum significantly occupied. In contrast, the activation energy for escape from the icosahedral funnel decreases as the temperature increases, because the system begins to occupy the many higher-energy minima in the funnel.

Although we are unable to use the master equation approach to examine the dynamics of relaxation from the liquid-like state because a statistical method that can account for the dynamical contribution from the huge number of liquid-like minima has not yet been developed, we are able to illustrate the type of processes that would operate by examining relaxation from the highest-energy minima in our sample (Fig. 1.14). The probability

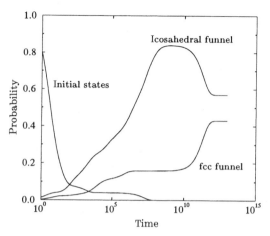

Figure 1.14. Relaxation of LJ_{38} from high-energy minima at a total energy of -160ϵ, showing the fast and slow contributions to the final probability of the fcc funnel. The time is in units of $(m\sigma^2/\epsilon)^{1/2}$.

rapidly flows out of the initial states down the PES toward the bottom of the two funnels. However, a nonequilibrium probability distribution results, which is stable for about four decades in time. Only then does the equilibration between the two funnels take place.

All the above analysis of the double-funnel energy landscape of LJ_{38} clearly shows how the icosahedral funnel acts as a kinetic trap hindering the system from reaching the global minimum. Therefore, global optimization methods that make use of the natural dynamics of the system, such as simulated annealing, are likely to fail or be very inefficient for this cluster. Annealing simulations illustrate this effect: for a 10^6 cycle cooling schedule the system never reached the global minimum and for a 10^7 cycle cooling schedule the global minimum was found only 2% of the time. Most of the global optimization methods that succeed for this difficult case [187,192,225–233] avoid some of these difficulties by transforming the energy landscape to a form for which global optimization is easier. However, even with such a transformation global optimization of LJ_{38} can still be difficult. For example, the time required to find the LJ_{38} global minimum with the basin-hopping algorithm is considerably longer than for adjacent sizes [188].

Although we have concentrated on LJ_{38} in this section, all the LJ clusters with nonicosahedral global minima illustrated in Figure 1.6 have double-funnel energy landscapes, and thus have similar thermodynamic and dynamic properties. In fact, these clusters represent a more extreme challenge for global optimization algorithms. Firstly, the search space is larger because of their size. Secondly, as a consequence of the larger size the pathway between the two funnels is likely to be longer, more complicated, and higher in energy than for LJ_{38}, thus making the interfunnel dynamics slower. This is certainly the case for LJ_{75} (Fig. 1.8b). Thirdly, the thermodynamic transition from the global minimum to the icosahedral funnel is located further below the melting temperature. For example, for LJ_{75} the Marks decahedron is most stable only below $T \approx 0.08 \epsilon k_B^{-1}$ [192] whereas the melting transition occurs at $T_m \approx 0.29 \epsilon k_B^{-1}$, and for LJ_{98} the tetrahedral global minimum is most stable only below $T \approx 0.0035 \epsilon k_B^{-1}$ [210]. This greater difficulty of global optimization is illustrated by the results for LJ_{75} and LJ_{98}. Only three unbiased algorithms [192,233,234] and two algorithms that involve seeding [229,235] have reported finding the Marks decahedron, and the LJ_{98} global minimum was first discovered only very recently, despite many applications of optimization algorithms to this cluster.

V. THIRTEEN-ATOM MORSE CLUSTERS

A. The Range of the Interatomic Potential

The range of the interatomic potential, relative to the pair equilibrium separation, has a profound influence on the physical properties of a chemical system. For example, the predicted lack of a liquid phase in C_{60} can be attributed to the short range of the interactions between the molecules [236]. Conversely, when the range of the potential is large, the critical temperature may lie well above the triple-point temperature, giving rise to a liquid phase that is stable over a wide range of temperatures [237,238], as for sodium. An energetic contribution to the destabilization of the liquid phase for short-ranged potentials has been predicted for both clusters and bulk matter by analyzing the structure and stability of liquid-like and solid-like minima [239].

For a pairwise potential, V_{ij}, the energy can be decomposed into contributions due to the n_{nn} nearest neighbor interactions, the "strain" due to deviations from the optimum pair separation in such pairs, and the non-nearest neighbor contribution, as represented, respectively, by the three terms in the expression [208]

$$V = -n_{nn}\epsilon + \sum_{\substack{i<j \\ r_{ij}<r_0}} [V_{ij} + \epsilon] + \sum_{\substack{i<j \\ r_{ij}\geq r_0}} V_{ij}, \qquad (1.57)$$

where ϵ is the pair well depth and r_0 is a criterion that defines nearest neighbors. The first two terms in Eq. (1.57) are most sensitive to the structure, and the balance between maximizing n_{nn} and minimizing the strain is the key factor in determining the relative stability of different structures. Since altering the range of the potential greatly affects the strain term for a given geometry, it shifts this balance, and can result in different morphologies being favored depending on the range of the potential [240]. For instance, in atomic clusters, decreasing the range favors decahedral and ultimately fcc over icosahedral packing because of the greater strain in the latter.

In addition to these thermodynamic and structural effects, the range of the potential can be expected to influence the dynamics of a system. This section summarizes a case study for the 13-atom Morse cluster, M_{13}, to investigate the connection between the global features of the energy landscape [159] and the dynamics [149] as a function of the range of the potential.

B. The Energy Landscape as a Function of Range

The Morse potential [241],

$$V = \sum_{i<j} e^{\rho(1-r_{ij}/r_e)} \left[e^{\rho(1-r_{ij}/r_e)} - 2 \right] \epsilon, \qquad (1.58)$$

is a convenient vehicle for the study of range effects because it contains just one adjustable parameter, ρ, which determines the width of the potential well with respect to the equilibrium pair separation r_e. r_e and ϵ serve as the units of length and energy. Physically meaningful values of ρ range from 3.15 and 3.17 for sodium and potassium [242] to 13.6 for C_{60} [80,243].

Using the methods described in Section III.B, we have generated databases of minima, transition states, and pathways for M_{13} at four values of the range parameter, ρ. In all four cases the global minimum is the icosahedron. As the first two lines of Table I show, the number of stationary points increases dramatically with ρ. While we can be confident that the searches for $\rho = 4$ and 6 are essentially complete, those for $\rho = 10$ and 14 are necessarily less so, and the number of Hessian eigenvectors searched from each minimum had to be reduced to limit the survey to a reasonable time. The larger number of minima for short-ranged potentials was noted some time ago [244]—as the range of the potential is increased, minima are "swallowed up" by the wider potential well [239].

As Table I also shows, the number $n_{min} - n_{pf}$ of minima lying outside the primary MSB (Section II.B) is a small fraction of the total. For $\rho = 4$, the energy landscape is a perfect funnel, and although the few minima lying outside this MSB at higher values of ρ technically constitute secondary funnels, they represent a very small proportion of phase space. The funnel-like nature of the landscapes for $\rho = 4$ and 6 is immediately visible from the

TABLE I
Some Properties of the Potential Energy Landscape of M_{13} at Four Values of the Range
Parameter ρ.

ρ	4	6	10	14
n_{min}	159	1439	9306	12760
n_{ts}	685	8376	37499	54439
$n_{\text{min}} - n_{\text{pf}}$	0	1	219	442
E_{gm}	-46.635	-42.440	-39.663	-37.259
ΔE_{gap}	3.024	2.864	2.245	0.468
$\langle \bar{v} \rangle_{\text{m}}$	1.187	1.625	2.615	3.660
$\langle |v^{\text{im}}| \rangle_{\text{ts}}$	0.396	0.473	0.637	0.628
$\langle b^{\text{up}} \rangle_{\text{p}}$	3.666	2.070	1.470	1.536
$\langle b^{\text{down}} \rangle_{\text{p}}$	0.461	0.543	0.583	0.784
$\langle \Delta E^{\text{con}} \rangle_{\text{p}}$	3.205	1.526	0.887	0.752
$\langle S \rangle_{\text{p}}$	2.457	1.735	1.030	0.971
$\langle D \rangle_{\text{p}}$	1.462	1.163	0.840	0.817
$\langle \tilde{N} \rangle_{\text{p}}$	6.673	5.939	6.093	5.918
$\langle n^{\text{gm}} \rangle_{\text{m}}$	1.525	2.447	3.744	3.885
$\langle S^{\text{gm}} \rangle_{\text{m}}$	2.579	3.534	3.573	3.357

All dimensioned quantities are tabulated in reduced units. n_{min} is the number of minima in the database, of which n_{pf} lie in the primary funnel. The number of transition states is n_{ts}. E_{gm} is the energy of the global minimum, with the next-lowest energy structure lying ΔE_{gap} higher. $\bar{v}_i$ is the geometric mean normal mode frequency at minimum i and v_i^{im} is the imaginary frequency at transition state i. b_i^{up} is the larger (uphill) barrier height between the two minima connected by transition state i, b_i^{down} is the smaller (downhill) barrier, and ΔE_i^{con} is the energy difference between the minima, so that $b_i^{\text{up}} = b_i^{\text{down}} + \Delta E_i^{\text{con}}$. S_i is the integrated path length between the two minima connected by transition state i, D_i is their separation in configuration space, and $\tilde{N}_i = N/\gamma$ is the cooperativity index of the rearrangement [Eq. (1.24)]. n_i^{gm} is the smallest number of steps from minimum i to the global minimum, and S_i^{gm} is the integrated length of this path. $\langle \cdots \rangle_{\text{m}}$, $\langle \cdots \rangle_{\text{ts}}$, and $\langle \cdots \rangle_{\text{p}}$ indicate averages where the index runs over minima, transition states, and nondegenerate pathways (i.e., pathways not merely connecting permutational isomers), respectively.

disconnectivity graphs in Figure 1.15. The upward shift of the $\rho = 6$ graph relative to $\rho = 4$ reflects the general increase in strain energy due to narrowing of the potential well. An increase in downhill barrier heights is also revealed by the somewhat longer branches at $\rho = 6$.

The rest of Table I lists some globally averaged properties of the energy landscapes. Some of the trends are straightforward to understand. For example, the average value of the geometric mean normal mode frequency, $\langle \bar{v} \rangle_{\text{m}}$, rises monotonically with ρ because of the increasing stiffness of shorter-ranged potentials.

Decreasing the range of the potential is accompanied by a general flattening of the landscape. This effect can be detected in the vicinity of the

58   DAVID J. WALES ET AL.

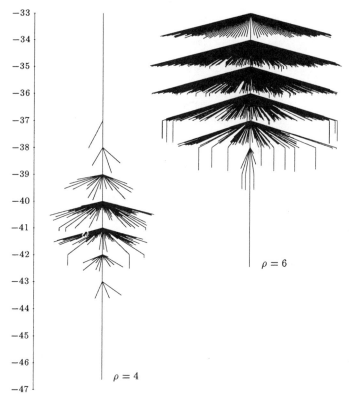

Figure 1.15. Disconnectivity graphs for M_{13} with $\rho = 4$ and $\rho = 6$ plotted on the same energy scale (in units of the pair well depth, ϵ).

global minimum by the decreasing energetic gap, ΔE_{gap}, to the second-lowest minimum. For $\rho < 13.90$, the latter is a defective icosahedron with one missing vertex and a capped face. The removal of a vertex allows the strained icosahedron to relax, causing a less steep increase in strain energy with ρ than for the perfect icosahedron. The striking drop in ΔE_{gap} between $\rho = 10$ and $\rho = 14$ is because the second-lowest minimum above $\rho = 13.90$ becomes the decahedron—previously a transition state—stabilized at high ρ by its lower strain [159].

The flattening of the landscape extends away from the global minimum. Table I shows that the average energy difference, $\langle \Delta E^{\text{con}} \rangle_{\text{p}}$, between connected minima decreases as ρ increases, indicating a lower energetic gradient of the energy landscape. At the same time, the average downhill barrier, $\langle b^{\text{down}} \rangle_{\text{p}}$, rises, suggesting that relaxation down the funnel to the global minimum will be kinetically hindered at high values of ρ. This prediction is borne out by the master equation dynamics presented in the next section.

The larger number of pathways on the PES for small ρ necessitates a shorter average path length, $\langle S \rangle_p$, and a closer separation of minima, $\langle D \rangle_p$ in configuration space. Since the average total path length, $\langle S^{gm} \rangle_m$, from a local minimum to the global minimum is relatively insensitive to ρ, these paths must involve a larger number, $\langle n^{gm} \rangle_m$, of individual rearrangements. However, the typical number of atoms involved in these rearrangements, as measured by the average of the cooperativity index, $\langle \tilde{N} \rangle_p$, does not vary much with ρ. This result contrasts with statistics for the larger clusters LJ$_{55}$ and $(C_{60})_{55}$, which showed that cooperative (high $\tilde{N}_i$) rearrangements are less likely for $(C_{60})_{55}$, where the potential range corresponds to $\rho \approx 14$ [72]. The 13-atom cluster is probably too small to support localized rearrangements in this way.

In concluding this section we note that, although the most obvious change in the Morse pair function with ρ is visible in the range of the attractive part of the potential, the core repulsion is also affected significantly. For small ρ, the repulsion is softer, and contributes to the increased width of the potential well. If one splices together a potential function using $\rho = 6$ for all separations $r_{ij} > r_e$ and a different value of ρ for $r_{ij} < r_e$, many of the effects described in this section are still observed. For example, the numbers of minima and transition states increase dramatically as ρ for the inner part of the potential is increased even though the range of the attraction is fixed at $\rho = 6$.

Many of the trends described in this section can be seen in the representative monotonic sequences in Figure 1.16. Relaxation to the global minimum is hindered when the range of the potential is short.

C.   Relaxation Dynamics

1.   Search Times

Previous studies of model potential energy landscapes [16] have shown that relaxation from high-energy states to the global minimum is most efficient when the PES has a large potential energy gradient with low downhill barriers and lacks secondary funnels that act as kinetic traps. The last of these conditions is satisfied for all the M$_{13}$ clusters described in Section V.B, but the remaining criteria indicate that relaxation should be easiest for long-ranged potentials.

This prediction is supported by the results of applying the master equation to the databases of Section V.B. As for LJ$_{38}$ in Section IV.E, the harmonic approximation was used in conjunction with the RRKM theory described in Section III.D.1, but for M$_{13}$ the complete databases were used

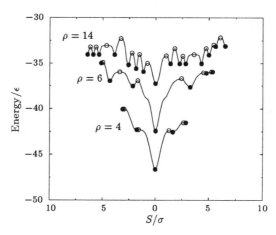

Figure 1.16. Representative monotonic sequences leading to the global minimum for three values of the range parameter ρ. S is the integrated distance along the reaction path from the global minimum. Minima are indicated by filled circles, and transition states by open circles. The plots demonstrate a number of features discussed in the text: the general increase in energy of the minima, the decreasing gap to the global minimum, the increasing downhill barrier heights, the shorter rearrangements, and the decreasing gradient toward the global minimum as the range of the potential decreases.

in the master equation, since we are interested in relaxation down the entire PES, not just the low-energy regions. Figure 1.17 shows, as a function of temperature, the "search time" taken for a strongly nonequilibrium distribution of minima to relax until the probability of the global minimum exceeds some threshold, here taken to be 0.4. The initial state at each value of ρ was an even distribution of probability among the minima that require the largest number of rearrangements to reach the global minimum. The qualitative shape of the curves is easily understood. As the temperature is increased from low values, the search time decreases because the thermal energy rises above the barriers between the minima. An optimal temperature is reached, where the search time is a minimum, above which the search time rises because the thermodynamic driving force toward the global minimum decreases. Eventually, the equilibrium probability of the global minimum falls below the threshold of 0.4, and the search time is no longer defined. Similar behavior has been observed for the same reasons in direct simulations of KCl clusters [245] and lattice protein models [11], as well as a master equation study of idealized energy landscapes [16].

As ρ increases, the search times become orders of magnitude longer, as expected from the decreasing overall slope of the landscape toward the global minimum and from the increasing downhill barrier heights. In

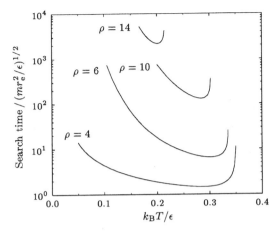

Figure 1.17. Search time as a function of temperature for M_{13} at four values of the range parameter, ρ. The search time is defined as the time taken for the probability of the global minimum to reach 0.4, starting from an even probability distribution among the minima in the layer furthest from the global minimum.

addition, the width of the curves decreases, so that the temperature range over which relaxation is reasonably fast (relative to the optimum time) is smaller when the range of the potential is short. This effect too can be rationalized in terms of the trends found in Section V.B. As the range of the potential is decreased, the energy gap between the global minimum and the other minima becomes smaller, and the energy range spanned by the minima narrows. Hence, these minima come into play at lower temperatures, and the temperature at which the global minimum ceases to dominate the equilibrium populations is lower when the potential is short-ranged. The sharp high temperature rise in the search time curve is therefore shifted to lower temperature when ρ is large. The higher downhill barriers at large ρ also cause interwell processes to slow down more rapidly as the temperature is lowered, causing the search time to rise sooner when the temperature is lowered below the optimum value. These effects combine to produce a narrower window of reasonable search times.

2. Relaxation of the Total Energy

An unambiguous way to define the relaxation time, τ_r, of a given property is to normalize its decay profile so that it starts at 1 and decays to 0, and then evaluate the total area under the curve [37]. For pure Debye relaxation, $\exp(-\lambda t)$, one simply obtains the inverse of the decay constant: $\tau_r = 1/\lambda$. For the multiexponential decay given by the master equation, the relaxation time is a weighted sum of the decay constants [149]. For simple kinetics, one would expect the relaxation time to follow the Arrhenius form, $\tau_r = \tau_0 \exp(A/k_B T)$. Figure 1.18b shows that when $\rho = 14$, M_{13} follows an Arrhenius law with $\tau_0 = 9.34 \times 10^{-3} m r_e^{1/2}$ and $A = 2.00\epsilon$. When $\rho = 4$, however, significant deviation from the linearized Arrhenius form is apparent as illustrated in Figure 1.18(a). The temperature dependence is

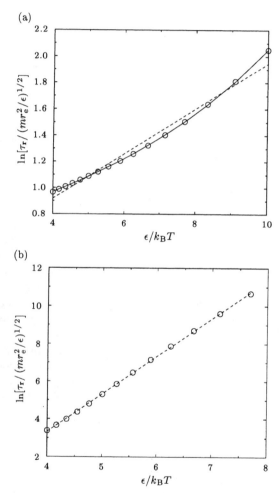

Figure 1.18. Arrhenius plots for the relaxation time of the total energy of M_{13} at (a) $\rho = 4$ and (b) $\rho = 14$. Circles are mean relaxation times from the master equation, dashed lines are fits to the Arrhenius form, and the solid line in (a) is a fit to the Vogel–Tammann–Fulcher (VTF) form.

better fitted by the VTF form [34]:

$$\tau_r = \tau_0 \exp\left[\frac{A}{k_B(T - T_0)}\right],\tag{1.59}$$

with $\tau_0 = 1.86(mr_e^2/\epsilon)^{1/2}$, $A = 0.0699\epsilon$, and $k_B T_0 = 0.051\epsilon$.

The above differences probably stem from the decreasing slope of the energy landscape as the range of the potential decreases. The energy intervals spanned by the minimum and transition state samples at higher ρ are narrower, making the landscape more uniform. This uniformity means that a change in the temperature has a similar effect on most of the individual interwell processes, each of which separately has an Arrhenius temperature dependence in the harmonic RRKM model [Eq. (1.48)]. On the steeper landscape of the $\rho = 4$ cluster, however, there is a greater spread of local minimum energies and barrier heights, so that as the temperature is lowered, some processes become "frozen out" before others, resulting in longer-relaxation times than expected by extrapolation of the high-temperature behavior.

In structural glasses, Arrhenius temperature dependence of relaxation times is associated with strong liquids, whereas VTF behavior is a signature of fragility [33]. If this classification can be applied to clusters, the results of this section suggest that increasing the range of the potential introduces a degree of fragility. Stillinger's picture of strong liquids having a "uniformly rough" energy landscape [37] is in line with the results of Section V.B: at low ρ paths between pairs of minima are organized into a larger-scale funnel, while for higher ρ the funnel feature is less prominent and a more uniform series of paths with significant barriers must be traversed to reach the global minimum.

3. Relaxation Modes

The analytic solution of the master equation decomposes the flow of probability into a series of exponentially decaying modes, each of which has a characteristic decay constant. It is instructive to look at how the contributions from these modes vary across the spectrum of time scales. From Eq. (1.41) mode j makes an important contribution to the probability evolution of minimum i if $u_i^{(j)} = \sqrt{P_i^{eq}}\tilde{u}_i^{(j)}$ is large in magnitude. The mode represents an overall flow *between* minima i and k if $u_i^{(j)}$ and $u_k^{(j)}$ have opposite signs. Inspection of the components of the eigenvectors for M_{13} at different values of ρ and different temperatures reveals some general trends. The slowest and fastest modes describe probability flow between a small number of minima, typically fewer than five. The fastest modes tend to be between minima that are directly connected by transition states. In contrast, the slowest modes are between unconnected minima, and probability flow involves intermediate minima. The slow modes tend to involve one highly populated minimum. Hence, if the initial probabilities of the other minima that feature in these slow modes (with eigenvector components of the opposite sign) are far from their equilibrium values, the slow modes may limit the overall relaxation. Modes with intermediate time scales may involve flow between a relatively large number of minima [149]. Hence, in a typical application of the master equation, the initial processes involve rapid equilibration between small groups of adjacent minima, followed by wider probability flow between larger groups, and finally slow adjustment of a few populations via processes involving multiple rearrangements.

VI. ANNEALING OF BUCKMINSTERFULLERENE

Annealing of C_{60} to Buckminsterfullerene (BF) in the graphite arc synthesis of Krätschmer et al. takes place on a time scale of around 0.001 seconds [246]. Isotopic substitution indicates that fullerene formation occurs from the condensation of atomic carbon vapor [247], and a number of mechanisms have been suggested for the formation of BF itself [248,249]. Kroto first proposed the "pentagon road" scheme based upon the growth of a nautilus-like shell, or "icospiral" [250]. He suggested that closure may occasionally occur as the shell curves around itself, preventing further growth. Clusters that avoid closure to C_{60} may grow into "buckyonions" [251]. The energetics of various pentagon road intermediates from C_{20} to C_{60} have been investigated by Bates and Scuseria [252].

Alternatively, in Heath's "fullerene road" model carbon radicals condense from the vapor to form graphite-like sheets, and the cage closes when it contains around 30 atoms [253]. In this size range closed fullerenes must contain unfavorable edge-sharing pentagons, and so Heath proposed that insertion of C_2 fragments [254] would continue until a deep minimum is reached at C_{60} or C_{70}. Calculations have subsequently suggested a connection between the C_2 extrusion process and annealing [255].

Manolopoulos and Fowler [256] have presented a model to examine the fullerene road mechanism. They used a "building game" approach [257] based on energetics and the number of adjacent pentagons at each nuclearity. They found that the fullerene road can explain the experimental abundance of C_{60} and C_{70} if the structural driving force is based only on energy minimization.

Another proposal is that cycloadditions may occur within large monocyclic rings, which progressively form loops and rearrange via 1,2-carbon shifts [258]. This model was developed to interpret ion chromatography experiments that yield arrival time spectra and require structural models to provide the collisional integrals that determine the drift rate [259, 260]. Calculations suggest that around 20 eV would be liberated in the formation of a fullerene by this route, which might assist the annealing process [258].

There are actually 1812 different C_{60} isomers with 20 hexagonal rings and 12 pentagons (excluding enantiomers) [261,262]. We have examined the annealing of these higher-energy isomers to BF using the master equation approach (Section III.D) [157]. To simplify the analysis we assume that C_{60} isomers containing faces other than five- and six-membered rings are not involved. We have also investigated only pathways involving one particular rearrangement mechanism, namely the "pyracylene" or "Stone–Wales" (SW) process [263].

A.   Connectivity of C_{60} Fullerene Isomers

There appears to be only one basic mechanism for the rearrangement of different C_{60} fullerene isomers, namely the SW or "pyracylene" process [263]. In this rearrangement, a pair of hexagons and a pair of pentagons effectively interchange their roles, providing a mechanism for pentagon migration (Fig. 1.19). Although there is an overall bond rotation of 90°, the

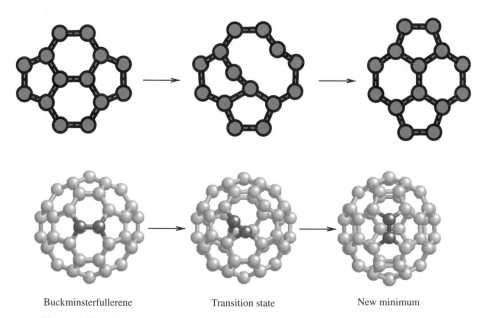

| Buckminsterfullerene | Transition state | New minimum |

Figure 1.19. The "pyracylene" or "Stone–Wales" (SW) rearrangement in C_{60}. Top: schematic view of the atoms in the SW patch. Bottom: pathway calculated with the density-functional tight-binding potential described in the text.

actual transition states are generally unsymmetrical, because the direct process is "forbidden" by orbital symmetry [255,263]. The arrangement of two pentagons and two hexagons illustrated at the top of Figure 1.19 has come to be known as a "Stone-Wales patch," and such patches can be identified in other fullerenes and in the terminal regions of carbon nanotubes [264]. The SW mechanism is similar to the hypothetical bond-switching mechanism proposed by Wooten, Winer, and Weaire for the preparation of amorphous silicon from the crystal [265].

Murry et al. [255] have previously located true transition states for the SW process, which they found to be nonconcerted and asymmetric, as expected. They have reported a two-step mechanism for the SW rearrangement via a single sp^3 transition state, and their best estimate of the corresponding barrier lies 0.5 eV lower than that of the symmetrical concerted rearrangement. These barriers are the results from single point DFT calculations with the BLYP [266,267] functional at MNDO [268] optimized geometries. Eggen et al [269] have further suggested that an additional carbon atom on the surface of the cluster may catalyze the annealing process by lowering the barrier for SW rearrangements [269]. "Stone–Wales maps" have been constructed for the connectivity of fullerene isomers linked by SW rearrangements [270], as in the present work.

Several MD studies of C_{60} annealing using a variety of potentials have previously failed to yield BF within the simulation times employed [271–278]. Given the observed experimental time scale these results are not surprising, although Xu and Scuseria were able to anneal the lowest energy defect structure into BF using a tight-binding potential [277]. Marcos et al. [278] have also investigated the transition from BF to the lowest energy pentagon defect structure using constant energy MD simulations of C_{60} and the empirical Tersoff potential [279]. They describe the process as a multistep, multiminimum path, in contrast to most other studies, but did not locate any true transition states.

More recently, Maruyama and Yamaguchi [280] have reported an MD simulation in which high-energy C_{60} minima were successfully annealed to BF. They used a novel temperature control method in their simulations and employed the Brenner potential [281]. Taking a semi-closed C_{60} pseudo-cage as the starting point, the authors found relaxation to BF after around 250 ns. This relaxation appears to proceed through a series of both ordinary and general (involving other polygons) SW rearrangements, mediated by what appear to be asymmetric transition states. From an Arrhenius plot an activation energy of $241 \, \text{kJ} \, \text{mol}^{-1}$ was deduced for the SW rearrangement, which is very low compared with other current estimates. The annealing to BF observed in this study is probably due to the low barriers obtained with the Brenner potential.

Energetic considerations alone cannot explain the experimental abundance of C_{60}, since C_{70} (D_{5h}) has far greater thermal stability [282–284]. Kinetic considerations must also be important, and a simple model suggests that closure at C_{60} should be expected if five- and six-membered rings are favored [257]. Furthermore, if C_{60} is formed through a series of exothermic reactions then the energy that is liberated may help to drive the annealing process to BF.

The spiral algorithm [262,285] was used in the present work to generate the required C_{60} isomers and their adjacency matrices, which may be used to prepare initial geometries for subsequent optimization [286]. The connectivity of the SW map for C_{60} has previously been examined by Austin et al. [270]. They found that the map contains 44 disjoint sets: the main component of 1710 connected isomers contains the global minimum BF, while the other disconnected graphs contain no more than 27 isomers. Austin and co-workers also used their graph theoretical representation to determine the smallest number of SW isomerizations (or "SW steps") required to reach BF from each local minimum. Each SW step away from the global minimum defines a set of structures, referred to collectively as a "Stone–Wales stack." BF itself is connected to only one other isomer with C_{2v} symmetry and two sets of fused pentagons. Hence, any relaxation process leading to BF must pass through this C_{2v} isomer. Austin et al. discovered that no more than 30 succesive SW rearrangements are required to convert any local minimum in the SW map to BF. The authors deduced the number of isomers in each SW stack, and the connectivities, both between successive stacks and within each stack.

We have confirmed the previous results for the SW map of C_{60} using a modified SPIRAL code [157, 262, 285]. Rearrangements were characterized up to and including the seventh stack, which includes a total of 197 minima and 547 transition states.

B.   Results

In our calculations we employed a density-functional tight-binding (DF-TB) potential for carbon based on the work of Frauenheim et al. [287–289]. Austin et al. have previously optimized all the 1812 local minima in the SW map using the semi-empirical QCFF/PI method [290]. We have verified that all 1812 candidate structures relax to true local minima for the DF-TB approach [157]. The DF-TB approach produces the smallest energy difference between the C_{2v} C_{60} isomer in stack 2 and BF, and lower barrier heights than have been found in other work [157]. A scatter plot of energies for all the minima as a function of stack number reveals a roughly linear correlation, in agreement with Austin et al. [270]. However, the potential energy gradient decreases somewhat as the SW stack number increases.

To obtain starting points for eigenvector-following transition state searches (Section III.A) the atoms in each SW patch were rotated by 22.5° about an axis corresponding to the radius vector of the midpoint of the SW bond. All the resulting transition states were found to be asymmetric, with an approximately sp^3 carbon atom.

The average uphill barrier height between SW stacks decreases somewhat with increasing SW step [157], in line with the decreasing average energy difference between stacks, which is one characteristic that we might expect of a funnel landscape (Section I). Austin et al. [270] previously noted that of the 1710 isomers contained in the main basin of the SW map, 898 isomers are connected by downhill paths to the global minimum. A typical monotonic sequence (Section II.B), corresponding to such a downhill path is shown in Figure 1.20.

The monotonic sequences for C_{60} are sawtoothed in character, as shown in Figure 1.20. The corresponding disconnectivity graph (Fig. 1.21) clearly has the "weeping willow" form described in Section II.C. The long, dangling branches are indicative of high barriers, and hence we expect the annealing efficiency to depend strongly on temperature (or total energy in the microcanonical ensemble). If the temperature (or energy) is high enough then the PES exhibits reasonable funneling properties: of the 3699 pathways leading from stack 7 to BF, nearly half (1789) correspond to monotonic sequences.

C. Relaxation Dynamics

All the thermodynamics and rate constant calculations were performed in the microcanonical ensemble according to the methods outlined in Section III.

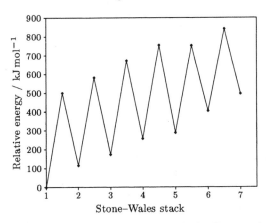

Figure 1.20. A representative monotonic sequence fo the C_{60} potential energy surface, showing the energies of minima and the transition states that connect them as a function of the Stone–Wales stack.

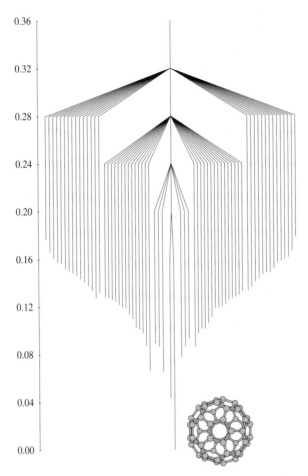

Figure 1.21. Disconnectivity graph for minima and transition states in the five lowest Stone–Wales stacks of C_{60}. Energy is in hartree relative to the global minimum Buckminsterfullerence structure. The graph including results up to stack seven has the same appearance.

Solutions to the master equation were obtained numerically using Bulirsh–Stoer [291] integration using an initial distribution consisting of the minima in stack 7 with equal weights. In this system the stack number provides a natural order parameter, and we sum the probabilities of all the minima in each stack in describing the relaxation dynamics. Equilibrium probabilities are shown in Figure 1.22 as a function of energy.

The occupation probability of BF as a function of time and total energy is shown in Figure 1.23. In the long time limit the maximum probability is

DAVID J. WALES ET AL.

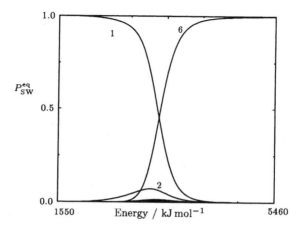

Figure 1.22. The equilibrium occupation probabilities for Stone–Wales stacks up to stack 7 as a function of the total energy relative to Buckminsterfullerene.

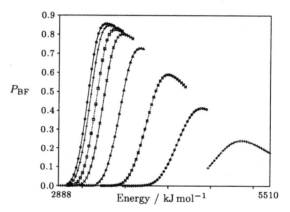

Figure 1.23. The occupation probability of Buckminsterfullerene (BF) as a function of energy (relative to the global minimum) and time, starting from an initial uniform distribution in Stone–Wales stack 7. The curves represent times (from left to right) of 3000, 2000, 1000, 500, 100, 1, and 0.1 μs.

large and the maximum is attained quite rapidly at higher energies. Figure 1.24 shows the stack populations as a function of time for relaxation at a total energy of 3360 kJ mol^{-1}. Transient populations develop in stacks 5 and 6 and BF becomes the dominant state after about 1.5 ms. These results agree with the form of the disconnectivity graph in Figure 1.21, namely that the PES has funneling properties if the total energy is sufficient to surmount the barriers. Above an energy of about 2800 kJ mol^{-1} (corresponding to an

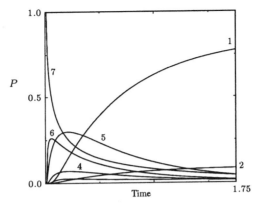

Figure 1.24. Evolution of the probability distribution among Stone–Wales stacks as a function of time for total energy $E = 3360\,\text{kJ}\,\text{mol}^{-1}$.

equipartition temperature of about 1900 K), relaxation to BF is predicted to occur on a millisecond time scale. However, this prediction should only be regarded as an order of magnitude estimate, since anharmonic corrections in RRKM theory can change the rate constant significantly. The contribution of structures containing heptagonal and square faces [292–294] could also be important.

VII. ALKALI HALIDE CLUSTERS

In Section IV, LJ_{38} and LJ_{75} provided examples of the trapping that can occur if there are a number of low-energy ordered structures that are well-separated in configuration space and lie at the bottom of their own funnels. An experimental example of this trapping has recently emerged for NaCl clusters. These clusters have only one energetically favorable morphology: the magic numbers that appear in mass spectra correspond to cuboidal fragments of the bulk crystal (rocksalt) lattice [295–297], hence they have been termed nanocrystals. However, in experiments that probed the mobility of size-selected cluster ions, multiple isomers were detected for most $(NaCl)_N Cl^-$ with $N > 30$ [298]. The populations in the different isomers were not initially equilibrated, but slowly evolved, allowing rates and activation energies for structural transitions between the nanocrystals to be obtained [299].

We were particularly interested in finding the mechanisms of the structural transitions between the different nanocrystals. Based on the small values of the activation energies, Hudgins et al. suggested that the rearrangement mechanisms might involve a sequence of surface diffusion steps [299]. Therefore, we chose to study one of the clusters probed by experiment, $(NaCl)_{35}Cl^-$, in detail. Our results will again show the utility of the energy landscape approach—the time scales of the transitions are longer than those easily accessible by conventional simulations—and provide another illustrative example of how the behavior of a system arises from the energy landscape.

To model the cluster we used the Tosi–Fumi parameterization of the Coulomb plus Born–Mayer form [300]. Although simple, this potential provides a reasonable description of the interactions. We then generated large (over 3000) samples of connected minima and transition states. Although our samples only represent a tiny fraction of the total number of stationary points, we have a good representation of the low-energy regions that are relevant to the structural transitions of $(NaCl)_{35}Cl^-$, and the network formed by this set of stationary points includes many pathways between the different $(NaCl)_{35}Cl^-$ nanocrystals.

A.   Low-Energy Nanocrystals for $(NaCl)_{35}Cl^-$

In the experiments on $(NaCl)_{35}Cl^-$ three peaks were resolved in the arrival time distribution [298,299]. However, the lowest-energy minima that we found for this clusters could be divided into four types (Fig. 1.25), namely an incomplete $5 \times 5 \times 3$ cuboid, a $6 \times 4 \times 3$ cuboid with a single vacancy, an incomplete $5 \times 4 \times 4$ cuboid, and an $8 \times 3 \times 3$ cuboid with a single vacancy. Based on calculations of the mobility of these structures using the exact hard-spheres scattering model [301], the three experimental peaks were reassigned to the $6 \times 4 \times 3$, $5 \times 5 \times 3$, and $8 \times 3 \times 3$ nanocrystals in order of increasing drift time. $5 \times 4 \times 4$ nanocrystals are probably not observed experimentally because of low equilibrium occupation probabilities (Section VII.D) and short residence times (Section VII.E).

B.   The $(NaCl)_{35}Cl^-$ PES

The disconnectivity graph for $(NaCl)_{35}Cl^-$ is shown in Figure 1.26. As the barriers between minima with the same cuboidal form are generally lower than those between minima that have different shapes, the disconnectivity graph splits into funnels corresponding to each cuboidal morphology. The separation is least clear for the $5 \times 4 \times 4$ minima because of the large number of different ways that the nine vacant sites can be arranged. The disconnectivity graph also shows that the barriers between the $5 \times 5 \times 3$, $6 \times 4 \times 3$, and $5 \times 4 \times 4$ nanocrystals are of similar magnitude, while the $8 \times 3 \times 3$ minima are separated from the rest by a considerably larger barrier.

The disconnectivity graph is also helpful for interpreting the $(NaCl)_{35}Cl^-$ dynamics observed in experiments [298,299]. In the formation process it is likely that a high-energy configuration is initially generated. The cluster then

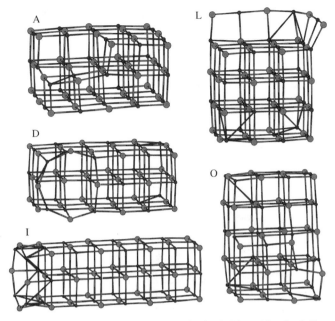

Figure 1.25.   The lowest-energy $5 \times 5 \times 3$ (A), $6 \times 4 \times 3$ (D), and $8 \times 3 \times 3$ (I) nanocrystals and the two lowest-energy $5 \times 4 \times 4$ nanocrystals (L and O). The sodium ions are represented by the smaller, darker circles and the chloride ions by the larger, lighter circles. The letter gives the energetic rank of the minimum when labeled alphabetically.

relaxes to one of the low-energy nanocrystals. Simulations for potassium chloride clusters indicate that this relaxation is particularly rapid for alkali halides because of the low barriers (relative to the energy difference between the minima) for downhill pathways [27,245]. However, there is a separation of time scales between this initial relaxation and the conversion of the metastable nanocrystals to the one with the lowest free energy. The large barriers between the different cuboids make them efficient traps.

C. Structural Transition Mechanisms

To understand the mechanisms for the structural transitions, we visualized many pathways connecting nanocrystals with different dimensions. In virtually all of them the major shape changes are achieved by the same type of mechanism. A typical example of this process occurs in the shortest path between the lowest energy $5 \times 5 \times 3$ and $6 \times 4 \times 3$ isomers (Fig. 1.27a). In this "glide" mechanism the two halves of the cluster slide past one another on a $\{110\}$ plane in a $\langle 1\bar{1}0 \rangle$ direction. Another example of this mechanism is illustrated in Figure 1.27b where two such rearrangements lead to the interconversion of the $8 \times 3 \times 3$ and $6 \times 4 \times 3$ nanocrystals.

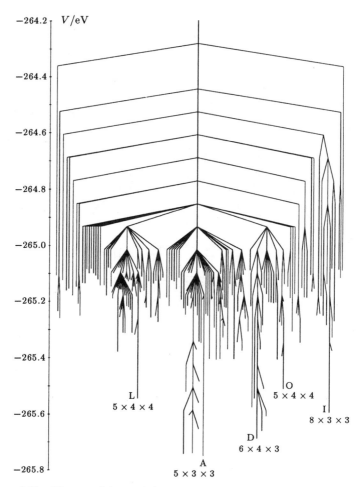

Figure 1.26.   Disconnectivity graph for $(NaCl)_{35}Cl^-$. Only branches leading to the lowest 200 minima are shown. The branches for the minima in Figure 1.25 have been labeled. Some of the funnels and subfunnels have also been labeled by their associated cuboidal shape.

We found one other mechanism by which major shape changes can occur. An example is shown in Figure 1.27c. In this "hinge" process the two halves of the cluster split apart and rotate about a common edge until they meet up again. Although visually appealing, this mechanism is unlikely to be dynamically relevant because the barriers involved are much larger than for the glide mechanism.

It is interesting to note that the plane and direction of the glide mechanism correspond to the primary slip system for NaCl [302]. Indeed,

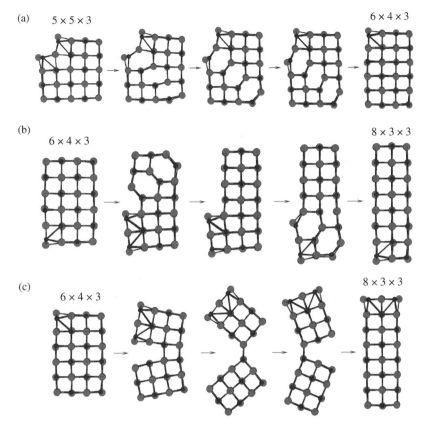

Figure 1.27. Three examples of the rearrangement mechanisms that lead to major shape changes in $(NaCl)_{35}Cl^-$. (a) A glide rearrangement that converts a $5 \times 5 \times 3$ into a $6 \times 4 \times 3$ isomer. (b) Two glide rearrangements that are part of the lowest energy path between the $6 \times 4 \times 3$ and $8 \times 3 \times 3$ isomers. (c) A higher energy "hinge" mechanism that converts a $6 \times 4 \times 3$ isomer directly to an $8 \times 3 \times 3$ isomer.

the structural transitions can be viewed as spontaneous plastic deformations to a thermodynamically more stable structure. We have observed similar ductility in simulations of NaCl nanowires under tension, with the same mechanism as for the clusters.

D. Thermodynamics of $(NaCl)_{35}Cl^-$

The large free energy barriers between the nanocrystals prevent an easy determination of the relative stabilities of the different nanocrystals by conventional simulations. Therefore, we use the harmonic superposition method (Section III.C) to examine this question. It can be seen from Figure 1.28 that

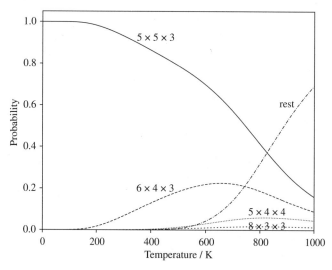

Figure 1.28. Equilibrium occupation probabilities of the $(NaCl)_{35}Cl^-$ nanocrystals computed using the harmonic superposition method.

the $5 \times 5 \times 3$ nanocrystal has the lowest free energy up until melting. This result is in agreement with experiments where the greater stability of the $5 \times 5 \times 3$ nanocrystals is indicated by the conversion of the other nanocrystals to the $5 \times 5 \times 3$ form with time. The $6 \times 4 \times 3$ nanocrystal also has a significant occupation probability, but the probabilities for the $8 \times 3 \times 3$ and $5 \times 4 \times 4$ nanocrystals are always small. The onset of melting is indicated by the rise of P_{rest} in Figure 1.28. However, this transition is too broad and occurs at too high a temperature (MC simulations indicate that melting occurs at $\sim 700\,K$) because the incompleteness of our sample of minima leads to an underestimation of the partition function for the liquid-like minima.

E.   Dynamics of $(NaCl)_{35}Cl^-$

The time scales of the structural transitions in $(NaCl)_{35}Cl^-$ mean that it is impossible to use conventional molecular dynamics to investigate the interfunnel dynamics. Instead we use the master equation method outlined in Section III.D. To reduce the computational expense and numerical difficulties we recursively removed from our sample those minima that are only connected to one other minimum—these "dead-end" minima do not contribute directly to the probability flow between different regions of the PES. The resulting pruned sample had 1624 minima and 2639 transition states. RRKM theory in the harmonic approximation was used to model the individual rate constants, $k_j^\dagger$.

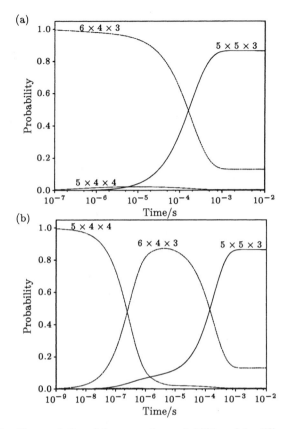

Figure 1.29. Time evolution of the occupation probabilities of the different nanocrystals at $T = 400$ K when the cluster is initially in minimum (a) D, and (b) O.

Some examples of the interfunnel dynamics are depicted in Figure 1.29. The results depend significantly on the starting configuration. For relaxation from the $6 \times 4 \times 3$ nanocrystal there is a small transient population of $5 \times 4 \times 4$ structures before equilibrium is established. In comparison the relaxation is much slower when the cluster starts from the lowest-energy $8 \times 3 \times 3$ minimum. This is a result of the large barrier to escape from this funnel (Fig. 1.26). When the system is initially in the $5 \times 4 \times 4$ minimum with lowest free energy (O), there is a large probability flow into the $6 \times 4 \times 3$ minima, which is then transferred to the $5 \times 5 \times 3$ funnel. Again, this behavior could have been predicted from the disconnectivity graph (Fig. 1.26), which shows that there is a lower barrier to go from minimum O to the $6 \times 4 \times 3$ funnel than to the funnel of the global minimum.

In the experiments, rate constants and activation energies were obtained for the conversion of the $6 \times 4 \times 3$ and $8 \times 3 \times 3$ nanocrystals into the $5 \times 5 \times 3$ nanocrystal [299]. It would, therefore, be useful if we could extract rate constants for the different interfunnel processes from the master equation dynamics. To achieve this we use the two methods outlined for LJ_{38} in Section IV.E. In Figure 1.30 we test the expression Eq. (1.52) by applying it to the interconversion of $6 \times 4 \times 3$ and $5 \times 5 \times 3$ nanocrystals. The two lines in the graph converge to the same plateau value, before both fall off beyond 0.001 s. This plateau corresponds to the time range for which the interfunnel passage dominates the evolution of the probabilities for the two funnels. At shorter times, when the occupation probabilities for the two funnels are still close to their initial values, there are many other contributing processes. At longer times the probabilities are both very close to their equilibrium values, and the slower equilibration with the $8 \times 3 \times 3$ funnel dominates the probability evolution. From the plateau in Figure 1.30 we obtain $k_+ + k_- = 5320 \, \mathrm{s}^{-1}$.

These rate constants can also be found by using the net-flow index to identify the eigenvalues of the transitions corresponding to the interfunnel processes. For example, at $T = 400 \, \mathrm{K}$ the mode with the most $5 \times 5 \times 3 \rightarrow 6 \times 4 \times 3$ character has $f^{5 \times 4 \times 3} = -0.339$, $f^{6 \times 4 \times 3} = 0.331$ and $\lambda = 5275 \, \mathrm{s}^{-1}$. The small difference from the value obtained above using Eq. (1.52) is probably due to mixing with a mode at $6388 \, \mathrm{s}^{-1}$. This mixing occurs because the eigenvalues of $\tilde{\mathbf{W}}$ (Section III.D) cannot cross as a function of temperature.

By calculating and diagonalizing $\tilde{\mathbf{W}}$ at a series of temperatures we were able to examine the temperature dependence of the interfunnel rate constants.

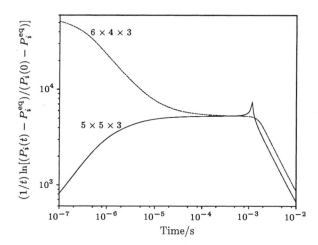

Figure 1.30. This graph tests the applicability of Eq. (1.52) to the interconversion of the $5 \times 5 \times 3$ and $6 \times 4 \times 3$ nanocrystals by following $(1/t) \ln [(P_i(t) - P_i^{\mathrm{eq}})/(P_i(0) - P_i^{\mathrm{eq}})]$ at $T = 400 \, \mathrm{K}$. Minimum D is the initial configuration.

TABLE II
Activation Energies, E_a, and Prefactors, A, for the Interfunnel Rate Constants
in $(NaCl)_{35}Cl^-$

From	To	Barrier/eV			A/s^{-1}
		E_a	ΔE	Expt.	
$6 \times 4 \times 3$	$5 \times 5 \times 3$	0.770	0.776	0.53 ± 0.05	1.43×10^{13}
$5 \times 5 \times 3$	$6 \times 4 \times 3$	0.846	0.840		2.26×10^{13}
$8 \times 3 \times 3$	$5 \times 5 \times 3$	1.055	1.055	0.57 ± 0.05	1.05×10^{15}
$5 \times 5 \times 3$	$8 \times 3 \times 3$	1.220	1.211		3.77×10^{14}

The activation energies are compared with ΔE, the difference in energy between the highest-energy transition state on the lowest-energy path directly between the two nanocrystals and the lowest-energy minimum of the starting nanocrystal, and experimental values for some of the processes [299].

For most of the interfunnel processes, the rate constants that we obtain fit well to the Arrhenius form, $k = A \exp(-E_a/k_B T)$, where E_a is the activation energy and A is the prefactor—values are given in Table II. In that table we also compare the activation energies to the differences in energy between the highest-energy transition state on the lowest-energy path between the relevant nanocrystals and the lowest-energy minimum of the starting nanocrystal. As for LJ$_{38}$ (Section IV.E) there is good agreement, providing further confirmation that activation energies for interfunnel processes can be related in a simple way to the PES. This finding also provides a firm link between dynamics and disconnectivity graphs, because these graphs provide a representation of just such barriers.

In Table II we also compare our activation energies to the two experimental values. Our values are too large by 0.24 and 0.49 eV. There are a number of possible sources of error—the incompleteness of our sampling of the PES, the harmonic approximation, the breakdown of RRKM theory, and the potential—but we believe that the potential is the major source. When polarization is included using the Welch potential [303], the estimated barriers become closer to the experimental values (the discrepancies are then 0.16 and 0.34 eV) but significant differences still remain [160]. For better agreement we probably need a potential that allows the properties of the ions to depend on the local environment, but such a potential [304,305] has not yet been developed for sodium chloride.

VIII. WATER CLUSTERS

Small water clusters have been the subject of intense theoretical and experimental interest over the last decade [118,306], due primarily to the advent of far-infrared vibration–rotation tunneling spectroscopy [307–310].

Here however, we are concerned with the effect of the hydrogen-bonding network upon the structure of the PES, and the associated thermodynamics and dynamics.

Bulk water was classified as a "strong" liquid by Angell in 1995 [38] (see Section I). However, this opinion has recently been revised on the basis of new experimental data. Ito et al. now suggest that bulk water changes character from fragile to strong on cooling below about 228 K [311]. Hence, around the usual melting point at 273 K water exhibits very fragile behavior but around the glass transition temperature of 136 K it appears to be a strong liquid. On warming above the glass transition temperature, amorphous solid water melts into a supercooled liquid until a temperature of between about 150 and 165 K is reached, when it devitrifies into crystalline ice [312]. However, before crystallizing, the viscosity of supercooled water exhibits non-Arrhenius behavior typical of fragile liquids [312]. Strong character is often associated with open network structures, but the hydrogen bonds that define the network structure of water are much weaker than the covalent bonds in materials such as SiO_2. However, as the temperature decreases the hydrogen bond strength increases relative to the available thermal energy, and we might speculate that this effect may account for the change in character observed by Ito et al. [311].

To begin with, it may be helpful simply to present some results without reference to the strong/fragile classification scheme. If water clusters exhibit qualitatively different behavior from systems such as atomic clusters, then this should be apparent from the disconnectivity graphs. We have generated graphs for $(H_2O)_6$ using both the TIP4P [313] and ASP-W4 [314,315] rigid body intermolecular potentials, and for $(H_2O)_{20}$ using TIP4P. The TIP4P potential is computationally undemanding compared to ASP-W4. It includes only effective pairwise terms, whereas ASP-W4 is a many-body potential. In fact, the TIP4P potential was not designed to model water clusters, but instead includes an enhanced dipole moment to account for the many-body terms present in the liquid. Nevertheless it has proved useful in the cluster regime, and should presumably be more accurate for larger systems. In our ASP-W4 calculations the induction energy was iterated to convergence. The ORIENT package [316], which includes eigenvector-following geometry optimization (Section III.A), was used to generate all the results reported in this section.

We first considered the water hexamer, which is the smallest cluster for which a noncyclic structure is found to be the global minimum, both theoretically [318–321] and experimentally [322]. For the TIP4P potential we have probably located the majority of the minima and transition states on the PES, while for ASP-W4 there are many more stationary points, and our sampling of the higher energy region of the surface is incomplete. For TIP4P our sample contains 133 minima and 540 transition states, while for ASP-W4 we found 370 minima and 1272 transition states. The smaller number of stationary points for the TIP4P potential is perhaps indicative of a longer effective range (Section V.A). Tsai and Jordan have previously identified ten classes of structure for $(H_2O)_6$, according to the number of 3-, 4-, 5-, and 6-membered rings [319]. The lowest-energy members of the four most important classes for the TIP4P potential are shown in Figure 1.31 in order of increasing energy, the cage being the global minimum.

The disconnectivity graph for TIP4P $(H_2O)_6$ in Figure 1.32 shows that members of the various structural classes do not occupy energetically separated regions of the PES. In fact, the surface is a good funnel, with only three minima not lying on monotonic sequences that lead to the global minimum. There is little side-branching to indicate larger barriers between families of structures, and this conclusion holds even if the separation of levels on the graph is halved (revealing more substructure). Barriers of similar heights exists for the interconversion of isomers between and within structural classes. Exceptions are the flipping motions of dangling hydrogens, which have low barriers and interconvert energetically similar structures within the same class, such as the two low-lying cages [118,306,323].

To a large extent the simple structure of the graph is a consequence of the small size of this system, where different structural classes are easily

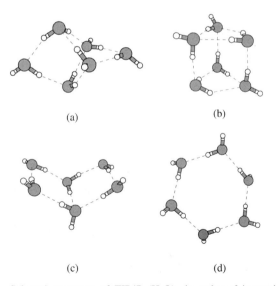

(a)

(b)

(c)

(d)

Figure 1.31. Selected structures of TIP4P $(H_2O)_6$ in order of increasing energy: the lowest-energy (a) cage, (b) prism, (c) book, and (d) ring. (Ball-and-stick representations produced using XMakemol [317].)

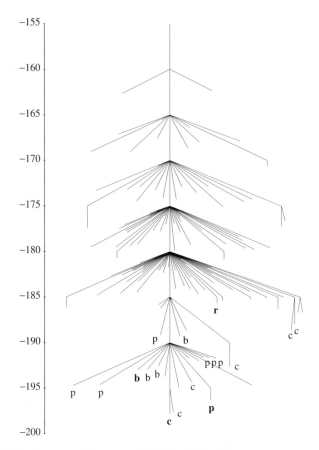

Figure 1.32. Disconnectivity graph for TIP4P $(H_2O)_6$. The energy scale is in units of kJ mol^{-1}. Some of the low-energy cage, prism, book, and ring structures are labeled c, p, b, and r, respectively, with the lowest-energy structure of each type in bold.

interconverted by a few rearrangements. The disconnectivity graph for ASP-W4 $(H_2O)_6$, illustrated in Figure 1.33, still exhibits relatively simple structure, despite the larger number of minima. The global minimum is a prism, rather than a cage, although the four cage isomers that have been invoked to explain the experimental tunneling splittings [118,306] are the lowest-lying local minima. In previous work, zero-point energy effects were suggested to explain why a cage structure might actually lie lowest [322].

For $(H_2O)_{20}$ we have only considered the TIP4P potential. This cluster can adopt a number of morphologies, depicted in Figure 1.34, and the global minimum for the TIP4P potential consists of face-sharing pentagonal prisms [324,325]. The box-kite and stacked-pentagonal-prism structures also lie

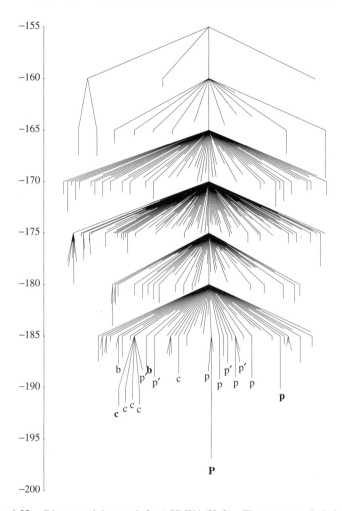

Figure 1.33. Disconnectivity graph for ASP-W4 $(H_2O)_6$. The energy scale is in units of kJ mol^{-1}. Some of the low-energy cage, prism, book, and ring structures are labeled c, p, b, and r, respectively, with the lowest-energy structure of each type in bold. p′ denotes a prism with one broken edge.

low in energy, while the lowest dodecahedral minima is separated from these minima by a significant energy gap. The energetics of these structures are very similar for the ASP-W4 potential, and are in reasonable agreement with recent *ab initio* results [326] that reversed previous claims [327].

It is only possible to obtain partial samples of the $(H_2O)_{20}$ surface, and to investigate the effects of different sampling schemes, we have compared three different surveys. To obtain an overall view of the PES we first tried

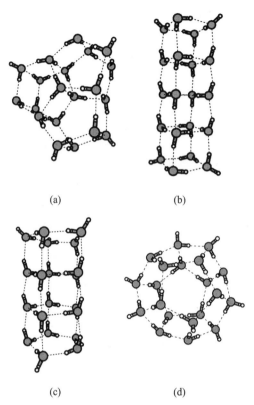

(a)   (b)

(c)   (d)

Figure 1.34. Structures of TIP4P $(H_2O)_{20}$ in order of increasing energy: (a) edge-sharing pentagonal prisms, $-872.99\,kJ\,mol^{-1}$; (b) fused cubes, $-869.44\,kJ\,mol^{-1}$; (c) face-sharing pentagonal prisms, $-867.30\,kJ\,mol^{-1}$; (d) dodecahedron, $-826.23\,kJ\,mol^{-1}$.

accepting all moves to newly found connected minima in step 3a of the sampling algorithm (Section III.B), with $n_{ev} = 2$. Starting from the global minimum this run (sample I) was terminated once 500 minima (and 513 transition states) had been found. The resulting disconnectivity graph is shown in Figure 1.35 [68]. Rather than a dominant funnel, there is repeated sub-branching, ultimately leading to minima of comparable energies that are separated by a range of barriers. Most of the barriers between directly connected pairs of minima lie below $20\,kJ\,mol^{-1}$, the energy of a typical hydrogen bond, and it is the hierarchical arrangement of these barriers that gives rise to the disconnectivity pattern. Minima first become connected in small groups, then at higher energy these groups become connected, and so

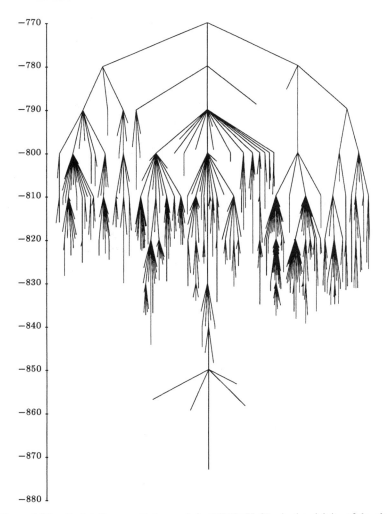

Figure 1.35.   Partial disconnectivity graph for TIP4P $(H_2O)_{20}$ in the vicinity of the global minimum. The energy scale is in units of kJ mol^{-1}.

on, leading to a "banyan tree" form above the lowest few minima. This result explains the relative difficulty of global optimization for water clusters compared, for example, with LJ clusters having the same number of degrees of freedom [325]. The difference arises partly from the coupling of angular and positional degrees of freedom. Isomers of a given morphology that differ only in the hydrogen-bonding pattern may span wide energy ranges [325] and be linked only by a long sequence of separate rearrangements. $(H_2O)_6$ is too small to show this effect.

We repeated the above survey starting from the other three minima in Figure 1.34. In each case we obtained disconnectivity graphs with similar features (not shown). The box-kite and stacked pentagonal prism minima are the lowest points on their respective graphs, but the dodecahedron is not because lower minima are immediately located.

These initial searches were designed to explore relatively wide regions of the PES without searching any particular energy range thoroughly. We therefore generated another sample (II) of 500 minima using $n_{ev} = 10$ and a Metropolis acceptance criterion [328]. Moves to new connected minima were accepted if a random number between 0 and 1 was greater than $\min[1, \exp(-\Delta V/k_B T)]$, where ΔV is the energy change associated with the step. The temperature parameter can be used to control the bias toward low-energy minima, and we employed a value of 293 K, which produced an acceptance ratio of 0.37. A total of 515 transition states had been found when the search was terminated at 500 minima, and the resulting disconnectivity graph is shown in Figure 1.36a. This search provides a better survey of the low energy region of the PES than the first one. The highest node where branching occurs is at -800 rather than $-770\,\mathrm{kJ\,mol^{-1}}$ for the first dataset, and the density of branch ends increases rapidly above around -850 rather than $-840\,\mathrm{kJ\,mol^{-1}}$. The effect of searching more modes from each minimum ($n_{ev} = 10$ rather than 2 for the previous run) is to increase the number of branches emerging from a typical node, because the region around each minimum is explored more thoroughly. However, despite these differences in detail, the hierarchical pattern of repeated sub-branching is still evident in Figure 1.36a.

A hierarchical disconnectivity graph could result if the lowest-energy pathways between minima in the sample have simply been overlooked. Unfortunately, we cannot prove that the lowest paths have been found. Instead, we took the original unbiased sample of 500 minima and used it to seed a fully biased $(T = 0)$ search with $n_{ev} = 5$ to generate a further 1000 minima (sample III). When this search was terminated, only the lowest four minima of the original sample had been used and transition state searches had been conducted from the lowest 125 minima in the new sample. If new transition states are discovered that provide lower energy pathways between these minima, then the corresponding nodes would move down the graph and branches would coalesce. In fact, the hierarchical pattern is preserved in the resulting disconnectivity graph (Fig. 1.36b). A number of new minima appear at low energy, highlighting the need for caution in the interpretation of graphs derived from incomplete samples. In the original unbiased sample the global minimum appears relatively isolated, but this result is an artifact

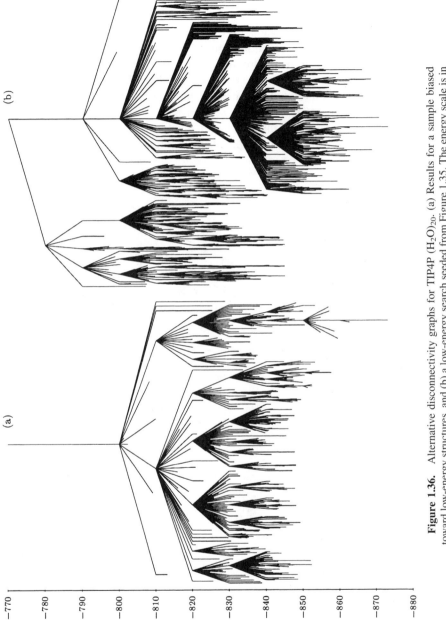

Figure 1.36. Alternative disconnectivity graphs for TIP4P $(H_2O)_{20}$. (a) Results for a sample biased toward low-energy structures, and (b) a low-energy search seeded from Figure 1.35. The energy scale is in units of kJ mol^{-1}.

87

of not exploring the low-energy regions of the PES thoroughly. In contrast, the searches biased toward low energy explore smaller regions of the PES. Nevertheless, our general conclusions about the nature of the PES are unaffected: if anything, the graph in Figure 1.36b looks more like a banyan tree than the original one in Figure 1.35.

Relaxation times for the total energy were calculated for each of the three samples as a function of temperature by integrating the scaled, shifted energy decay curve as described in Section V.C.2. The master equation was solved analytically by matrix diagonalization (Section III.D) using two different starting distributions. The results in Figure 1.37a were obtained

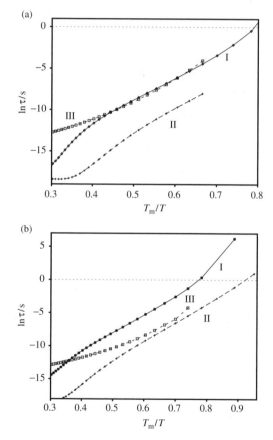

Figure 1.37. Energy relaxation time as a function of temperature for $(H_2O)_{20}$. (a) The starting distribution corresponds to a very high temperature $\sim 10^4$ K for each point. (b) The equilibrium distribution at double the temperature of interest is used as the starting point in every case. The three data sets correspond to samples, I, II, and III described in the text.

by starting from the equilibrium distribution at around 10^4 K for every point. Figure 1.37b was obtained by starting from the equilibrium distribution at a temperature equal to twice the temperature of interest. The most important feature is the same for the different temperature jumps, namely that sample III exhibits significant non-Arrhenius behavior. Higher temperature results are not shown, since we do not expect the Markovian assumption inherent in the master equation approach to be valid in this limit. The precise values of the relaxation time are also unlikely to be accurate due to all the approximations involved. A melting temperature around $T_m = 160$ K was found in previous work for this cluster [324,329], and we have used this value of T_m to scale the temperature axis in Figure 1.37.

The relaxation behavior in Figure 1.37 is quite sensitive to the quality of the sample of stationary points. Only the largest sample III exhibits pronounced non-Arrhenius curvature. In the other two samples the results are probably different because the global minimum is separated by a large energy gap from almost all the other minima. The curvature for sample III could be interpreted as an increasing "barrier height" with decreasing temperature. In fact, the barriers corresponding to single-step pathways do not increase much in the lower regions of the PES. However, because of the hierarchical structure, relaxation may involve surmounting a number of barriers linking minima of progressively higher energy. Hence, if we think of the barrier between two minima in terms of the highest point on the lowest-energy path, which may involve several transition states, then this quantity does increase on average with decreasing temperature.

Hierarchical surfaces may also provide a realisation of the Adam–Gibbs model [178,179], which assumes that relaxation is due to "cooperatively rearranging regions" whose size increases with decreasing temperature. On a hierarchical landscape, structural change involves longer pathways with higher barriers as the temperature decreases.

Stillinger has suggested that the PES of a strong liquid would be "uniformly rough" [37]. This description is only true for the $(H_2O)_{20}$ cluster in the sense that the barriers between directly connected minima rarely exceed $20\,\text{kJ mol}^{-1}$; the distribution of transition state energies is actually quite broad, and the sub-branching of the disconnectivity graph suggests "cratering" of the surface. Finally, we should be cautious about extrapolating from the PES of the $(H_2O)_{20}$ "nanodroplet" to draw conclusions about bulk water, since all the water molecules in the cluster are really in surface environments. Nevertheless, the results are clearly qualitatively different from those obtained for most of the other systems discussed in this review.

IX. BIOMOLECULES

Simulation of biomolecules must face two principal difficulties—system size and the potential. Simplified representations of the interatomic interactions are unavoidable, especially for large proteins. However, such approximations make quantitative comparison with experiment problematic. On the other hand, insight may still be gained by pursuing simplified models that capture the essential (or at least relevant) properties of real molecules. Indeed, much understanding has emerged from numerous studies of lattice protein models, which not only take a highly coarse-grained view of molecular structure, but also restrict the configuration space to a grid. The absence of a continuous configuration space produces a discrete PES where concepts such as catchment basins for local minima and transition states are no longer defined. However, a free energy landscape can still be obtained in terms of order parameters such as the radius of gyration.

The next stage of realism beyond a lattice model is a continuum bead model, where atoms are grouped together into beads, and the beads are allowed to move continuously in space. Typically, one bead represents a single amino acid in a protein. All-atom representations of biomolecules constitute a further level of sophistication. In this final section of our review we investigate the potential energy surfaces of a continuum bead heteropolymer and a tetrapeptide modeled by an all atom potential. Disconnectivity graphs immediately provide a visual distinction between "frustrated" and "unfrustrated" systems in both cases.

A. A Model Polypeptide

In this section we present disconnectivity graphs for an off-lattice model protein. The original model exhibits frustration [330] and does not fold efficiently to the global potential energy minimum. However, when the frustration is removed by constructing the corresponding "Gō-like" model, the resulting surface resembles a funnel [158].

1. The Model Potential

The model heteropolymer that we have considered was introduced by Honeycutt and Thirumalai to illustrate their proposed "metastability hypothesis" that a polypeptide may adopt a variety of metastable folded conformations with similar structural characteristics but different energies [331,332]. The model has $N = 46$ beads linked by stiff bonds, and there are three types of bead: hydrophobic (B), hydrophilic (L), and neutral (N). The sequence is

$$B_9N_3(LB)_4N_3B_9N_3(LB)_5L.$$

The potential energy is given by [332]

$$V = \frac{1}{2}K_r \sum_{i=1}^{N-1}(r_{i,i+1} - r_e)^2 + \frac{1}{2}K_\theta \sum_{i}^{N-2}(\theta_i - \theta_e)^2$$

$$+ \epsilon \sum_{i}^{N-3}[A_i(1 + \cos\varphi_i) + B_i(1 + \cos 3\varphi_i)]$$

$$+ 4\epsilon \sum_{i=1}^{N-2}\sum_{j=i+2}^{N} C_{ij}\left[\left(\frac{\sigma}{r_{ij}}\right)^{12} - D_{ij}\left(\frac{\sigma}{r_{ij}}\right)^6\right], \qquad (1.60)$$

where r_{ij} is the distance between beads i and j. The units of distance and energy are σ and ϵ. The first term represents the bonds linking successive beads with $K_r = 231.2\epsilon\sigma^{-2}$ [333], while the second term is a sum over the bond angles, θ_i, defined by the triplets of atomic positions $\mathbf{r}_i$ to $\mathbf{r}_{i+2}$, with $K_\theta = 20\epsilon\,\mathrm{rad}^{-2}$ and $\theta_e = 105°$. The third term is a sum over the dihedral angles, φ_i, defined by the quartets $\mathbf{r}_i$ to $\mathbf{r}_{i+3}$. If the quartet involves no more than one N monomer then $A_i = B_i = 1.2$, generating a preference for the *trans* conformation ($\varphi_i = 180°$), whereas if two or three N monomers are involved then $A_i = 0$ and $B_i = 0.2$. This choice makes the three neutral segments of the chain flexible and likely to accommodate turns. The last term in Eq. (1.60) represents the nonbonded interactions, and σ is set equal to r_e.

The coefficients for different combinations of monomer types are

$$
\begin{array}{llll}
i, j \in B & C_{ij} = 1 & D_{ij} = 1 \\
i \in L, j \in L, B & C_{ij} = \frac{2}{3} & D_{ij} = -1 \\
i \in N, j \in L, B, N & C_{ij} = 1 & D_{ij} = 0
\end{array}
$$

with $C_{ij} = C_{ji}$ and $D_{ij} = D_{ji}$. Hence, hydrophobic monomers experience a mutual Van der Waals attraction, and all other combinations are purely repulsive, with interactions involving a hydrophilic monomer being of longer range.

The global minimum of this BLN model is the four-stranded β-barrel [332] shown in Figure 1.38. The hydrophobic beads prefer the core environment, while the neutral beads support the turns. Berry and co-workers have previously investigated the self-assembly of this system from separated strands and have performed a principal coordinate analysis [334,335].

2. *Exploring the Landscape*

Conjugate-gradient minimization was used to calculate the pathways following small displacements from each transition state. Every CG optimization was followed by eigenvector-following minimization to ensure

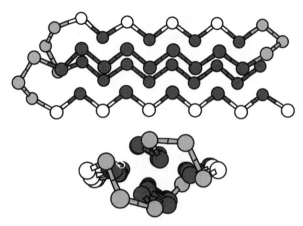

Figure 1.38. Side and end views of the global minimum of the BLN model. Hydrophobic, hydrophilic, and neutral beads are shaded dark gray, white, and light gray, respectively.

that true minima were obtained. The sampling scheme of Section III.B was employed with $n_{ev} = 20$ to explore the low-energy part of the PES. The search was terminated when 500 minima and 636 transition states had been found.

The resulting disconnectivity graph for the low-energy part of the BLN model landscape is shown in Figure 1.39. The PES is clearly not a single funnel, but instead exhibits some hierarchical structure with repeated splitting at successive nodes and long descending branches. A number of low-energy minima are present with different arrangements of the four strands found in the global minimum (Fig. 1.38), and they are separated by high effective barriers. The shortest paths between these minima generally involve at least ten separate rearrangements, and there is little thermodynamic driving force toward the global minimum.

Berry et al. [333] have previously presented monotonic sequences leading to the global minimum of the BLN model. They observed that the barriers are relatively low compared to the energy gradient defined by the connected minima, and categorized the BLN model as a "structure seeker." However, only 67 of our 500 minima lie on monotonic sequences leading to the global minimum, while the other β-barrel structures find themselves at the bottom of different monotonic sequences containing comparable or even larger numbers of minima. Hence, this system "seeks" only a general β-barrel structure, and interconversion of the low-energy minima is predicted to be slow. We therefore expect to see a separation of time scales, one for relaxation to a β-barrel minimum, and another for relaxation to the global minimum.

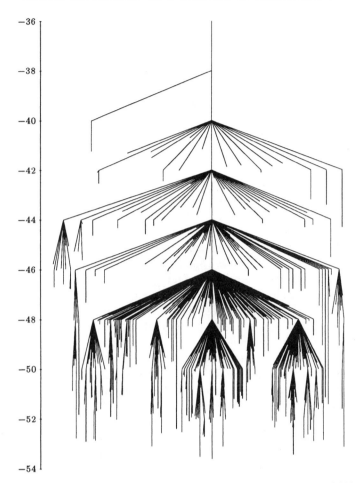

Figure 1.39.   Disconnectivity graph for the BLN model, based on a sample of 500 minima and 636 transition states. The energy is in units of ϵ.

Guo and Brooks [336] applied MD simulation to the BLN model and identified a collapse transition to compact states with an associated latent heat. They were able to discriminate between collapse and folding to the global minimum using suitable order parameters, and concluded that collapse occurred before any appreciable native structure was attained.

Nymeyer et al. [337] concluded that the BLN model exhibits a "rough" free energy landscape from its thermodynamic and dynamic behavior. They compared these results with those for a modified potential in which frustration is largely eliminated, and we examine this system in the following section.

Our conclusion that the original model is not a good folder is in good agreement with the above simulations, and with annealing studies conducted by Berry et al. [333].

3. Frustration

Following Nymeyer et al. [337], we removed the effects of frustration by leaving out all the attractive interactions between pairs of monomers that are not in contact in the native state (global minimum). This transformation was achieved by setting $D_{ij} = 0$ in Eq. (1.60) for nonbonded pairs of hydrophobic monomers that are separated by more than 1.167σ in the global minimum. The remaining attractive forces are more specific. The modified potential is referred to as a Gō-model, following Gō and co-workers [338], who constructed model potentials by considering the native contacts in this way.

Five hundred low-lying minima and 805 transition states were located for this Gō-model using the same sampling scheme as above. The resulting disconnectivity graph is shown in Figure 1.40; it is clearly much more funnel-like, with no low-energy minima separated from the global minimum by large barriers. The energy range spanned by the 500 minima is larger than for the original model because the Gō-model has a significantly lower number of minima per unit energy. The highest-energy minima in our sample for the original model are still quite compact, while those for the Gō-model exhibit considerable unfolding of the β-barrel.

Plots of potential energy versus shortest integrated path length to the global minimum are shown in Figure 1.42, and reveal far less correlation for the unmodified potential. The number of rearrangements on the shortest path to the global minimum is shown for both models in Figure 1.41, and exhibits a broader distribution for the original model with a higher average number of steps. Both these features confirm the reorganization of the Gō-model PES into a funnel-like structure. The average properties of pathways summarized in Table III provide further evidence. Uphill barriers are generally higher and downhill barriers lower for the Gō-model, revealing a larger downhill energy gradient. Not surprisingly, Nymeyer et al. [337] observed cooperative collapse and folding to the global minimum in their simulations with the Gō-model [337], accompanied by a single narrow peak in the heat capacity.

The disconnectivity graphs provide a rather clear way to distinguish the original BLN model and the Gō-model. Frustration in the original model is signaled by the hierarchical structure with a number of low-energy minima each lying at the bottom of its own funnel. When the frustration is removed in the Gō-model the graph resembles the "palm tree" form discussed in Section II.C. Actually, the PES for the Gō-model is still far from an ideal

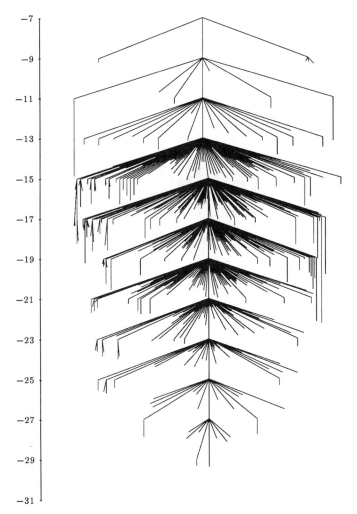

Figure 1.40. Disconnectivity graph for the BLN Gō-model, based on a sample of 500 minima and 805 transition states. The energy is in units of ϵ.

funnel. Only 124 of the 500 minima lie in the primary MSB, so that relaxation from a high-energy minimum is likely to require several uphill steps. Nevertheless, the global minimum will be located much more efficiently than in the original model.

B. A Tetrapeptide

Here we present disconnectivity graphs for an all-atom model of a short polypeptide, iso-butyryl-(ala)$_3$-NH-methyl (IAN). This particular peptide

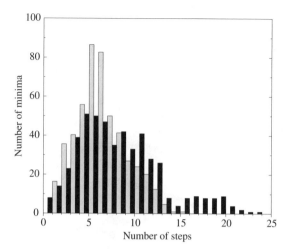

Figure 1.41. Distribution of the number of rearrangement along the shortest path from a given minimum to the global minimum for the BLN model (black) and the Gō-like model (gray).

has been studied before [59,61] using a different potential. Two slightly differing force fields are used, leading to two rather different graphs, one funnel-like and one more complicated.

1. The Potential

The force field used was the AMBER [339] potential of Cornell et al. The potential energy is given by

$$V = \sum_{\text{bonds}} k_r(r - r_0)^2 + \sum_{\text{angles}} k_\theta(\theta - \theta_0) + \sum_{\text{dihedrals}} k_\phi(1 + \cos(n\phi + \delta))$$
$$+ \sum_{i<j} \left[\frac{A_{ij}}{r_{ij}^{12}} - \frac{B_{ij}}{r_{ij}^6} + \frac{q_i q_j}{\epsilon_{ij} r_{ij}} \right] \tag{1.61}$$

where r, r_0, θ, and θ_0 are the bond length, equilibrium bond length, bond angle, and equilibrium bond angle, respectively. Bond lengths and angles are held near to their equilibrium values by a simple harmonic potential. There is an explicit dihedral angle term, taking the form of a truncated Fourier series. For the majority of dihedral angles this series consists of a single term as in Eq. (1.61), with periodicity n, and phase shifted by δ. ϕ is the actual dihedral angle. For the main backbone dihedrals there are up to three such terms each with different n and δ. This term also covers "improper dihedrals," which are groups of three atoms bonded to a central atom, such

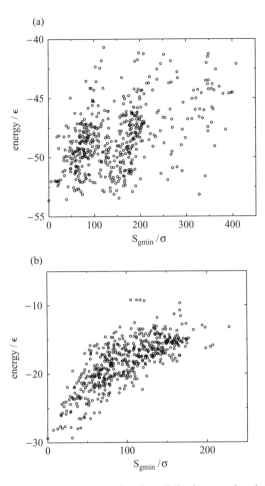

Figure 1.42. Energy of minima as a function of the integrated path length along the shortest path to the global minimum. Upper panel: the BLN model; lower panel: the Gō-like model.

as a carbon sp^2 center. In this case the angle is defined to produce an energetic penalty for the system to move away from planarity. The last term in Eq. (1.61) covers the nonbonded interactions for two atoms i and j separated by at least three bonds and a distance r_{ij} apart. The dispersion energy is represented by a 6–12 potential, with coefficients A_{ij} and B_{ij} that depend on the atom types i and j. The electrostatic energy is modeled by a Coulombic interaction of atom-centered point charges q_i and q_j with dielectric constant ϵ_{ij}. The difference between the potentials used to generate

DAVID J. WALES ET AL.

TABLE III
Properties of Individual Pathways for the BLN and Gō Models

Model	BLN	Gō-like
$\langle b^{\text{up}} \rangle_{\text{p}}$	2.59	3.07
$\langle b^{\text{down}} \rangle_{\text{p}}$	0.862	0.635
$\langle \Delta E^{\text{con}} \rangle_{\text{p}}$	1.73	2.43

b_i^{up} is the uphill barrier height between the two minima connected by transition state i, and b_i^{down} is the downhill barrier. $\Delta E_i^{\text{con}} = b_i^{\text{up}} - b_i^{\text{down}}$ is the energy difference between the two minima. The angle brackets indicate averaging over the sample of pathways. The units of energy are ϵ.

our two graphs lies in ϵ_{ij}. For the first graph we used a constant dielectric $\epsilon_{ij} = 1$ and for the second graph we used a distance-dependent dielectric $\epsilon_{ij} = r_{ij}$. The latter choice introduces dielectric screening and is commonly used as an implicit representation of solvent [340]. It therefore provides an approximate description of the aqueous peptide for comparison with the behavior in vacuo.

2. Results

The graph in Figure 1.43 for $\epsilon_{ij} = 1$ shows an almost perfect funnel-like landscape, although at the bottom there are three distinct branches. A previous study [59] on this molecule led to a much simpler graph (139 minima, as compared to 657 here) as the model employed was not all-atom, and only considered the backbone torsion angles as degrees of freedom. We have included all degrees of freedom. Nevertheless, Becker and Karplus also concluded that the landscape possesses funneling properties, with small features outside of the main funnel. We would expect this system to be a good "folder" as there are no significant kinetic traps.

In contrast, the graph in Figure 1.44 for $\epsilon_{ij} = r_{ij}$ is much more complicated. Four main branches can be identified, all of which lead to relatively low-lying minima. These branches all represent kinetic traps, and are separated from the global minimum by energy barriers of the order of 20 kcal mol^{-1}. We expect this system to exhibit frustration (Section IX.A). The other major difference between this graph and the first is that it contains 2548 minima, roughly four times as many as for $\epsilon_{ij} = 1$. We believe that in both cases we have found the vast majority of the low-lying minima on the surface. From Section V.B we expect the number of minima to increase for $\epsilon_{ij} = r_{ij}$ because the screened Coulomb term in the potential energy is shorter ranged.

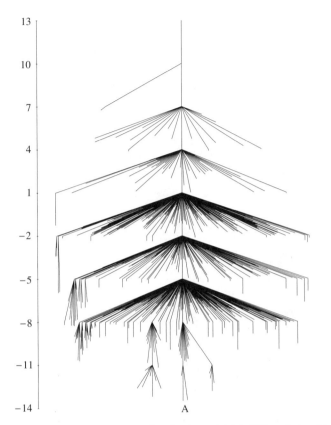

Figure 1.43. Disconnectivity graph for iso-butyryl-(ala)$_3$-NH-methyl with $\epsilon_{ij} = 1$. Energies are in kcal mol^{-1}. The global minimum is at $E = -13.146$ kcal mol^{-1} and the sample contains 657 minima and 2391 transition states.

Figure 1.45 shows the conformations labeled A through E on Figure 1.43 and Figure 1.44. The two global minima (A and C) both have single turns of a helix, though *not* an α-helix. The other three structures (B, D, and E) represent the kinetic traps from Figure 1.44. B and D each have one *cis* peptide bond, and E has two. A *cis* peptide bond is one where the carbonyl oxygen and the amide hydrogen are on the same side of the bond, forcing the chain back on itself. This arrangement is energetically unfavorable with respect to the *trans* conformation, and so these structures are several kcal mol^{-1} higher in energy than C (structures B and D are about 4 kcal mol^{-1} higher, E is about 8 kcal mol^{-1} higher).

Our results for IAN in vacuo and with implicit solvent are clearly rather different. For a native biomolecule we would expect the PES to have

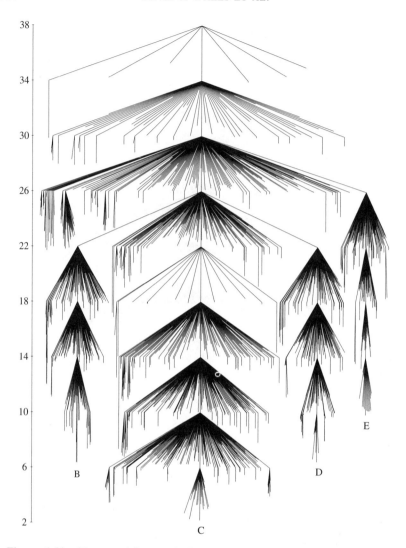

Figure 1.44. Disconnectivity graph for iso-butyryl-(ala)$_3$-NH-methyl with $\epsilon_{ij} = r_{ij}$. Energies are in kcal mol^{-1}. The global minimum is at $E = 2.195$ kcal mol^{-1} and the sample contains 2548 minima and 7311 transition states.

funneling characteristics when solvent is present, assuming that the active state corresponds to the global free energy minimum under ambient conditions. In contrast, our results for IAN suggest that the global minimum, A (Fig. 1.45), should be readily located in vacuo, but in solution the system should be frustrated. Of course, this is a rather small peptide, and it may not

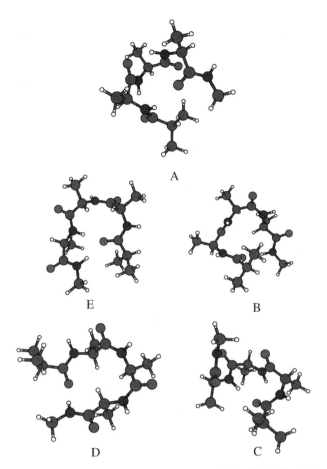

Figure 1.45. Iso-butyryl-(ala)$_3$-NH-methyl minima A, B, C, D, E. A is the global minimum for $\epsilon_{ij} = 1$. C for $\epsilon_{ij} = r_{ij}$. B, D, and E are low-lying kinetic traps for the latter potential.

be representative of larger biomolecules. Alternatively, the empirical potential, especially the distance-dependent dielectric term, may simply be inadequate. Using explicit solvent would increase the number of degrees of freedom significantly, but may be necessary in future work.

Acknowledgments

DJW is grateful to the Royal Society and the EPSRC for financial support. JPKD is grateful to Emmanuel College, Cambridge, for the award of the Sir Alan Wilson Research Fellowship.

REFERENCES

1. M. Born and J. R. Oppenheimer, *Ann. Physik* **84**, 457 (1927).

2. J. N. Murrell, S. Carter, S. C. Farantos, P. Huxley, and A. J. C. Varandas, *Molecular Potential Energy Functions* (Wiley, Chichester, 1984).

3. P. G. Mezey, *Potential Energy Hypersurfaces* (Elsevier, Amsterdam, 1987).

4. D. J. Wales and J. P. K. Doye, in *Large Clusters of Atoms and Molecules*, edited by T. P. Martin, pp. 241–280 (Kluwer, Dordrecht, 1996).

5. C. Levinthal, in *Mössbauer Spectroscopy in Biological Systems, Proceedings of a Meeting Held at Allerton House, Monticello, Illinois*, edited by P. DeBrunner, J. Tsibris and E. Munck, p. 22 (University of Illinois Press, Urbana 1969).

6. C. B. Anfinsen, in *New Perspectives in Biology*, edited by M. Sela, pp. 42–50 (Elsevier, Amsterdam, 1964).

7. C. Levinthal, *J. Chim. Phys.* **65**, 44 (1968).

8. C. J. Camacho and D. Thirumalai, *Phys. Rev. Lett.* **71**, 2505 (1993).

9. M. R. Ejtehadi, N. Hamedani, and V. Shahrezaei, *Phys. Rev. Lett.* **82**, 4723 (1999).

10. A. Šali, E. Shakhnovich, and M. Karplus, *Nature* **369**, 248 (1994).

11. N. D. Socci and J. N. Onuchic, *J. Chem. Phys.* **101**, 1519 (1994).

12. T. R. Sosnick, L. Mayne, and S. W. Englander, *Proteins: Struct., Func. and Gen.* **24**, 413 (1996).

13. J. D. Bryngelson, J. N. Onuchic, N. D. Socci, and P. G. Wolynes, *Proteins: Struct., Func. and Gen.* **21**, 167 (1995).

14. R. Zwanzig, A. Szabo, and B. Bagchi, *Proc. Natl. Acad. Sci. USA* **89**, 20 (1992).

15. R. Zwanzig, *Proc. Natl. Acad. Sci. USA* **92**, 9801 (1995).

16. J. P. K. Doye and D. J. Wales, *J. Chem. Phys.* **105**, 8428 (1996).

17. P. E. Leopold, M. Montal, and J. N. Onuchic, *Proc. Natl. Acad. Sci. USA* **89**, 8271 (1992).

18. H. Frauenfelder, S. G. Sligar, and P. G. Wolynes, *Science* **254**, 1598 (1991).

19. P. G. Wolynes, J. N. Onuchic, and D. Thirumalai, *Science* **267**, 1619 (1995).

20. N. D. Socci, J. N. Onuchic, and P. G. Wolynes, *Proteins: Struct., Func. and Gen.* **32**, 136 (1998).

21. T. Lazaridis and M. Karplus, *Science* **278**, 1928 (1997).

22. N. D. Socci, J. N. Onuchic, and P. G. Wolynes, *J. Chem. Phys.* **104**, 5860 (1996).

23. A. Šali, E. Shakhnovich, and M. Karplus, *J. Mol. Biol.* **235**, 1614 (1994).

24. H. S. Chan, *Nature* **373**, 664 (1995).

25. A. Gutin, A. Sali, V. Abkevich, M. Karplus, and E. I. Shakhnovich, *J. Chem. Phys.* **108**, 6466 (1998).

26. W. H. Zachariasen, *J. Am. Chem. Soc.* **54**, 3841 (1932).

27. K. D. Ball, R. S. Berry, R. E. Kunz, F.-Y. Li, A. Proykova, and D. J. Wales, *Science* **271**, 963 (1996).

28. M. Goldstein, *J. Chem. Phys.* **51**, 3728 (1969).

29. G. P. Johari and M. Goldstein, *J. Chem. Phys.* **53**, 2372 (1970).

30. F. H. Stillinger and T. A. Weber, *Phys. Rev. A* **25**, 978 (1982).

31. F. H. Stillinger and T. A. Weber, *Phys. Rev. A* **28**, 2408 (1983).

32. F. H. Stillinger and T. A. Weber, *Science* **225**, 983 (1984).

33. C. A. Angell, *J. Non-Cryst. Solids* **131–133**, 13 (1991).
34. G. W. Scherer, *J. Am. Ceram. Soc.* **75**, 1060 (1992).
35. G. Williams and D. C. Watts, *J. Chem. Soc.*, Faraday Trans. **66**, 80 (1970).
36. R. Böhmer, K. L. Ngai, C. A. Angell, and D. J. Plazek, *J. Chem. Phys.* **99**, 4201 (1993).
37. F. H. Stillinger, *Science* **267**, 1935 (1995).
38. C. A. Angell, *Science* **267**, 1924 (1995).
39. S. Sastry, P. G. Debenedetti, and F. H. Stillinger, *Nature* **393**, 554 (1998).
40. B. Doliwa and A. Heuer, *Phys. Rev. Lett.* **80**, 4915 (1998).
41. D. J. Lacks, *Phys. Rev. Lett.* **80**, 5385 (1998).
42. D. L. Malandro and D. J. Lacks, *J. Chem. Phys.* **107**, 5804 (1997).
43. A. Heuer, *Phys. Rev. Lett.* **78**, 4051 (1997).
44. L. Angelani, G. Parisi, G. Ruocco, and G. Viliani, *Phys. Rev. Lett.* **81**, 4648 (1998).
45. G. Daldoss, O. Pilla, G. Viliani, C. Brangian, and G. Ruocco, *Phys. Rev. B* **60**, 3200 (1999).
46. M. Morse, *Amer. Math. Soc. Colloq. Publ.* **18** (1934).
47. M. Morse and S. S. Cairns, *Critical Point Theory in Global Analysis and Differential Topology* (Academic Press, New York, 1969).
48. P. G. Mezey, *Chem. Phys. Lett.* **82**, 100 (1981).
49. F. H Stillinger, *Phys. Rev. E* **59**, 48 (1999).
50. F. G. Amar and R. S. Berry, *J. Chem. Phys.* **85**, 5943 (1986).
51. J. N. Murrell and K. J. Laidler, *J. Chem. Soc.*, Faraday Trans. **64**, 371 (1968).
52. R. E. Leone and P. v. R. Schleyer, *Angew. Chem. Int. Ed. Engl.* **9**, 860 (1970).
53. J. W. McIver and R. E. Stanton, *J. Am. Chem. Soc.* **94**, 8618 (1972).
54. R. E. Stanton and J. W. McIver, *J. Am. Chem. Soc.* **97**, 3632 (1975).
55. P. Pechukas, *J. Chem. Phys.* **64**, 1516 (1976).
56. J. G. Nourse, *J. Am. Chem. Soc.* **102**, 4883 (1980).
57. R. S. Berry and R. Breitengraser-Kunz, *Phys. Rev. Lett.* **74**, 3951 (1995).
58. R. E. Kunz and R. S. Berry, *J. Chem. Phys.* **103**, 1904 (1995).
59. O. M. Becker and M. Karplus, *J. Chem. Phys.* **106**, 1495 (1997).
60. S. Saito, M. Matsumoto, and I. Ohmine, *Adv. Class. Traj. Meth.* **4**, 105 (1999).
61. R. Czerminski and R. Elber, *J. Chem. Phys.* **92**, 5580 (1990).
62. Y. Levy and O. M. Becker, *Phys. Rev. Lett.* **81**, 1126 (1998).
63. K. M. Westerberg and C. A. Floudas, *J. Chem. Phys.* **110**, 9259 (1999).
64. P. Garstecki, T. X. Hoang, and M. Cieplak, *Phys. Rev. E* **60**, 3219 (1999).
65. P. Sibani, J. C. Schön, P. Salamon, and J. Andersson, *Europhys. Lett.* **22**, 479 (1993).
66. J. C. Schön, *Ber. Bunsenges. Phys. Chem.* **100**, 1388 (1996).
67. J. C. Schön, H. Putz, and M. Jansen, *J. Phys. Condensed Matter.* **8**, 143 (1996).
68. D. J. Wales, M. A. Miller, and T. R. Walsh, *Nature* **394**, 758 (1998).
69. H. Frauenfelder and D. T. Leeson, *Nature Structural Biology* **5**, 757 (1998).
70. R. G. Palmer, D. L. Stein, E. Abrahams, and P. W. Anderson, *Phys. Rev. Lett.* **53**, 958 (1984).
71. B. A. Huberman and M. Kerszberg, *J. Phys. A: Math. Gen.* **18**, L331 (1985).

72. D. J. Wales, *J. Chem. Phys.* **101**, 3750 (1994).

73. G. M. Crippen and H. A. Scheraga, *Archives of Biochemistry and Biophysics* **144**, 462 (1971).

74. J. Pancíř, *Coll, Czech. Chem. Comm.* **40**, 1112 (1974).

75. R. L. Hilderbrandt, *Computers and Chemistry* **1**, 179 (1977).

76. C. J. Cerjan and W. H. Miller, *J. Chem. Phys.* **75**, 2800 (1981).

77. J. Simons, P. Jørgenson, H. Taylor, and J. Ozment, *J. Phys. Chem.* **87**, 2745 (1983).

78. A. Banerjee, N. Adams, J. Simons, and R. Shepard, *J. Phys. Chem.* **89**, 52 (1985).

79. J. Baker, *J. Comp. Chem.* **7**, 385 (1986).

80. D. J. Wales and J. Uppenbrink, *Phys. Rev. B* **50**, 12342 (1994).

81. L. J. Munro and D. J. Wales, Faraday Discuss. **106**, 409 (1997).

82. L. J. Munro and D. J. Wales, *Phys. Rev. B* **59**, 3969 (1999).

83. M. Page and J. W. McIver, *J. Chem. Phys.* **88**, 922 (1988).

84. W. H. Press, S. A. Teukolsky, W. T. Vetterling, and B. P. Flannery, *Numerical Recipes in FORTRAN* (Cambridge University Press, Cambridge, 2nd edn., 1992).

85. B. Friedrich, Z. Herman, R. Zahradnik, and Z. Havlas, *Adv. Quant. Chem.* **19**, 257 (1988).

86. A. Banerjee and N. P. Adams, *Int. J. Quant. Chem.* **43**, 855 (1992).

87. J. W. McIver and A. Komornicki, *J. Am. Chem. Soc.* **94**, 2625 (1972).

88. D. O'Neal, H. Taylor, and J. Simons, *J. Phys. Chem.* **88**, 1510 (1984).

89. L. R. Pratt, *J. Chem. Phys.* **85**, 5045 (1986).

90. R. Elber and M. Karplus, *Chem. Phys. Lett.* **139**, 375 (1987).

91. J. Baker, *J. Comp. Chem.* **8**, 563 (1987).

92. R. S. Berry, H. L. Davis, and T. L. Beck, *Chem. Phys. Lett.* **147**, 13 (1988).

93. J. Nichols, H. Taylor, P. Schmidt, and J. Simons, *J. Chem. Phys.* **92**, 340 (1990).

94. M. C. Smith, *Int. J. Quant. Chem.* **37**, 773 (1990).

95. J. Baker and W. H. Hehre, *J. Comp. Chem.* **12**, 606 (1991).

96. J. Baker, *J. Comp. Chem.* **13**, 240 (1992).

97. I. V. Ionova and E. A. Carter, *J. Chem. Phys.* **98**, 6377 (1993).

98. C. J. Tsai and K. D. Jordan, *J. Phys. Chem.* **97**, 11227 (1993).

99. A. Matro, D. L. Freeman, and J. D. Doll, *J. Chem. Phys.* **101**, 10458 (1994).

100. S. F. Chekmarev, *Chem. Phys. Lett.* **227**, 354 (1994).

101. O. S. Smart, *Chem. Phys. Lett.* **222**, 503 (1994).

102. J.-Q. Sun and K. Ruedenberg, *J. Chem. Phys.* **101**, 2157 (1994).

103. J.-Q. Sun and K. Ruedenberg, *J. Chem. Phys.* **101**, 2168 (1994).

104. J. M. Bofill and M. Comajuan, *J. Comp. Chem.* **16**, 1326 (1995).

105. I. V. Ionova and E. A. Cater, *J. Chem. Phys.* **103**, 5437 (1995).

106. F. Jensen, *J. Chem. Phys.* **102**, 6706 (1995).

107. C. Peng, P. Y. Ayala, H. B. Schlegel, and M. J. Frisch, *J. Comp. Chem.* **17**, 49 (1996).

108. J. Baker and F. Chan, *J. Comp. Chem.* **17**, 888 (1996).

109. Q. Zhao and J. B. Nicholas, *J. Chem. Phys.* **104**, 767 (1996).

110. P. Y. Ayala and H. B. Schlegel, *J. Chem. Phys.* **107**, 375 (1997).

111. A. Ulitsky and D. Shalloway, *J. Chem. Phys.* **106**, 10099 (1997).

112. W. Quapp, M. Hirsch, O. Imig, and D. Heidrich, *J. Comp. Chem.* **19**, 1087 (1998).

113. H. Goto, *Chem. Phys. Lett.* **292**, 254 (1998).

114. D. J. Wales, *J. Am. Chem. Soc.* **115**, 11180 (1993).

115. D. J. Wales and T. R. Walsh, *J. Chem. Phys.* **105**, 6957 (1996).

116. T. R. Walsh and D. J. Wales, *J. Chem. Soc.*, Faraday Trans. **92**, 2505 (1996).

117. D. J. Wales and T. R. Walsh, *J. Chem. Phys.* **106**, 7193 (1997).

118. D. J. Wales, *J. Chem. Phys.* **110**, 10403 (1999).

119. D. J. Wales, in *Theory of Atomic and Molecular Clusters*, edited by J. Jellinek, pp. 86–110 (Springer-Verlag, Heidelberg, 1999).

120. B. Paizs, G. Fogarasi, and P. Pulay, *J. Chem. Phys.* **109**, 6571 (1998).

121. O. Farkas and H. B. Schlegal, *J. Chem. Phys.* **109**, 7100 (1998).

122. N. Mousseau and G. T. Barkema, *Phys. Rev. E* **57**, 2419 (1998).

123. J. Baker, D. Kinghorn, and P. Pulay, *J. Chem. Phys.* **110**, 4986 (1999).

124. E. B. Wilson, J. C. Decius, and P. C. Cross, *Molecular Vibrations* (Dover, New York, 1980).

125. D. J. Wales, *J. Chem. Soc.*, Faraday Trans. **89**, 1305 (1993).

126. F. S. Acton, *Numerical Methods that Work*, p. 204 (Washington, Mathematical Association of America, 1990).

127. F. B. Hildebrand, *Methods of Applied Mathematics* (Dover, New York, 1992).

128. A. F. Voter, *J. Chem. Phys.* **106**, 4665 (1997).

129. G. Henkelman and H. Jónsson, *J. Chem. Phys.* **111**, 7010 (1999).

130. J. P. K. Doye and D. J. Wales, *Z. Phys. D* **40**, 194 (1997).

131. G. T. Barkema and N. Mousseau, *Phys. Rev. Lett.* **77**, 4358 (1996).

132. M. R. Hoare, *Adv. Chem. Phys.* **40**, 49 (1979).

133. D. J. McGinty, *J. Chem. Phys.* **55**, 580 (1971).

134. J. J. Burton, *J. Chem. Phys.* **56**, 3133 (1972).

135. G. Franke, E. R. Hilf, and P. Borrmann, *J. Chem. Phys.* **98**, 3496 (1993).

136. D. J. Wales, *Mol. Phys.* **78**, 151 (1993).

137. J. P. K. Doye and D. J. Wales, *J. Chem. Phys.* **102**, 9659 (1995).

138. G. M. Torrie and J. P. Valleau, *J. Comp. Phys.* **23**, 187 (1977).

139. D. D. Frantz, D. L. Freeman, and J. D. Doll, *J. Chem. Phys.* **93**, 2769 (1990).

140. K. D. Ball and R. S. Berry, *J. Chem. Phys.* **111**, 2060 (1999).

141. A. M. Ferrenberg and R. H. Swendsen, *Phys. Rev. Lett.* **61**, 2635 (1988).

142. I. R. McDonald and K. Singer, Discuss. Faraday Soc. **43**, 40 (1967).

143. A. M. Ferrenberg and R. H. Swendsen, *Phys. Rev. Lett.* **63**, 1195 (1989).

144. F. Sciortino, W. Kob, and P. Tartaglia, *Phys. Rev. Lett.* **83**, 3214 (1999).

145. S. Büchner and A. Heuer, *Phys. Rev. E* **60**, 6507 (1999).

146. W. Forst, *Theory of Unimolecular Reactions* (Academic Press, New York, 1973).

147. S. F. Chekmarev and I. H. Umirzakov, *Z. Phys. D* **26**, 373 (1993).

148. K. D. Ball and R. S. Berry, *J. Chem. Phys.* **109**, 8557 (1998).

149. M. A. Miller, J. P. K. Doye, and D. J. Wales, *Phys. Rev. E* **60**, 3701 (1999).

150. P. C. Haarhoff, *Mol. Phys.* **7**, 101 (1963).

151. K. D. Ball and R. S. Berry, *J. Chem. Phys.* **109**, 8541 (1998).
152. J. P. K. Doye and D. J. Wales, *J. Chem. Phys.* **102**, 9673 (1995).
153. N. G. van Kampen, *Stochastic Processes in Physics and Chemistry* (North-Holland, Amsterdam, 1981).
154. R. E. Kunz, *Dynamics of First-Order Phase Transitions* (Deutsch, Thun, 1995).
155. B. Kunz, R. S. Berry, and T. Astakhova, *Surf. Rev. Lett.* **3**, 307 (1996).
156. R. E. Kunz, P. Blaudeck, K. H. Hoffmann, and R. S. Berry, *J. Chem. Phys.* **108**, 2576 (1998).
157. T. R. Walsh and D. J. Wales, *J. Chem. Phys.* **109**, 6691 (1998).
158. M. A. Miller and D. J. Wales, *J. Chem. Phys.* **111**, 6610 (1999).
159. M. A. Miller, J. P. K. Doye, and D. J. Wales, *J. Chem. Phys.* **110**, 328 (1999).
160. J. P. K. Doye and D. J. Wales, *Phys. Rev. B* **59**, 2292 (1999).
161. M. Cieplak, M. Henkel, J. Karbowski, and J. R. Banavar, *Phys. Rev. Lett.* **80**, 3654 (1998).
162. J. Wang, J. N. Onuchic, and P. G. Wolynes, *Phys. Rev. Lett.* **76**, 4861 (1996).
163. W. Götze, in *Liquids, Freezing and the Glass Transition*, edited by J.-P. Hansen, D. Levesque, and J. Zinn-Justin, pp. 287–499 (North-Holland, Amsterdam, 1991).
164. W. Götze, *J. Phys.*: Condens. Matter **11**, A1 (1999).
165. W. Kob, *ACS Symp. Ser.* **676**, 28 (1997).
166. R. G. Gilbert and S. C. Smith, *Theory of Unimolecular and Recombination Reactions* (Blackwell, Oxford, 1990).
167. C. Leon and K. L. Ngai, *J. Phys. Chem. B* **103**, 4045 (1999).
168. J. Sabelko, J. Ervin, and M. Gruebele, *Proc. Natl. Acad. Sci. USA* **96**, 6031 (1999).
169. R. Zwanzig, *J. Chem. Phys.* **103**, 9397 (1995).
170. J. G. Saven, J. Wang, and P. G. Wolynes, *J. Chem. Phys.* **101**, 11037 (1994).
171. D. J. Watts and S. H. Strogatz, *Nature* **393**, 440 (1998).
172. G. Diezemann, H. Sillescu, G. Hinze, and R. Bohmer, *Phys. Rev. E* **57**, 4398 (1998).
173. G. Diezemann and K. Nelson, *J. Phys. Chem. B* **103**, 4089 (1999).
174. S. A. Brawer, *J. Chem. Phys.* **81**, 954 (1984).
175. J. C. Dyre, *Phys. Rev. Lett.* **58**, 792 (1987).
176. V. I. Arkhipov and H. Bässler, *J. Phys. Chem.* **98**, 662 (1994).
177. J. C. Dyre, *Phys. Rev. B* **51**, 12276 (1995).
178. G. Adam and J. H. Gibbs, *J. Chem. Phys.* **43**, 139 (1965).
179. J. H. Gibbs and E. A. DiMarzio, *J. Chem. Phys.* **28**, 373 (1958).
180. K. L. Ngai, *J. Phys. Chem. B* **103**, 5895 (1999).
181. D. J. Wales, *Science* **271**, 925 (1996).
182. J. P. K. Doye and D. J. Wales, *Phys. Rev. Lett.* **80**, 1357 (1998).
183. C. B. Anfinsen, *Science* **181**, 223 (1973).
184. D. Baker and D. A. Agard, *Biochemistry* **33**, 7505 (1994).
185. S. Kirkpatrick, C. D. Gelatt, and M. P. Vecchi, *Science* **220**, 671 (1983).
186. L. Piela, J. Kostrowicki, and H. A. Scheraga, *J. Phys. Chem.* **93**, 3339 (1989).
187. J. Pillardy and L. Piela, *J. Phys. Chem.* **99**, 11805 (1995).
188. D. J. Wales and H. A. Scheraga, *Science* **285**, 1368 (1999).

189. J. P. K. Doye, D. J. Wales, and R. S. Berry, *J. Chem. Phys.* **103**, 4234 (1995).

190. Z. Li and H. A. Scheraga, *Proc. Natl. Acad. Sci. USA* **84**, 6611 (1987).

191. Z. Li and H. A. Scheraga, *J. Mol. Struct.* **179**, 333 (1988).

192. D. J. Wales and J. P. K. Doye, *J. Phys. Chem. A* **101**, 5111 (1997).

193. A simple basin-hopping fortran program may be downloaded from URL http://brian.ch.cam.ac.uk/software.html.

194. R. P. White and H. R. Mayne, *Chem. Phys. Lett.* **289**, 463 (1998).

195. J. P. K. Doye, D. J. Wales, and M. A. Miller, *J. Chem. Phys.* **109**, 8143 (1998).

196. J. E. Jones and A. E. Ingham, *Proc. R. Soc. A* **107**, 636 (1925).

197. L. T. Wille, in *Annual Reviews of Computational Physics VII*, edited by D. Stauffer, p. 25 (World Scientific, Singapore, 2000).

198. D. Romero, C. Barron, and S. Gomez, *Comp. Phys. Comm.* **123**, 87 (1999).

199. D. J. Wales, J. P. K. Doye, A. Dullweber, and F. Y. Naumkin, The Cambridge Cluster Database, URL http://brian.ch.cam.ac.uk/CCD.html.

200. A. L. Mackay, *Acta Cryst.* **15**, 916 (1962).

201. J. Farges, M. F. de Feraudy, B. Raoult, and G. Torchet, *Adv. Chem. Phys.* **70**, 45 (1988).

202. O. Echt, K. Sattler, and E. Recknagel, *Phys. Rev. Lett.* **47**, 1121 (1981).

203. I. A. Harris, R. S. Kidwell, and J. A. Northby, *Phys. Rev. Lett.* **53**, 2390 (1984).

204. I. A. Harris, K. A. Norman, R. V. Mulkern, and J. A. Northby, *Chem. Phys. Lett.* **130**, 316 (1986).

205. O. Echt, O. Kandler, T. Leisner, W. Miehle, and E. Recknagel, *J. Chem. Soc.*, Faraday Trans. **86**, 2411 (1990).

206. S. Kakar, O. Björneholm, J. Weigelt, A. R. B. de Castro, L. Tröger, R. Frahm, T. Möller, A. Knop, and E. Rühl, *Phys. Rev. Lett.* **78**, 1675 (1997).

207. S. Gomez and D. Romero, in *Proceedings of the First European Congress of Mathematics*, vol. III, pp. 503–509 (Birkhauser, Basel, 1994).

208. J. P. K. Doye and D. J. Wales, *Chem. Phys. Lett.* **247**, 339 (1995).

209. L. D. Marks, *Phil. Mag. A* **49**, 81 (1984).

210. R. H. Leary and J. P. K. Doye, *Phys. Rev. E* **60**, R6320 (1999).

211. J. P. K. Doye, M. A. Miller, and D. J. Wales, *J. Chem. Phys.* **111**, 8417 (1999).

212. F. H. Stillinger and D. K. Stillinger, *J. Chem. Phys.* **93**, 6013 (1990).

213. J. A. Northby, *J. Chem. Phys.* **87**, 6166 (1987).

214. M. A. Miller, *Energy Landscapes and Dynamics of Model Clusters*, PhD thesis, (University of Cambridge, Cambridge, UK, 1999).

215. C. L. Cleveland, W. D. Luedtke, and U. Landman, *Phys. Rev. Lett.* **81**, 2036 (1998).

216. C. L. Cleveland, W. D. Luedtke, and U. Landman, *Phys. Rev. B* **60**, 5065 (1999).

217. P. Labastie and R. L. Whetten, *Phys. Rev. Lett.* **65**, 1567 (1990).

218. R. M. Lynden-Bell and D. J. Wales, *J. Chem. Phys.* **101**, 1460 (1994).

219. R. E. Kunz and R. S. Berry, *Phys. Rev. Lett.* **71**, 3987 (1993).

220. R. E. Kunz and R. S. Berry, *Phys. Rev. E* **49**, 1895 (1994).

221. P. J. Steinhardt, D. R. Nelson, and M. Ronchetti, *Phys. Rev. B* **28**, 784 (1983).

222. J. S. van Duijneveldt and D. Frenkel, *J. Chem. Phys.* **96**, 4655 (1992).

223. J. P. K. Doye, M. A. Miller, and D. J. Wales, *J. Chem. Phys.* **110**, 6896 (1999).

224. J. P. K. Doye and D. J. Wales, *J. Chem. Phys.* **111**, 11070 (1999).

225. C. Barrón, S. Gómez, and D. Romero, *Appl. Math. Lett.* **9** 75 (1996).

226. D. M. Deaven, N. Tit, J. R. Morris, and K. M. Ho, *Chem. Phys. Lett.* **256**, 195 (1996).

227. J. A. Niesse and H. R. Mayne, *J. Chem. Phys.* **105**, 4700 (1996).

228. R. H. Leary, *J. Global Optimization* **11**, 35 (1997).

229. M. D. Wolf and U. Landman, *J. Phys. Chem. A* **102**, 6129 (1998).

230. R. P. White, J. A. Niesse, and H. R. Mayne, *J. Chem. Phys.* **108**, 2208 (1998).

231. R. V. Pappu, R. K. Hart, and J. W. Ponder, *J. Phys. Chem. B* **102**, 9725 (1998).

232. K. Michaelian, *Chem. Phys. Lett.* **293**, 202 (1998).

233. F. Schoen and M. Locatelli, Abstract for the *International Workshop on Global Optimization* (1999).

234. R. H. Leary, Abstract for the *International Workshop on Global Optimization* (1999).

235. C. Barrón, S. Gómez, D. Romero, and A. Saavedra, *Appl. Math. Lett.* **12**, 85 (1999).

236. M. H. J. Hagen, E. J. Meijer, G. C. A. M. Mooij, D. Frenkel, and H. N. W. Lekkerkerker, *Nature* **365**, 425 (1993).

237. N. W. Ashcroft, *Nature* **365**, 387 (1993).

238. E. Lomba and N. G. Almarza, *J. Chem. Phys.* **100**, 8367 (1994).

239. J. P. K. Doye and D. J. Wales, *J. Phys. B* **29**, 4859 (1996).

240. J. P. K. Doye and D. J. Wales, *J. Chem. Soc.*, Faraday Trans. **93**, 4233 (1997).

241. P. M. Morse, *Phys. Rev.* **34**, 57 (1929).

242. L. A. Girifalco and V. G. Weizer, *Phys. Rev.* **114**, 687 (1959).

243. L. A. Girifalco, *J. Phys. Chem.* **96**, 858 (1992).

244. M. R. Hoare and J. McInnes, Faraday Discuss. **61**, 12 (1976).

245. J. P. Rose and R. S. Berry, *J. Chem. Phys.* **98**, 3262 (1993).

246. W. Krätschmer, L. D. Lamb, K. Fostiropoulos, and D. R. Huffman, *Nature* **347**, 354 (1990).

247. G. Meijer and D. S. Bethune, *J. Chem. Phys.* **93**, 7800 (1990).

248. H. W. Kroto, J. R. Heath, S. C. O'Brien, R. F. Curl, and R. E. Smalley, *Nature* **318**, 162 (1985).

249. J. R. Heath, S. C. O'Brien, Q. Zhang, Y. Liu, R. F. Curl, H. W. Kroto, F. K. Tillel, and R. E. Smalley, *J. Am. Chem. Soc.* **107**, 7779 (1985).

250. H. W. Kroto, *Science* **242**, 1139 (1988).

251. K. G. McKay, H. W. Kroto, and D. J. Wales, *J. Chem. Soc.*, Faraday Trans. **88**, 2815 (1992).

252. K. R. Bates and G. E. Scuseria, *J. Phys. Chem. A* **101**, 3038 (1997).

253. J. R. Heath, in *Fullerenes-Synthesis, Properties, and Chemistry of Large Carbon Clusters*, edited by G. S. Hammond and V. J. Kuck, no. 481 in ACS Symposium Series, p. 1 (Washington, DC, American Chemical Society, 1992).

254. M. Endo and H. W. Kroto, *J. Phys. Chem.* **96**, 220 (1992).

255. R. L. Murry, D. L. Strout, G. K. Odom, and G. E. Scuseria, *Nature* **366**, 665 (1993).

256. D. E. Manolopoulos and P. W. Fowler, in *The Far Reaching Impact of the Discovery of C_{60}, edited by* W. Andreoni, vol. 316 of *NATO ASI Series, Series E.*, p. 51, Dordrecht, Netherlands (NATO, Kluwer Academic Publ., 1996).

257. D. J. Wales, *Chem. Phys. Lett.* **141**, 478 (1987).

258. D. L. Strout and G. E. Scuseria, *J. Phys. Chem.* **100**, 6492 (1996).

259. J. M. Hunter and M. F. Jarrold, *J. Am. Chem. Soc.* **117**, 10317 (1995).

260. G. von Helden, M. T. Hsu, N. Gotts, and M. T. Bowers, *J. Phys. Chem.* **97**, 8182 (1993).

261. X. Liu, D. J. Klein, W. A. Seitz, and T. G. Schmalz, *J. Comp. Chem.* **12**, 1265 (1991).

262. D. E. Manolopoulos, J. C. May, and S. E. Down, *Chem. Phys. Lett.* **181**, 105 (1991).

263. A. J. Stone and D. J. Wales, *Chem. Phys. Lett.* **128**, 501 (1986).

264. D. Mitchell, P. W. Fowler, and F. Zerbetto, *J. Phys. B.* **29**, 4895 (1996).

265. F. Wooten, K. Winer, and D. Weaire, *Phys. Rev. Lett.* **54**, 1392 (1985).

266. A. D. Becke, *Phys. Rev. A* **38**, 3098 (1988).

267. C. Lee, W. Yang, and R. G. Parr, *Phys. Rev. B* **37**, 785 (1988).

268. M. J. S. Dewar and W. Thiel, *J. Am. Chem. Soc.* **99**, 4899 (1977).

269. B. R. Eggen, M. I. Heggie, G. Jungnickel, C. D. Latham, R. Jones, and P. R. Briddon, *Science* **272**, 87 (1996).

270. S. J. Austin, P. W. Fowler, D. E. Manolopoulos, and F. Zerbetto, *Chem. Phys. Lett.* **235**, 146 (1995).

271. P. Ballone and P. Milani, *Phys. Rev. B* **42**, 3201 (1990).

272. J. R. Chelikowsky, *Phys. Rev. Lett.* **67**, 2970 (1991).

273. J. R. Chelikowsky, *Phys. Rev. Lett.* **45**, 12062 (1992).

274. C. Z. Wang, C. H. Xu, C. T. Chan, and K. M. Ho, *J. Phys. Chem.* **96**, 3563 (1992).

275. X. D. Jing and J. R. Chelikowsky, *Phys. Rev. B* **46**, 5028 (1992).

276. J. Y. Yi and J. Bernholc, *Phys. Rev. B* **48**, 5724 (1993).

277. C. H. Xu and G. E. Scuseria, *Phys. Rev. Lett.* **72**, 669 (1994).

278. P. A. Marcos, M. J. López, A. Rubio, and J. A. Alonso, *Chem. Phys. Lett.* **273**, 367 (1997).

279. J. Tersoff, *Phys. Rev. B* **38**, 9902 (1988).

280. S. Maruyama and Y. Yamaguchi, *Chem. Phys. Lett.* **286**, 343 (1998).

281. D. W. Brenner, *Phys. Rev. B* **42**, 9458 (1990).

282. P. W. Fowler, *Chem. Phys. Lett.* **131**, 444 (1986).

283. M. D. Newton and R. E. Stanton, *J. Am. Chem. Soc.* **108**, 2469 (1986).

284. P. W. Fowler and J. Woolrich, *Chem. Phys. Lett.* **127**, 78 (1986).

285. P. W. Fowler and D. E. Manolopoulos, *An Atlas of Fullerenes*, vol. 30 of *International Series of Monographs on Chemistry* (OUP, Clarendon Press, Oxford, 1995).

286. D. E. Manolopoulos and P. W. Fowler, *J. Chem. Phys.* **96**, 7603 (1992).

287. J. Widany, T. Frauenheim, T. Koehler, M. Sternberg, D. Porezag, G. Jungnickel, and G. Seifert, *Phys. Rev. B* **53**, 4493 (1996).

288. D. Porezag, T. Frauenheim, T. Kohler, G. Seifert, and R. Kaschner, *Phys. Rev. B* **51**, 12947 (1995).

289. G. Seifert, D. Porezag, and T. Frauenheim, *Int. J. Quant. Chem.* **58**, 185 (1996).

290. S. J. Austin, P. W. Fowler, D. E. Manolopoulos, G. Orlandi, and F. Zerbetto, *J. Phys. Chem.* **99**, 8076 (1995).

291. J. Stoer and R. Bulirsh, *Introduction to Numerical Analysis* (Springer-Verlag, New York, 1980).

292. P. W. Fowler, T. Heine, D. Mitchell, G. Orlandi, R. Schmidt, G. Seifert, and F. Zerbetto, *J. Chem. Soc.*, Faraday Trans. **92**, 2203 (1996).

293. A. Ayuela, P. W. Fowler, D. Mitchell, R. Schmidt, G. Seifert, and F. Zerbetto, *J. Phys. Chem.* **100**, 15634 (1996).

294. P. W. Fowler, T. Heine, D. E. Manolopoulos, D. Mitchell, G. Orlandi, R. Schmidt, G. Seifert, and F. Zerbetto, *J. Phys. Chem.* **100**, 6984 (1996).

295. J. E. Campana, T. M. Barlak, R. J. Colton, J. J. DeCorpo, J. R. Wyatt, and B. I. Dunlap, *Phys. Rev. Lett.* **47**, 1046 (1981).

296. R. Pflaum, P. Pfau, K. Sattler, and E. Recknagel, *Surf. Sci.* **156**, 165 (1985).

297. Y. J. Twu, C. W. S. Conover, Y. A. Yang, and L. A. Bloomfield, *Phys. Rev. B* **42**, 5306 (1990).

298. P. Dugourd, R. R. Hudgins, and M. F. Jarrold, *Chem. Phys. Lett.* **267**, 186 (1997).

299. R. R. Hudgins, P. Dugourd, J. M. Tenenbaum, and M. F. Jarrold, *Phys. Rev. Lett.* **78**, 4213 (1997).

300. M. P. Tosi and F. G. Fumi, *J. Phys. Chem. Solids* **25**, 45 (1964).

301. A. A. Shvartsburg and M. F. Jarrold, *Chem. Phys. Lett.* **86**, 261 (1996).

302. M. T. Sprackling, *The Plastic Deformation of Simple Ionic Crystals* (Academic Press, London, 1976).

303. D. O. Welch, O. W. Lazareth, and G. J. Diens, *J. Chem. Phys.* **94**, 835 (1976).

304. M. Wilson, P. A. Madden, N. C. Pyper, and J. H. Harding, *J. Chem. Phys.* **104**, 8068 (1996).

305. N. T. Wilson, M. Wilson, P. A. Madden, and N. C. Pyper, *J. Chem. Phys.* **105**, 11209 (1996).

306. D. J. Wales, in *Advances in Molecular Vibrations and Collision Dynamics*, edited by J. M. Bowman and Z. Bačič, vol. 3, pp. 365–396 (JAI Press, Stamford, 1998).

307. R. C. Cohen and R. J. Saykally, *J. Phys. Chem.* **94**, 7991 (1990).

308. N. Pugliano and R. J. Saykally, *Science* **257**, 1937 (1992).

309. R. J. Saykally and G. A. Blake, *Science* **259**, 1570 (1993).

310. K. Liu, J. D. Cruzan, and R. J. Saykally, *Science* **271**, 929 (1996).

311. K. Ito, C. T. Moynihan, and C. A. Angell, *Nature* **398**, 492 (1999).

312. C. A. Angell, *J. Phys. Chem.* **97**, 6339 (1993).

313. W. L. Jorgensen, J. Chandrasekhar, J. D. Madura, R. W. Impey, and M. L. Klein, *J. Chem. Phys.* **79**, 926 (1983).

314. C. Millot and A. J. Stone, *Mol. Phys.* **77**, 439 (1992).

315. C. Millot, J. C. Soetens, M. T. C. M. Costa, M. P. Hodges, and A. J. Stone, *J. Phys. Chem. A* **102**, 754 (1998).

316. A. J. Stone, A. Dullweber, M. P. Hodges, P. L. A. Popelier, and D. J. Wales, *Orient: A Program for Studying Interactions Between Molecules*, Version 3.2 (University of Cambridge, Cambridge, 1995).

317. M. P. Hodges, *XMakemol: A Program for Visualising Atomic and Molecular Systems*, version 4.09 (1998).

318. K. S. Kim, M. Dupuis, G. C. Lie, and E. Clementi, *Chem. Phys. Lett.* **131**, 451 (1986).

319. C. J. Tsai and K. D. Jordan, *Chem. Phys. Lett.* **213**, 181 (1993).

320. J. K. Gregory and D. C. Clary, *J. Phys. Chem.* **100**, 18014 (1996).

321. J. Kim and K. S. Kim, *J. Chem. Phys.* **109**, 5886 (1998).

322. K. Liu, M. G. Brown, C. Carter, R. J. Saykally, J. K. Gregory, and D. C. Clary, *Nature* **381**, 501 (1996).

323. J. K. Gregory and D. C. Clary, *J. Phys. Chem. A* **101**, 6813 (1997).

324. D. J. Wales and I. Ohmine, *J. Chem. Phys.* **98**, 7245 (1993).

325. D. J. Wales and M. P. Hodges, *Chem. Phys. Lett.* **286**, 65 (1998).

326. A. Khan, *J. Chem. Phys.* **110**, 11884 (1999).

327. A. Khan, *J. Phys. Chem.* **99**, 12450 (1995).

328. N. Metropolis, A. W. Rosenbluth, M. N. Rosenbluth, A. H. Teller, and E. Teller, *J. Chem. Phys.* **21**, 1087 (1953).

329. A. Baba, Y. Hirata, S. Saito, I. Ohmine, and D. J. Wales, *J. Chem. Phys.* **106**, 3329 (1997).

330. G. Toulouse, *Comm. Phys.* **2**, 115 (1977).

331. J. D. Honeycutt and D. Thirumalai, *Proc. Natl. Acad. Sci. USA* **87**, 3526 (1990).

332. J. D. Honeycutt and D. Thirumalai, *Biopolymers* **32**, 695 (1992).

333. R. S. Berry, N. Elmaci, J. P. Rose, and B. Vekhter, *Proc. Natl Acad. Sci. USA* **94**, 9520 (1997).

334. B. Vekhter and R. S. Berry, *J. Chem. Phys.* **110**, 2195 (1999).

335. N. Elmaci and R. S. Berry, *J. Chem. Phys.* **110**, 10606 (1999).

336. Z. Guo and C. L. Brooks III, *Biopolymers* **42**, 745 (1997).

337. H. Nymeyer, A. E. García, and J. N. Onuchic, *Proc. Natl. Acad. Sci. USA* **95**, 5921 (1998).

338. Y. Ueda, H. Taketomi, and N. Gō, *Biopolymers* **17**, 1531 (1978).

339. W. D. Cornell, P. Cieplak, C. I. Bayly, I. R. Gould, K. M. M. Jr., D. M. Ferguson, D. C. Spellmeyer, T. Fox, J. W. Caldwell, and P. A. Kollman, *J. Am. Chem. Soc.* **117**, 5179 (1995).

340. B. R. Gelin and M. Karplus, *Proc. Natl. Acad Sci. USA* **72**, 2002 (1975).

341. R. J. Speedy, *J. Chem. Phys.* **110**, 4559 (1999).

342. D. Liu and J. Nocedal, *Math. Prog. B* **45**, 503 (1989).

343. N. K. Kopsias and D. N. Theodoros, *J. Chem. Phys.* **109**, 8573 (1998).

SOLID-FLUID EQUILIBRIUM: INSIGHTS FROM SIMPLE MOLECULAR MODELS

P. A. MONSON

Department of Chemical Engineering
University of Massachusetts Amherst, MA 01003

D. A. KOFKE

Department of Chemical Engineering, State University of New York
at Buffalo, Buffalo, NY 14260

I. INTRODUCTION

One of the most remarkable phenomena in nature is the collective organization of atoms and molecules into crystalline solids. That molecules arrange themselves in this fashion—with ordering that extends distances several orders of magnitude greater than the molecular size—is an underappreciated wonder of physics. Moreover, the molecular physics that leads to crystallization is shared by many other phenomena including liquid crystallinity [1] and the behavior of dense colloids [2,3], as well as a variety of so-called self-assembly phenomena that have been of so much recent interest in the materials science community [4,5]. Most remarkable is that the molecular physics that leads to the formation of solids can often be explained qualitatively using surprisingly simple molecular models. One might even venture to claim that such phenomena can be understood fundamentally only through the application of such models. Indeed, it is fair to say that the existence of the crystalline phase—the question of how it can ever be thermodynamically more stable than the fluid—could not be explained prior to the 1950s, when its presence was unambiguously demonstrated in the hard-sphere model.

This article is a review of some of the things that have been learned about the phase equilibrium between solids and liquids (SLE) or solids and fluids

Advances in Chemical Physics, Volume 115, edited by I. Prigogine and Stuart A. Rice.
ISBN 0-471-39331-2 © 2000 John Wiley & Sons, Inc.

(SFE) from the perspective of molecular theory. We seek to provide an insight into solid-fluid phase diagrams in terms of the underlying molecular interactions, using as a basis the simplest molecular models that will exhibit the phenomena seen in nature. Our primary interest will be in van der Waals solids, although some of the ideas about the effect of molecular size and shape on packing in van der Waals solids must surely be relevant to covalent and ionic solids. By the term "van der Waals solid," in general we mean one where the phase diagram can be understood by treating the system as a collection of molecules interacting via short-range repulsive and long-range attractive (dispersion) forces, although we will expand this definition to incorporate intermolecular potentials involving electrostatic (dipole-dipole, quadrupole-quadrupole, etc.) interactions or hydrogen bonding.

The study of solid phases using sophisticated models of the intermolecular forces is an active area for research [6–9]. Such work is important for making a very close connection with experiments and for developing quantitative models for intermolecular forces. On the other hand, we believe that simple models are more useful in gaining a more global perspective of the link between phase behavior and molecular interactions. So what constitutes a simple model? Our guide is Occam's razor. The research we review here is concerned with either finding the simplest model that can explain a given set of phenomena or investigating the phase behavior that can emerge from a simple model. Thus, the hard-sphere model and its extensions feature prominently. In addition to hard spheres we will discuss extensions of the hard-sphere model to mixtures, nonspherical molecules, and electrolytes. We will also consider, although to a lesser extent, the Lennard-Jones 12-6 potential and mixtures of Lennard-Jones particles. We hope that readers unfamiliar with this field will be impressed by the complexity of the phase behavior that can emerge from these simple models. We also firmly believe that understanding the behavior of these simple models offers considerable insights into the behavior of real systems and provides a much firmer foundation for pursuing studies of SFE with more complex models.

There are three subject areas that fall naturally under the umbrella of the present article:

- A review of Monte Carlo and molecular dynamics simulation methods for studying solid-fluid equilibrium;
- A review of theoretical methods for calculating solid-fluid equilibrium;
- A discussion of the properties of various molecular models and what they teach us about the connections between intermolecular forces and phase diagrams;

All three areas will be addressed here. The application of classical density functional theory has led to some of the most important recent theoretical advances in SFE and these have been the subject of several authoritative review articles [10–16]. On the other hand, we know of no recent comprehensive review addressing theoretical approaches other than density functional theories (DFT) and the other two subject areas, particularly the last one, and it was this that motivated us to write this chapter. We hope that the somewhat broader coverage of molecular modeling research in SFE given in this chapter will be of benefit to researchers new to the field. We should mention that this Chapter is written from a perspective that is more strongly influenced by liquid-state statistical mechanics than by solid-state theory. The interests of the authors in the problem at hand are an outgrowth of their previous work on phase equilibrium in fluids and fluid mixtures.

In understanding the solid-fluid phase diagrams of van der Waals solids, the most fundamental feature of the interactions are the short-range repulsions between the molecules. The simplest intermolecular potential that describes this behavior is the hard-sphere model. We therefore begin with a discussion of what is known about the hard-sphere phase diagram and how we can use this phase diagram to develop a phenomenological description of the phase behavior of simple molecules in the context of mean field or generalized van der Waals theory.

A. The Phase Diagram of the Hard-Sphere Model

The solid-fluid transition in the hard-sphere system has been the subject of great interest for about 50 years now. Although it was originally only a computer model, researchers in colloid science have come close to a complete experimental realization of the hard-sphere model [2,3]. The hard-sphere system is one with an intermolecular potential of the form

$$u(r) = \infty \qquad r < \sigma$$
$$= 0 \qquad r > \sigma \qquad (2.1)$$

Since the intermolecular potential energy of a configuration of hard spheres is either zero or infinite, the Boltzman factor, $\exp(-\beta U_N)$, is either one or zero and the configurational partition function is independent of temperature. Thus, the full behavior of this model is described by a single isotherm.

The possibility of a fluid-to-solid transition in the hard-sphere model was first predicted by Kirkwood and his co-workers [17–19]. This prediction was part of the stimulus for the celebrated studies of hard spheres by Alder and Wainwright [20] at the Lawrence-Livermore National Laboratory using the molecular dynamics (MD) method and by Wood and Jacobson [21] at the

Los Alamos National Laboratory using the Monte Carlo (MC) method. The work of Wood followed the slightly earlier seminal work of Metropolis et al. [22] in which the MC simulation method was first described. These early computer simulation studies established that the hard-sphere system has a high-density solid-like phase and a lower-density, fluid-like phase. Subsequently, Hoover and Ree [23,24] were able to make a rigorous calculation of the free energy of the hard-sphere model in both the fluid and solid states and show that the thermodynamic properties were consistent with a first-order fluid-to-solid transition with a density increase of about 10% on freezing. The stable solid phase structure is face-centered cubic (fcc), although recent calculations show that the difference in free energy between the fcc, and hcp structures is very small [25–28].

The phase diagram of the hard-sphere model is of enormous importance in developing a fundamental understanding of the solid-fluid transition; this diagram is shown in Figure 2.1. First of all, the hard-sphere model provides a fundamental geometric picture of the process of freezing by isothermal compression. The hard-sphere pressure-density isotherm is similar to high-temperature isotherms of simple fluids such as the rare gases or methane or nitrogen. Freezing by isothermal compression in such systems was demonstrated in pioneering work on phase equilibrium at high pressures,

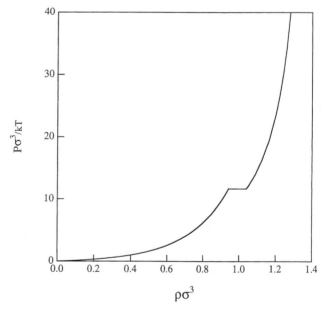

Figure 2.1. The phase diagram of the hard-sphere system as a plot of reduced pressure versus reduced density based on the results of Hoover and Ree [24].

such as that by Bridgman [29]. On the other hand, for hard spheres, increased ordering by compression in the solid phase is analogous to the effect of decreasing temperature in a system with attractive forces. This last point can be made more clearly by replotting Figure 2.1 using a reciprocal pressure axis, as shown in Figure 2.2. The quantity $kT/(P\sigma^3)$ may be interpreted as a dimensionless temperature, and we see that the hard-sphere phase diagram can be represented as an isobaric temperature-versus-density diagram. Figure 2.3 shows computer graphics visualizations from MC simulations of the hard-sphere system in the coexisting solid and fluid states; there is no doubting the crystallinity of the solid phase.

A worthwhile question to ask is why there should be a phase transition from a fluid to a solid in the hard-sphere model. We can formulate an initial perspective by considering the configurational Helmholtz free energy of the hard-sphere system, which is related to the configurational partition function via

$$\frac{A}{NkT} = -\frac{1}{N} \ln Z_N \qquad (2.2)$$

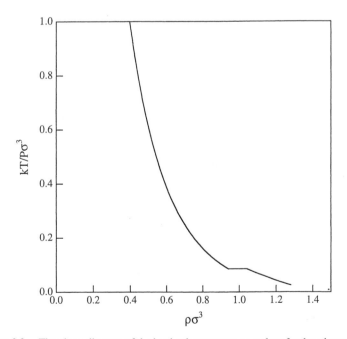

Figure 2.2. The phase diagram of the hard-sphere system as a plot of reduced temperature (inverse reduced pressure) versus reduced density based on the results of Hoover and Ree [24].

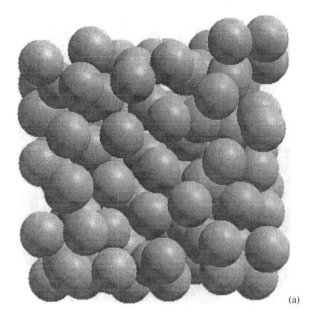

(a)

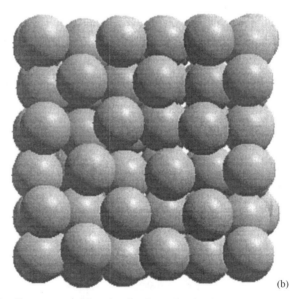

(b)

Figure 2.3. Computer graphics visualizations of states from a 108-particle Monte Carlo simulation of the hard-sphere system: (a) coexisting fluid state; (b) coexisting solid state.

Here Z_N is the configurational partition function given by

$$Z_N = \int \cdots \int \exp[-\beta U_N(\mathbf{r}_1 \ldots \mathbf{r}_N)]d\mathbf{r}_1 \ldots d\mathbf{r}_N \qquad (2.3)$$

where $U_N(\mathbf{r}_1 \ldots \mathbf{r}_N)$ is the N-particle potential energy for the system. For hard spheres this integral is simply the hypervolume of configuration space for which there are no hard-sphere overlaps. Now consider the contribution to Z_N from configurations with a high degree of crystalline order. Such configurations can be distinguished by an appropriate order parameter, such as one defined in terms of a Fourier sum of the particle position vectors. Contributions to Z_N from ordered configurations are present at all densities below the close-packed density, $\rho_{cp} = \sqrt{2}/\sigma^3$. However, for most of the density range (densities for which the average nearest neighbor distance is significantly greater than σ) the contribution from such configurations is very small, the integral being dominated completely by contributions from disordered configurations of the spheres. An estimate of the ratio of the number of disordered configurations to the number of ordered configurations in the limit of low density can be made by comparing the configurational partition function for an ideal gas ($Z_N = V^N/N!$) with that for an ideal gas in which the molecules are confined to Wigner-Seitz cells of volume V/N (in this case $Z_N = (V/N)^N$). The ratio of these for large N is e^N, a huge number. This ratio yields a free-energy difference (per molecule) of kT between the ordered and disordered configurations. This value represents the "ordering penalty" that we understand intuitively (and correctly) to disfavor the spontaneous formation of a crystal in a low-density system.

Now, as the density increases, the value of the configurational integral decreases and the free energy increases. More significantly, for high densities the disordered configurations of spheres come to represent a smaller fraction of the accessible configuration space—understandably, fewer ordered than disordered configurations are lost as the volume is decreased. For a macroscopic system the decrease in the disordered configuration fraction with increasing density is infinitesimal but persistent. Then, at some point the fraction drops precipitously to zero, and the overwhelming contribution to the integral in Eq. (2.3) comes from configurations with a high degree of crystalline order. Of all possible crystalline structures the one that provides the most configuration space at any given density is also the one that allows the highest close-packing density. This will be the fcc or hcp structure. Thus it is easy to understand that the hard-sphere system should be disordered at low density and ordered in either the fcc or hcp structures at high density (recognizing that the very slight free energy difference between the fcc and hcp structures of hard spheres requires a more sophisticated analysis).

In as much as the configurational internal energy of a hard sphere system is zero, we can also write

$$\frac{S}{Nk} = \frac{1}{N} \ln Z_N \qquad (2.4)$$

Thus, for hard spheres the entropy is determined by the hypervolume of configuration space for which there are no hard-sphere overlaps. At high density the translationally ordered configurations in the hard-sphere system are of greater entropy than the translationally disordered configurations by virtue of the greater fraction of accessible configuration space that they represent. This is another instance where the popular idea that there is a direct relationship between the entropy and the "degree of disorder" in a system fails. In fact, the thermodynamics of hard spheres can be discussed completely in terms of the free energy and without recourse to any link between entropy and disorder. To us this seems a more worthwhile approach in as much as Eq. (2.2) remains correct for any type of interactions, whereas Eq. (2.4) is correct only for hard spheres.

B.   Generalized van der Waals Theory

We have seen that by interpreting the pressure in a hard-sphere system as an inverse temperature we may also use the hard-sphere model to give a phenomenological model of freezing or melting by isobaric cooling or heating, respectively. Thermodynamic perturbation theory allows us to relate the properties of systems with attractive intermolecular forces to hard-sphere systems [30]. In its simplest form, the mean field or van der Waals approximation, perturbation theory gives the following expression for the Helmholtz free energy in the system

$$A = A_{HS} - a\rho \qquad (2.5)$$

where A_{HS} is the Helmholtz energy of the hard-sphere reference system, a is a parameter giving the strength of the attractive interactions in the system, and ρ is the density. This equation was first used to calculate SFE by Longuet-Higgins and Widom [31]. When an accurate description of the free energy of hard spheres in both the fluid and solid phase is used, this equation yields the phase diagram shown in Figure 2.4. Thus, the phenomenological theory of vapor-liquid equilibrium (VLE) represented by the van der Waals equation of state may be extended to yield such a theory for the entire phase diagram by including the full phase behavior of the hard-sphere reference system. On the other hand, this is not really a theory of SLE in the sense that

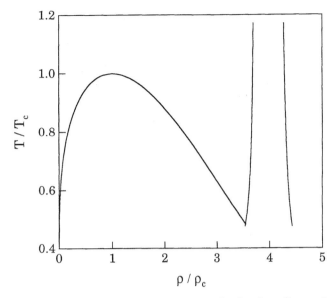

Figure 2.4. Corresponding states temperature versus density phase diagram for spherical molecules calculated from the generalized van der Waals theory of Longuet-Higgins and Widom [31].

it is a theory of VLE, because the transition behavior must be incorporated into the hard-sphere properties—the existence of SLE is not an outcome of the treatment.

Our purpose in these last two subsections has been to show how the simplest fundamental description of SFE for van der Waals solids can emerge from the hard-sphere model and mean field theory. Much of the remainder of the chapter deals with how we extend this kind of approach using simple molecular models to describe more complex solid-fluid and solid-solid phase diagrams. In the next two sections, we discuss the numerical techniques that allow us to calculate SFE phase diagrams for molecular models via computer simulation and theoretical methods. In Section IV we then survey the results of these calculations for a range of molecular models. We offer some concluding remarks in Section V.

II. COMPUTER SIMULATION METHODS
FOR SOLID-FLUID EQUILIBRIUM

Computer simulations of molecular models by MC and MD techniques have revolutionized our understanding of the solid-fluid phase transition.

The hard-sphere phase transition was only the first of many important discoveries that have emerged from these techniques. As has been described frequently in the past, these simulations can serve us in three ways. First, they provide essentially exact results for a given molecular model, which can be used to test the accuracy of approximate theories. Second, through comparison with experiment they can be used to assess the quantitative accuracy of a given molecular model. Third, they can be used for exploring the phenomenological behavior of molecular models in a parallel manner to that in which an experimentalist studies the behavior of a real system. We do not review the techniques themselves because there are now excellent textbooks on the subject [32–34]. We focus instead on the specific aspects of these methods associated with the calculation of SFE.

A. Free-Energy Calculations

Thermodynamic equilibria are evaluated by analysis of the free energy. When a thermodynamic system can exist in one of several distinct phases —differing perhaps in density, molecular composition, symmetry, and so on—the most stable of them is the one that minimizes the appropriate free energy for the choice of independent thermodynamic variables. The same criterion is invoked (directly or indirectly) when one attempts to analyze phase equilibria by molecular simulation. For molecular simulation to be broadly useful it is important then that it can be applied to the evaluation of free energies of model systems.

Molecular simulation never provides a direct measurement of the free energy, in the sense that it never performs a numerical quadrature of the partition function for the model system. Instead, simulation provides the *difference* in free energy between the system of interest (the *target* system) and a reference system, for which the free energy may be known. The reference system may differ from the target system in its thermodynamic state (e.g., temperature or density), or it may represent an entirely different molecular model. The free energy of the reference may be known from a separate molecular simulation study, or it may form a system simple enough that its free energy can be determined analytically. The free-energy difference may be an end in itself, since this difference is all that is needed to ascertain the relative stability of two candidate phases.

Before going further we should point out that many free-energy calculations on solids are performed approximately, usually involving the application of some type of harmonic treatment [35]. These approximate methods are not within the scope of this review.

I. Methods

There exist many ways to evaluate free energy differences by molecular simulation [34,36–39]. Free-energy perturbation (FEP) methods compute the difference as an ensemble average [40]

$$\exp[-((\beta A)_1 - (\beta A)_0)] = \langle \exp[-((\beta U)_1 - (\beta U)_0]\rangle_0 \qquad (2.6)$$

The right-hand side of this equation describes an ensemble average over configurations of the reference system (subscript 0), in which the quantity being averaged is the Boltzmann factor of the difference in energy (divided by kT for the respective system) of the systems. The formula is very easily programmed, and its measurement may be performed without compromising the calculation of other ensemble averages. Unfortunately, in practice the formula as written is almost never useful for computing solid-phase free energies. The method breaks down if the configurations important to the target system are inadequately sampled during the simulation of the reference. The breakdown is difficult to circumvent when studying the solid state. The situation is somewhat different for fluids, and the formula is very useful for these systems. With fluids one can turn to the Euler equation [41], which expresses the Helmholtz free energy (for example) in terms of the pressure and the chemical potential

$$A = -pV + N\mu \qquad (2.7)$$

The pressure is easily evaluated via the virial theorem, and the chemical potential can be measured by application of Eq. (2.6). This application, known as Widom's insertion method [42,43], involves measuring the energy change accompanying the insertion of a molecule at a randomly selected point in the simulation volume. This energy change is an *intensive* quantity; it is independent of the size of the system. Connected to this, and more important, is the fact that the degree of overlap of the portion of phase space relevant to the target and reference systems, expressed as a fraction of the phase space explored in the reference-system simulation, is also independent of the system size. In contrast, when the energy difference $U_1 - U_0$ is *extensive*—increasing with system size—the probability of encountering a configuration important to the target system, while simulating the reference, vanishes exponentially with system size. The route to the free energy via the chemical potential, which makes direct simulation of fluid-phase free energies viable, is not available to solid-phase simulations. The periodic boundary conditions routinely used in simulation do not permit the insertion of a molecule without the introduction of a defect on the crystal lattice. Of course real crystals do exhibit defects, but typically not at densities of 1 in 100 or 1000 that would be implied by accepting them as a consequence of a particle insertion measurement of the chemical potential. Thus, very large system sizes are needed to permit a particle insertion method to be viable for a solid-phase simulation.

Staging methods can be applied to remedy the sampling problems encountered with FEP calculations [44–46]. Here the free-energy difference of interest is broken into a set of smaller differences, each of which is evaluated by an FEP calculation performed at each intermediate stage. A small free-energy difference usually implies a greater overlap of target and reference-system configurations, so with a sufficient number of stages each FEP intermediate can be made tractable. An important consideration is the direction in which the FEP calculation is performed. In principle, either system (the reference of known free energy or the target) may play the role of the "0" or "1" system in Eq. (2.6), however, in practice the performance of the calculation may differ greatly with the direction in which the FEP calculation is performed. A good rule of thumb is always to select the "0" system as the one with the greater entropy (e.g., the higher temperature) [47]. Regardless, FEP methods—even with staging—usually are not viable for solid-phase calculations, particularly if the intermediate stages have no intrinsic interest beyond their role as a computational aid. Because the free-energy difference is extensive, too many stages are required to produce a result efficiently.

Thermodynamic integration (TI) is the most widely used method for computing solid-phase free energy differences. In a way, TI represents a generalization of FEP, in that multistage FEP may be viewed as a very high-order TI scheme. High-order methods are very accurate, but tend to be unstable, which is certainly the case with FEP. With TI, the reference and target systems are joined by a reversible path, and the free-energy difference is expressed

$$A_1 - A_0 = \int_{\lambda_0}^{\lambda_1} \frac{\partial A}{\partial \lambda} d\lambda \qquad (2.8)$$

where λ describes movement along the reversible path. λ may be a state variable, in which case the integrand is the conjugate thermodynamic property (e.g., the volume and the pressure, respectively), or λ may be a parameter of the Hamiltonian, in which case the integrand may be formulated as a less familiar ensemble average [48]. A series of simulations is performed for various values of λ between λ_0 and λ_1. At each λ the derivative $A'(\lambda) = \partial A/\partial \lambda$ is measured and the integral is evaluated via a standard quadrature formula. Evaluation of the derivative can in most cases be accomplished without difficulty, and TI on the whole is very reliable. Unlike FEP, it does not require or expect overlap of the important regions of configuration space from one quadrature point to the next. Instead, it makes some assumption (embodied in the quadrature formula) about the smoothness of the derivative $A'(\lambda)$ along the integration path. The method fails if this assumption is in error, as happens when the integration path crosses a thermodynamic phase transition.

Closely related to TI is the adiabatic switching method of Watanabe and Reinhardt [49]. This approach involves following a thermodynamic path via a molecular dynamics simulation in which the parameter λ varies slowly with time. For a sufficiently slow variation, the process is adiabatic and the entropy of the initial and final states will be the same; extensions of the basic idea permit evaluation of isothermal and other free-energy changes. The question of how slow the variation must be to remain an adiabatic process has been the subject of some study and analysis [50,51]. De Koning and Antonelli [52] have demonstrated its use by calculating the difference in free energy between two Einstein crystals.

Between TI and FEP are several methods that evaluate the free-energy difference by examination of cumulants [53] or histograms [54–56] of the integrand of Eq. (2.8), $A'(\lambda)$. The idea is that this additional information can be obtained from a simulation with almost no additional cost, and it can be used to extract the maximum information available to compute the free-energy difference. These methods can be effective, particularly when applied in a staging scheme. As a caveat, when applying these methods, one should be careful not to be lured into the assumption that the distribution of $A'(\lambda)$ follows a Gaussian form. It is only near the peak of the distribution that this approximation holds, and the utility of these methods comes from the extra information they provide about the wings of this distribution. For example, the decay at one of the wings (say, $A' > 0$) is commonly of the form $\exp(-aA')$ rather than the Gaussian $\exp(-aA'^2)$; the distinction is important, because the contribution to the free-energy difference here may be of the form $\exp(+bA')$ (here a and b are positive constants).

Expanded ensembles and related "parameter-hopping" techniques (see citations in Ref. 34) describe another class of methods for computing free-energy differences. The basic feature of these methods is the treatment of the path variable λ as a dynamical or ensemble variable to be sampled during the simulation. The free-energy difference between systems defined by λ_0 and λ_1 may be obtained simply from the observed probabilities $f(\lambda)$ (taken in the form of histograms) of finding the system in each state. Usually the system greatly favors one state over the other, and if given complete freedom to move between them it may never visit one of the states during a simulation of reasonable length. To remedy this a weighting function $w(\lambda)$ may be applied to bias the system toward one or the other state. Then the (residual) free-energy difference is given by

$$A_1 - A_0 = kT \ln \frac{f(\lambda_1)w(\lambda_0)}{f(\lambda_0)w(\lambda_1)} \qquad (2.9)$$

Often it is difficult to construct a useful weighting function that spans the reference and target systems. In this case a staging approach may be applied to join the two systems via a series of simulations, as described above for the FEP method. In this context, histogram reweighting methods [54–56] can be introduced to extract the greatest amount of information from the collective histogram data. Like the FEP methods, the performance of parameter-hopping techniques greatly suffers if used for an extensive rather than an intensive free-energy difference. The ability of the system to move from one value of λ to another depends on the overlap of their important-configuration spaces. For extensive free-energy differences this overlap is small, so it may require long simulations for λ to sample well (many times back and forth) a substantial range of values. Also, in these situations the performance of the technique can be exquisitely sensitive to the choice of the weighting function.

2. Reference Systems

If one wishes to establish a value for an absolute free energy, and not just a free-energy difference, ultimately the target system must be joined to a reference for which the free energy can be evaluated analytically. Very few such systems have been formulated. For solids, the choices are (a) a non-interacting single-occupancy gas; (b) a noninteracting cell model; (c) a noninteracting Einstein crystal; and (d) a harmonic lattice (noninteracting phonon gas). These systems are depicted in Figure 2.5. References (a) and (b) can be cast as the same system, but it is worthwhile to keep them conceptually distinct. One would like the reference to share many common features with the target system, if only to minimize the accumulated errors involved in traversing the path that yields their free-energy difference. Usually one wants both systems to exhibit the same crystal symmetry. It is especially important that the path joining the systems be reversible, and that there exists no phase transitions along the way.

The first rigorous free-energy calculation performed for a solid was completed for the hard-sphere model by Hoover and Ree [24] using thermodynamic integration from a noninteracting single-occupancy gas. In this reference, each sphere center is artificially confined to its own Wigner–Seitz cell (other cell shapes may be used), and the density is low enough that

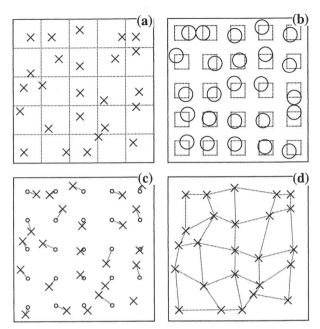

Figure 2.5. Analytically solvable reference systems for calculation of the free energy of solid phases. Simple square lattice is shown for illustration. (a) Noninteracting single-occupancy gas. Dotted lines indicate boundaries that constrain each particle to a single-occupancy cell, and "×" marks positions of the noninteracting particles. This system is obtained by increasing the volume to infinite size. (b) Noninteracting cell model. Particle centers are constrained to the regions described by the dotted-line squares. If the sphere size indicates the range of pair interactions, the cell size shown is just at the point where the spheres become noninteracting. This system is reached by decreasing the cell size. (c) Noninteracting Einstein system. "×" marks positions of the noninteracting particles, which are attached to sites (indicated by the small circles) via Hookean springs (indicated by the dotted lines). (d) Harmonic lattice. Neighboring particles interact via Hookean springs, which are indicated by the dotted lines. Periodic-boundary interactions are not shown.

an ideal-gas or virial treatment may be applied to yield a value for the free energy. Thermodynamic integration proceeds along a path of increasing density. At sufficiently high density the single-occupancy constraint becomes redundant, as the particle interactions naturally keep each sphere within its cell. At these densities the free-energy difference between the artificially confined and naturally unconfined systems is zero. Hoover and Ree observed a cusp in the pressure-density isotherm at the point where the single-occupancy system becomes mechanically stable, but they did not detect a thermodynamic phase transition. Subsequent studies [57] have since indicated the possibility of a very weak first-order transition along this path, yet the results reported by Hoover and Ree have been tested against more accurate studies based on a different reference system, and they hold up quite well. Regardless, integration from a single-occupancy gas is not considered a reliable approach to calculation of solid-phase free energies, owing to the uncertainties associated with phase transitions that are likely to exist along the integration path.

The single-occupancy cell model approach of Hoover and Ree can be formulated in a different, and safer, direction. In the single-occupancy gas, the cell volume increases with the system volume as the density is lowered. Instead one can formulate a path in which the cell constraint becomes tighter while the system volume (density) remains fixed. As the path is traversed, each molecule is confined to an increasingly small region centered on its lattice position. For hard potentials a point is reached in which the molecules become incapable of excursions that lead to overlap with other molecules, and the system becomes noninteracting (for spheres this occurs when the diameter of the constraint cell D satisfies $D \leq a - \sigma$, where a is the nearest-neighbor lattice distance and σ is the sphere diameter). The free energy is then easily obtained for the noninteracting system. For longer-range potentials the molecules might never reach a point of noninteraction, but for sufficiently small cells the free energy can be computed by applying an Einstein-crystal treatment (discussed below). Rickman and Philpot [58] employed this reference with a staged FEP scheme to compute the free energy of the Lennard-Jones model. With this approach one computes the free-energy difference $A(\lambda) - A(\lambda')$ by merely counting the fraction of the time that *all* particles are simultaneously within a distance λ of their lattice sites, given that they are all constrained to be within a distance $\lambda' > \lambda$ during the simulation. However, unless $\lambda' - \lambda$ is very small (of order $0.001\,\sigma$ for $\lambda < 0.07\,\sigma$), this probability is too small to measure. Consequently, many stages (on the order of 20 or so) are needed to obtain the free energy. Perhaps other free-energy difference methods (e.g., thermodynamic integration) could yield a result with less effort.

The noninteracting Einstein crystal has each molecule bound to its cell center by a Hookean (linear) spring, with no interactions or any sort between the molecules. Thus the configuration-space Hamiltonian for this system is

$$H = \omega \sum_{i=1}^{N} (\mathbf{r}_i - \mathbf{r}_i^0)^2 \qquad (2.10)$$

where $\mathbf{r}_i$ is the center-of-mass position of molecule i, and $\mathbf{r}_i^0$ is the corresponding lattice site for the molecule. The free energy of this system, minus any temperature-dependent kinetic contributions is simply

$$A = -\tfrac{3}{2} NkT \ \ln(\pi kT/\omega) \tag{2.11}$$

Two types of paths can be constructed to connect a target system to an Einstein reference. In the first, the target-system Hamiltonian is maintained while the Einstein spring constant ω is continuously increased. For sufficiently large values of ω the molecules remain very close to their lattice sites, so the target Hamiltonian is essentially a constant, and the reference (plus the static lattice energy of the target) is obtained. However, it is more practical to invoke either a FEP calculation or a virial treatment before this (arbitrary) limit is reached [25]. Alternatively, one may construct a path in which the target Hamiltonian is turned off as the Einstein springs are turned on. The noninteracting Einstein crystal is then obtained at a well-defined value of the integration parameter. The noninteracting Einstein crystal forms the reference for the Frenkel-Ladd technique [25], which is presently the method of choice for free-energy calculations [59,60–66]. Important details of its application vary with the nature of the system being studied, including whether the potential is continuous or discontinuous, and whether the molecules are monatomic or multiatomic. A thorough discussion is presented in the text by Frenkel and Smit [34].

The last reference system we discuss is the lattice of interacting harmonic oscillators. In this system each atom is connected to its neighbors by a Hookean spring. By diagonalizing the quadratic form of the Hamiltonian, the system may be transformed into a collection of independent harmonic oscillators, for which the free energy is easily obtained. This reference system is the basis for lattice-dynamics treatments of the solid phase [67]. If $\mathbf{D}$ is the dynamical matrix for the harmonic system (such that element D_{ij} describes the force constant for atoms i and j), then the free energy is

$$A = \tfrac{1}{2} kT \ \ln \det(\mathbf{D}/2\pi kT) + \Phi_0 \tag{2.12}$$

where Φ_0 is the static-lattice energy. Contributions due to uniform translation of the system in each direction must be removed before evaluating the free energy this way. This adjustment can be accomplished by evaluating the determinant from the eigenvalues of $\mathbf{D}$, removing the zero values, or by holding one atom fixed and removing the corresponding rows and columns (one for each coordinate direction) from the dynamical matrix [68]. Elimination of these contributions adds to the finite-size effects of the measurement, and for precise calculations these translational contributions can be added back into the free energy. The appropriate means to apply this correction was advanced very recently [274].

The force constants between the atoms may be selected to optimize the free-energy calculation. Usually they are given in terms of the second derivatives of the target-system potential with respect to the molecular coordinates, but this need not be the case. A typical path for connecting a target system to a harmonic lattice is by decreasing the temperature [69,70]. At sufficiently low temperature the particles experience only small excursions from their minimum energy configuration, so the energy may be expanded in terms of their coordinates. The first-order terms vanish, leaving only the static part plus the second-order (harmonic) contributions. This approach fails, however, when applied to systems with discontinuous potentials, or to systems that exhibit a polymorphic transition to another crystalline phase as the temperature is lowered.

There is one caveat to note in relation to these reference systems: Molecular dynamics simulations of harmonic and nearly harmonic models are known to exhibit very poor ergodicity behavior, leading to very poor sampling of phase space. Special measures should be taken if applying molecular dynamics simulation in a free-energy calculation involving such a model reference system [71]. If MC simulation is employed, there is no intrinsic problem with ergodicity and the issue is of no concern.

3. Free-Energy Differences Between Crystal Structures

Absolute free energies give more information than needed to determine the relative stability of two crystal structures, since only the difference in free energy between them is necessary. If the free-energy difference is small but finite (as for polymorphic transitions between similar phases), the errors accompanying the measurement of the free energy of each phase can overwhelm the overall measurement of their free-energy difference; this can also be a problem when locating weak first-order transitions, where small errors in measuring the difference translate into large errors in locating the coexistence conditions. A better strategy is to formulate a reversible path that joins one phase to another, thereby facilitating direct measurements of their free-energy difference. Bruce et al. [26] recently proposed such a scheme, and demonstrated its use by calculating the very small free-energy difference between fcc and hcp hard spheres at melting.

In their approach, particle positions are expressed in terms of the hcp/fcc lattice vectors and displacements of each sphere from its associated lattice site. Configurations are sampled via the displacement vectors. For each configuration (set of displacement vectors), the number of sphere overlaps that would be present for the two choices of the lattice vectors is observed. Configurations are biased according to the these overlap numbers, with substantial weighting given to those configurations in which both lattice structures have no overlapping spheres. Like other parameter-hopping methods, the actual free-energy difference is recovered by counting the number of non-overlap configurations for each lattice structure, adjusted appropriately to remove the sampling bias. The error analysis for the calculation indicates that it provides a free-energy difference that is at least an order of magnitude more precise than those measured in previous studies. Remarkably, Pronk and Frenkel [28] were able to further apply the method to measure the hard-sphere hcp/fcc interfacial free energy. It should be noted that Pronk and Frenkel's application of the lattice-switch method could not reproduce the result of Bruce et al. to within statistical precision. Pronk and Frenkel verified their own calculation by separately performing an integration from an Einstein crystal. They concluded that the Einstein-integration is as precise as that produced by the lattice-switch method, for the same amount of computation.

The approach developed by Bruce et al. is particular to the fcc/hcp problem, and while it may not be trivial to extend it to other situations, it is fair to view it as a method capable of general application. However, it should be noted that the method does measure an *extensive* free-energy difference. It worked in this case, despite the difficulties discussed in Section II.A.1, for two reasons: (1) the investigators are clever and patient and (2) the free-energy difference is not completely extensive, but has some intensive character—the difference between the lattices involves shifting of planes, which can lead to overlaps only in the one dimension perpendicular to the planes.

4. Free-Energy Bounds

In some situations it is sufficient to know only bounds on the free energy. For instance, establishment of a lower bound on the free energy of a particular phase may permit the conclusion that it is not the stable phase at a particular state, if another phase can be found with free energy below this bound. Free-energy bounds also play a central role in theories that apply variational methods to approximate the free energy. The idea is to form a free-energy bound as a function of some variational parameter(s) (e.g., the elements of the dynamical matrix in a lattice-dynamics approach), and to minimize the free-energy bound with respect to the parameter(s).

Free energy bounds can be established via the Gibbs–Bogoliubov inequality [72], which follows from Eq. (2.6) by considering the convexity of the exponential function

$$(\beta A)_1 \leq (\beta A)_0 + \langle (\beta U)_1 - (\beta U)_0 \rangle_0 \qquad (2.13)$$

An upper bound follows from a simple reassignment of the indexes (note the change in the system governing the sampling of the average)

$$(\beta A)_1 \geq (\beta A)_0 + \langle((\beta U)_1 - (\beta U)_0)\rangle_1 \qquad (2.14)$$

Self-consistent phonon theory (SCPT) is based on Eq. (2.13), which may be applied without using molecular simulation since the average is taken over the reference system. Morris and Ho [73] describe application of the complementary formula, Eq. (2.14) in the same context, which does require molecular simulation to evaluate the ensemble average. In both SCPT and its extension the reference system is selected to minimize or maximize the free energy, as appropriate.

Frenkel (personal communication) has described a formula that should yield a tighter bound on the free energy

$$(\beta A)_1 \leq (\beta A)_0 + 2\ln\langle\exp[-((\beta U)_1 - (\beta U)_0)/2]\rangle_0 \qquad (2.15)$$

This result may be viewed as an application of the Schwartz inequality. The problem with this formula is that it suffers from the same numerical troubles that plague the exact FEP perturbation formula: the integrand might have significant contributions from configurations that are poorly sampled by the simulation.

B.   Determining the Optimum Solid Phase Geometry

One of the key problems in calculating the free energy of solids is the determination of the most stable crystal structure. The simulation methods that we have just described are carried out for an assumed crystal structure. Thus it might be necessary to consider a variety of crystal structures in order to locate the structure of lowest free energy. Alternatively, one can incorporate structural relaxations into MD and MC simulations as was first suggested by Parrinello and Rahman [74]. Ordinarily such structural relaxations are suppressed by the use of a fixed simulation box shape and the periodic boundaries. The basic idea in the Parinello–Rahman approach is simply to allow the simulation cell to change shape during the calculation. Such a shape change can be incorporated into the equations of motion in isobaric molecular dynamics simulations. Parrinello and Rahman showed how their method could be used to study the dynamics of a change in crystal structure brought about by a change in the intermolecular potential. Najafabadi and Yip [75] have generalized the approach to allow implementation in MC simulations in an isostress-isothermal ensemble. This allows the study of solid phases under an arbitrary stress and was used by them to investigate a stress induced transition between fcc to body centered cubic bcc phases in a model of solid iron. The formulation for an isotropic stress in MC simulations was developed independently by Yashoneth and Rao [76]. Incorporation of box shape changes is now a feature of most simulation studies of solid phases, especially when there is some question about whether the crystal structure and/or cell shape is optimal. The method of Parrinello and Rahman is perhaps the closest we have come to a computational method for finding the solid phase geometry that minimizes the free energy of the system.

C. Phase Equilibrium Calculations

One can mimic experiments to construct the most straightforward approaches to measuring solid-fluid coexistence conditions by molecular simulation. One might take a cold solid, gradually warm it, and observe the melting temperature, or proceed similarly by cooling a hot liquid. The problem with this approach is that seen also in experiment: metastability might permit the system to remain in the initial phase well past the true thermodynamic freezing point; moreover, too-rapid cooling may lead to a preponderance of defects in the crystalline phase [77]. The problem is even more severe in simulation than experiment, owing to the extremely short time scales accessible to simulation. Melting the crystal is less difficult, particularly if defects are present to act as nucleation sites for the formation of the liquid. Still, it is difficult to pin down the melting temperature as the region surrounding the defect may undergo premelting behavior before the thermodynamic melting point is reached. Generally, approaches based on the direct observation of bulk melting or freezing are useful only as means for putting bounds on the transition. The problems with applying these techniques to solid-fluid studies are attenuated, but still persist, when applied to polymorphic or plastic-crystal transitions [78].

Alternatively, thermodynamic phase equilibrium in a model system can be evaluated by beginning the simulation with two (or more) phases in the same simulation volume, in direct physical contact (i.e., with a solid-fluid interface). This approach has succeeded [79], but its application can be problematic. Some of the issues have been reviewed by Frenkel and McTague [80]. Certainly the system must be large (recent studies [79,81,82] have employed from 1000 up to 65,000 particles) to permit the bulk nature of both phases to be represented. This is not as difficult for solid-liquid equilibrium as it is for vapor-liquid, because the solid and liquid densities are much more alike (it is a weaker first-order transition) and the interfacial free energy is smaller. However, the weakness of the transition also implies that a system out of equilibrium experiences a smaller driving force to the equilibrium condition. Consequently, equilibration of the system, particularly at the interface, may be slow.

The most commonly used approach to the measurement of phase coexistence points involves the simulation of each phase separately, in its own simulation volume out of direct contact with the other phase. The coexistence conditions are then established by invoking one of the methods reviewed above for free-energy measurement. At the condition of mutual equilibrium the chemical potential or Gibbs free energy per molecule is the same in both phases, while away from the equilibrium state the stable phase is the one with the lower free energy. The only unattractive feature of this methodology is that the measurement of the free energy can be computationally intensive, requiring simulations at several state points that may themselves not be of any intrinsic interest. Nevertheless, the approach is viable, and forms the foundation for the vast majority of the solid-fluid phase coexistence data reported in the literature for model systems.

The method is shown schematically in Figure 2.6. In this example, at a single temperature the equation of state (pressure versus density) of each phase is evaluated over their corresponding regions of stability, beginning in each case from a limit in which the free energy is known. The free energy at

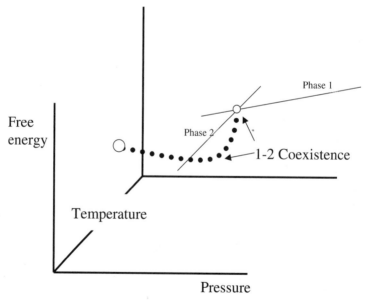

Figure 2.6. A schematic diagram of Gibbs free energy per molecule versus pressure for a single-component system exhibiting a first-order phase transition. The lines give the free energy versus pressure relationship for each single phase and the crossing point gives the condition of phase equilibrium. The dotted line traversing the temperature axis describes the path of a Gibbs–Duhem integration performed along the pressure-temperature coexistence line.

any point is evaluated by integrating these data; the result is shown as the two solid curves displayed in the figure. In the range of pressures in which the measurements overlap, both phases are at least metastable, and the transition is found at that point where the chemical potentials match. A great many variations are possible within this basic approach by considering integrations in variables other than the pressure at constant temperature. Such extensions are necessary, for example, with mixtures, where the composition of the two phases is unknown at the equilibrium condition; here an approach rooted in a semigrand formalism is appropriate, with the ratio of fugacities (or chemical-potential differences) being the relevant field variable [48,83,84]. One may also consider integrations in which a parameter of the intermolecular potential serves as the field variable. Thus one may evaluate the free energy of a Yukawa model by integrating along some suitable parameter that "morphs" it into a Lennard-Jones model. This may be a much more efficient and convenient means to establish an absolute free energy of an unstudied model system, as compared to an integration path that connects it to one of the analytically solvable models discussed in Section II.A.2.

Once a state point of coexistence is established, additional state points can be determined expeditiously through application of the Gibbs–Duhem integration method [48,85,86]. In this approach a differential equation for the coexistence line is used to guide the establishment of state points away from the known coexistence point. The most well known such formula is the Clapeyron equation [41]

$$\left(\frac{dP}{d\beta}\right)_\sigma = -\frac{\Delta h}{\beta \Delta v} \qquad (2.16)$$

where $\beta = 1/kT$ and Δh and Δv are the difference in molar enthalpy and molar volume, respectively, between the two phases. Clearly this equation applies to integrations in the pressure-temperature plane. Continuing the example shown in Figure 2.6, one traces out the coexistence line in the pressure-temperature plane (shown as the dotted line) by integrating this first-order differential equation. The right-hand side of the equation must be evaluated by molecular simulation of the two phases, but the method is efficient in that each pair of simulations yields a point on the coexistence line. Again, many useful variations are possible by considering integration paths other than in the pressure-temperature plane [48]; some of these applications are reviewed below.

D. Defects

Point defects are amenable to analysis by equilibrium statistical thermo-dynamics. The simplest formula for estimating the void concentration in a crystalline array considers only the energy penalty for creation of the void, which is offset by the entropy gain of making the void site distinguishable from the occupied sites. Then the void fraction is estimated by

$$\tilde{n} = e^{-\beta U_0} \tag{2.17}$$

The energy of formation U_0 is to a first approximation given by the energy of sublimation. Of course this treatment completely neglects the configurational free energy changes connected to the formation of the defect. Better treatments incorporate this effect via quasiharmonic [35], local harmonic [87], or other free-energy minimization [88] methods. However, studies of the validity of these approaches have found them to be inadequate, significantly underestimating the defect free-energies for temperatures above half the melting point [89]. This outcome is difficult to interpret because the studies have been applied to complex, realistic models. The information needed to improve these theories must come from the study of simpler model potentials, where the qualitative features can be isolated and understood. This is especially true when considering multiatomic molecular crystals, for which there have been no molecular simulation studies examining vacancy free energies.

Line and planar defects arise in crystals during the growth process. Unlike point defects, these features scale with the size of the system (to some power), and their unfavorable energetics cannot be overcome by the additional configuration space they introduce, so they are not present in the equilibrium crystal. Nevertheless, they are sufficiently (meta)stable that they can be examined in the context of equilibrium thermodynamics. Many studies of this form apply some variant of harmonic analysis, rather than employ the rigorous (and time-consuming) methods of molecular simulation.

The appropriate thermodynamic and statistical-mechanical formalism for the application of molecular simulation to the study of point defects has been given only recently, by Swope and Andersen [90]. These workers identified the number of lattice sites M as a key thermodynamic variable in the characterization of these systems. A real solid phase is free to adopt a value for M that minimizes the system free energy, because it can in principle create or destroy lattice sites through the migration of molecules to and from the surface of the crystal. The resulting bulk crystal can thus disconnect the molecule number N from the lattice-site number M, and thereby achieve an equilibrium of lattice defects in the form of vacancies and interstitials.

Application of periodic boundary conditions in molecular simulation leads to the introduction of a constraint on the number of lattice sites present in a simulated system. Without distorting the underlying lattice, a crystalline phase can adopt only specific numbers of lattice sites consistent with the periodic boundaries (e.g., $M = 32$, 108, 256, etc., for a fcc lattice in cubic periodic boundaries). This constraint applies even if the number of molecules is permitted to fluctuate, such as in a grand-canonical simulation, because the lattice sites are distinct from the molecules. Consequently, application of the usual simulation methods yields results that are in principle not representative of the behavior of the bulk phase being modeled. Swope and Andersen provide an illuminating example. A simulation in the grand-canonical ensemble is conducted at fixed T, V, and μ. Implicitly, M is fixed by the choice of V and the geometry of the system. In fact, there is a nonzero range of values of the volume that correspond to a particular value of M (say, 108). Thus, variation of V within this range corresponds to variation of the volume per lattice site, and one should expect that the ensemble averages (of, for example, the pressure) taken in the simulation will depend on the volume V. But in a real pure crystal, as in a pure fluid, the intensive properties (such as the pressure) do not vary with the size of the system (V) when T and μ are fixed. In the real crystal, the intensive variable M/V is free to adopt a value to minimize the free energy, which implies that instead the thermodynamic conjugate to M, and not M itself, is "imposed." This property, the "chemical potential of a lattice site" has been identified by Swope and Andersen and designated with the symbol ν_c. They show that an equilibrated crystalline system has $\nu_c = 0$.

The lattice thermodynamics described by Swope and Andersen are formulated as the thermodynamics of a constrained system. A system with a fixed value of M exhibits a free energy A_c, pressure p_c, and chemical potential μ_c appropriate to a constrained system. Moreover, a Euler equation exists to relate the principal thermodynamic quantities

$$\nu_c = \tilde{a}_c + p_c \tilde{v} - \mu_c \tilde{n} \tag{2.18}$$

where the tilde indicates a quantity that has been made intensive by dividing by the number of lattice sites M. A Gibbs-Duhem equation relates changes in the constraint-system field variables

$$d\nu_c = \tilde{v} dp_c - \tilde{n} d\mu_c \tag{2.19}$$

Swope and Andersen describe how this formalism can be used to construct a simulation protocol for study of crystalline phases. In essence, it is necessary to explore values of $\tilde{v}$ and $\tilde{n}$ in search of values that make the lattice-site chemical potential ν_c equal to zero. The pressure p_c and chemical potential μ_c can be evaluated using existing techniques (although μ_c is tricky for the lattice system). With $\tilde{a}_c$ known at one point (say by Frenkel–Ladd integration) to establish a single value of v_c, subsequent values v_c can be measured through application of the Gibbs–Duhem formula above.

We are aware of no studies that actually apply the formalism presented by Swope and Andersen. However, Squire and Hoover [91] decades earlier performed a well-conceived molecular simulation study of vacancies, addressing the issue raised by Swope and Anderson without invoking their formalism. Their approach has the same basis as many current theoretical treatments (such as those using Eq. 2.17) as well as that used in more recent simulation studies [71]. Key to Squire and Hoover's calculation is the (justifiable) assumption that the defects are independent and tightly localized, so that their cumulative effect can be gauged by performing a simulation of a system with only one of them. The vacancy concentration is then obtained by minimizing a low-vacancy model free energy. For the triple-point of the Lennard-Jones model, the state examined by Squire and Hoover, this concentration was found to be 1 in 3000. Notable is that this result is obtained from a simulation of only 108 particles. Other systems, at other state conditions, may not adhere to the independent-defect approximation employed by Squire and Hoover, in which case it becomes necessary to invoke the Swope–Andersen formalism.

III. THEORIES OF SOLID-FLUID EQUILIBRIUM

As a prelude to discussing theories of SFE it is worthwhile to make a brief digression to contrast the calculation of SFE with the calculation of vapor-liquid equilibrium (VLE). The problem of developing a phenomenological theory of VLE (away from the critical region) was, in a sense, solved by van der Waals with the formulation of his equation of state, although, of course, quantitatively accurate statistical mechanical calculations were not possible until the development of computer simulations, integral equation theories, and thermodynamic perturbation theories over the last few decades. In formulating a theory of VLE we can anticipate that since the liquid and vapor states are continuous, a single theoretical formulation will be applicable for both phases. On the other hand, there is no such continuity between solid and fluid states. The transition is fundamentally one where, in the language of condensed matter physics, the symmetry is broken during the change of phase. Nevertheless, two of the most influential theoretical approaches to the calculation of the SFE are based, on the one hand, upon treating the liquid phase as a disordered solid and, on the other hand, upon treating the solid phase as a highly ordered fluid. In this section we give a review of these approaches and some others. Our goal here has been to offer perspective rather than detail since some of these topics have been reviewed in detail elsewhere and by authors actively involved in their development and application [10–16, 92].

A. Cell Theories for the Free Energies of Solids and for SFE

There is an extensive literature spanning a period from the mid-1930s to the mid-1960s that describes efforts to develop theories for the thermodynamics of liquids by assuming that the volume of the system was divided into cells on a lattice. At the simplest level the partition function can then be calculated by assuming independent motion of the molecules in a potential field created by the neighboring molecules assumed fixed to their lattice sites. Much of this work is reviewed in an excellent monograph by Barker [92]. We now know that such efforts were of limited utility in developing the theory of liquids, which is now firmly based in the language of the radial distribution function [30]. In fact, the cell theories are usually a more appropriate description of solids rather than fluids and the recognition of this fact leads to some very useful results for the thermodynamic properties of solids.

Barker [92–94] has presented a general formulation of the cell theory and we give a brief review of his approach here. We will restrict our discussion to single-component atomic solids and discuss the application to mixtures and nonspherical molecules later. Suppose we have a system of N molecules in the canonical ensemble. The configurational partition function, Eq. (2.205), may be rewritten by breaking the volume into N identical subvolumes or cells so that

$$Z = \frac{1}{N!} {\sum_{m_1 \ldots m_N}}' Z^{(m_1 \ldots m_N)} \left(\frac{N!}{\prod_s m_s!} \right) \qquad (2.20)$$

Here the quantity $Z^{(m_1 \ldots m_N)}$ is a configurational partition function evaluated only for configurations in which m_1 particular molecules lie in cell 1, m_2 particular molecules lie in cell 2, etc. The multinomial coefficient gives the number of ways of assigning molecules to the cells for a given set of occupation numbers, $\{m_i\}$, the summation is over all $\{m_i\}$, and the prime on the summation indicates that the sum of occupation numbers in any assignment of molecules to cells must equal N. A useful approximation for a defect-free solid is to assume that only single occupancy of each cell occurs so that

$$Z_N = Z^{(1 \ldots 1)} \qquad (2.21)$$

Cell theories seek a further approximation for $Z^{(1 \ldots 1)}$ as a product of single-cell partition functions. The starting point is to develop an effective one-molecule potential energy that describes the interaction of any molecule with its environment. One way to do this is to write a decomposition of the intermolecular pair potential, $u(\mathbf{r}_i, \mathbf{r}_j)$, via

$$u(\mathbf{r}_i, \mathbf{r}_j) = w_{ij}(\mathbf{r}_j) + w_{ji}(\mathbf{r}_i) + \Delta(\mathbf{r}_i, \mathbf{r}_j) \qquad (2.22)$$

where the $w_{ij}(\mathbf{r}_j)$ are as yet unspecified one-particle functions and $\Delta(\mathbf{r}_i, \mathbf{r}_j)$ is a perturbation term. If we define an effective one-particle potential energy function via

$$u_0(\mathbf{r}_i) = \sum_j w_{ji}(\mathbf{r}_i) \qquad (2.23)$$

then the configurational partition function may be written as

$$Z = Z_0 \left\langle \prod_{i<j}(1 + f_{ij}) \right\rangle_0 \qquad (2.24)$$

where

$$f_{ij} = \exp[-\Delta(\mathbf{r}_i, \mathbf{r}_j)/kT] - 1 \qquad (2.25)$$

and the angled bracket denotes an ensemble average in a system with potential energy given by summing Eq. (2.23) over all molecules and Z_0 is the partition function for this system. By expansion of the product in Eq. (2.24), a perturbation series or cell cluster expansion is obtained. Different theories can be obtained by making different prescriptions for the $w_{ij}(\mathbf{r}_j)$. One possibility is to write

$$w_{ij}(\mathbf{r}_j) = \frac{1}{2}u(\mathbf{r}_i^{(o)}, \mathbf{r}_j^{(o)}) + [u(\mathbf{r}_i^{(o)}, \mathbf{r}_j) - u(\mathbf{r}_i^{(o)}, \mathbf{r}_j^{(o)})] \qquad (2.26)$$

where the superscript (o) on a position vector denotes the location of the center of the cell or lattice site. With this prescription one obtains an expansion for Z in which Z_0 is the partition function in the Lennard–Jones and Devonshire (LJD) approximation [95–98], or free volume approximation as it is sometimes called. In this case we have

$$Z_0 = \left[\int \exp[-u_0(\mathbf{r})/kT]d\mathbf{r} \right]^N \qquad (2.27)$$

For hard spheres using a dodecahedral Wigner–Seitz cell, the one-body integral in Eq. (2.27) can be calculated analytically [99], although the resulting expressions are quite complex. For other potentials a numerical evaluation is required. In early work with the cell theory it was common to approximate the cell shape by an effective spherical cavity for which the integral could be evaluated analytically, a procedure referred to as the "smearing" approximation [92]. For hard spheres a very simple expression for the free energy and equation of state is obtained in this way and this agrees quite well with the MC simulation results. However, the results obtained without using the smearing approximation are even more accurate. In recent work on applying the cell theory to nonspherical molecules and mixtures, MC integration methods have been used to evaluate the one-body integrals [100,101] without making any assumptions about the cell shape. These methods can be executed extremely quickly on currently available computer workstations.

To go beyond the LJD approximation without evaluating higher-order corrections, Barker [93] proposed choosing $w_{ij}(\mathbf{r}_j)$ such that the value of f_{ij} averaged over $\mathbf{r}_j$ is zero for all values of $\mathbf{r}_i$. This results in a series expansion for Z in which Z_0 is the partition function in an approximation corresponding to the well known Bethe–Guggenheim or quasi-chemical approximation for lattice gases or binary alloys. Barker called this approximation the "self-consistent field" (SCF) theory because it is an approximation in the spirit of the Hartree–Fock approximation in quantum mechanics. This approximation is much more complicated to implement than the LJD approximation in as much as the $w_{ij}(\mathbf{r}_j)$ are obtained by the solution of an integral equation that must generally be done numerically. On the other hand, many of the terms in the cell cluster expansion that are present when Eq. (2.22) is used now vanish and the convergence of the cell cluster expansion should be superior with this choice of reference system.

The LJD cell theory was originally proposed to yield a theory of gases and liquids, and for the Lennard-Jones 12-6 potential Eq. (2.27) yields a van der Waals loop and a critical point. However, Barker [92] gives a careful analysis of why the LJD theory is essentially a theory of solids and that the low-density behavior of the model does not give a physically realistic description of fluids. In the third and fourth in their series of papers Lennard-Jones and Devonshire [97,98] set forth a theory of SFE using the cell theory based on a generalization of the order-disorder theory of binary alloys [102] combined with the LJD theory. In this theory the system is pictured as a simple cubic lattice composed of two interpenetrating fcc sublattices. The ordered state or solid corresponds to all the sites on one sublattice being occupied with all the sites on the other sublattice empty. The fluid phase corresponds to partial occupancy of both sublattices with no long-range order among the occupied sites. Viewed from a more modern perspective, the flaws in this approach are legion. Perhaps the most serious problem is that it predicts a critical point for the solid-liquid transition. However, the initial results obtained were regarded as quite promising and the theory was subsequently extended to SFE of nonspherical molecules [103,104].

Barker and Cowley have applied the SCF theory to single-occupancy cell models of both hard spheres and hard disks [94,105]. They obtained very accurate solid phase properties at high density. When single or zero occupancy of the cells is allowed for, the theory yields solutions corresponding to a lower-density phase in addition to the solid phase. Maxwell constructions using the free energies from the two solutions yield phase equilibrium results for hard spheres and disks that are in good agreement with the MC simulation results. The fluid phase equations of state obtained from the theory as implemented are significantly less accurate than the solid phase equation of state. However, this is due to the neglect of multiple cell occupancy and higher order terms in the cell cluster expansion. It is important to recognize that this approach has a much firmer theoretical basis than the LJD order-disorder theory. In Barker's formulation it is clear that the SCF theory could, in principle, accurately describe the fluid state provided that a sufficient number of terms in the cell cluster expansion could be included to account for the correlation effects that are so important in the fluid phase, and if multiple occupancy of the cells could be considered. On the other hand, to do this successfully may require prohibitively complex and computationally expensive calculations. The same may also be true of the original LJD theory, although the convergence of the cell cluster expansion will be slower in that case. Leaving aside the issue of whether the cell theory can give an accurate theory of solid-fluid coexistence, it does provide a useful approach to calculating the properties of solids. We return to this point later in this section where we discuss theoretical approaches to SFE that use the cell theory for the solid phase and liquid state theory for the fluid phase.

Recently, Cottin and Monson [101] have shown how the LJD theory can be applied to solid phase mixtures. The basic approximation they make (in addition to single-cell occupancy) is to write the configurational partition function for an n-component system as a product of cell partition functions expressed as

$$Z_{N_1 \ldots N_n} = \frac{1}{\prod\limits_{j=1}^{n} N_j!} \prod_i z_i^{N p_i} \qquad (2.28)$$

where the product over i is 1 over all possible types of cell, p_i is the fraction (or cell probability) of cells of type i, and N is the total number of cells (equal to the total number of molecules). In this work a cell consists of the central molecule and its nearest neighbors, so that the type of a cell is determined by the species of molecule at its center and the composition and arrangements of the neighboring molecules. The cell probability, p_i, is readily obtained by using the approximation that the probability that any nearest neighbor site is occupied by a molecule of a particular species is equal to the mole fraction of that species. This approximation can be shown to be similar to the Bragg–Williams or mean field approximation. Cottin and Monson have developed a simple algorithm that allows the large number of cell partition functions that contribute to the product in Eq. (2.27) to be evaluated simultaneously, resulting in a calculation time comparable to that for a single-component system. Moreover, the structure of the approximation means that once the set of Z_i's are determined at a given density, the partition function and hence the Helmholtz free energy are known at all compositions for that density. In previous implementations of the cell theory for mixtures, which were to liquids rather than solids, a single cell type was used for each component [106,107]. Cottin and Monson have applied this approach to binary hard sphere mixtures [101,108] and to Lennard–Jones 12-6 mixtures [109].

B. Density Functional Theories

Over the last two decades most of the theoretical activity in the field of SFE has been devoted to development and application of classical density functional theories (DFTs) of inhomogeneous systems. The basic physical idea underlying this approach is that the influence of pair correlations upon the molecular density distribution in the melting solid is similar to that seen in the freezing liquid. Thus, information about the structure of one equilibrium phase could be used to calculate the free energy of the other. In as much as there are some tractable routes to the pair correlations in liquids [30], it makes sense to use the liquid as the reference state for the calculation. DFTs of SFE treat the solid as a highly inhomogenous fluid. Several reviews of the DFT have appeared over recent years [10–16] and we refer the reader to these sources for more thorough and authoritative presentations than we can provide here.

The physical ideas underlying the application of DFT to SFE first emerged in the work of Kirkwood and Monroe [17,18], who sought an alternative to the LJD order-disorder theory expressed in the language of the radial distribution function. They considered the solution of the Kirkwood integral equation [110] relating the one-body distribution function, $\rho^{(1)}(\mathbf{r})$, and the two-body distribution function, $\rho^{(2)}(\mathbf{r}, \mathbf{r}')$. They applied the approach to the freezing of liquid argon, for which they approximated $\rho^{(2)}(\mathbf{r}, \mathbf{r}')$ by the experimentally determined radial distribution function of the liquid phase at a single density. They looked for solutions to the integral equation for which

$\rho^{(1)}(\mathbf{r})$ would exhibit the periodicity of the crystal lattice. The utility of their results was limited by their approximation for $\rho^{(2)}(\mathbf{r}, \mathbf{r}')$ and the use of a truncated Fourier series for $\rho(\mathbf{r})$. Nevertheless, the idea of treating the solid as an inhomogeneous fluid was first set out in this work. The nature of the solutions to the Kirkwood and Born–Green–Yvon [30] integral equations, as well as their counterpart expressed in terms of the one-body and two-body direct correlation functions [111], in the neighborhood of the solid-fluid transition has been the subject of several more recent studies [112–114]. Much of this work is focused on the question of whether solid-fluid coexistence can be identified with a bifurcation point in the solution space of the integral equation. This route does not appear to yield quantitatively accurate information about the solid-fluid transition.

1. Functional Taylor Expansion Theories

The first application of what would now be called density functional theory to the calculation of SFE was given by Ramakrishnan and Yussouff [115,116], although they did not use this terminology. A rederivation of their theory within the density functional context and a generalization to the theory of the solid-liquid interface was presented subsequently by Haymet and Oxtoby [117]. During the following decade this approach has been extensively developed and applied to a variety of problems.

We make take as a starting point for such theories the following expression for the grand potential of an inhomogeneous system of spherical molecules

$$\Omega[\rho] = A[\rho] + \int \rho(\mathbf{r})[\phi(\mathbf{r}) - \mu]d\mathbf{r} \qquad (2.29)$$

In this expression $\rho(\mathbf{r})$ is the single molecule density. $A[\rho]$ is the intrinsic Helmholtz free energy, μ is the chemical potential, and $\phi(\mathbf{r})$ is an external field, which we can ignore in the present application. Eq. (2.29) expresses the dependence of the grand potential upon the functional form of the molecular density distribution in the system. For a homogeneous fluid phase $\rho(\mathbf{r})$ will be uniform and for a solid it will have the periodicity of the crystal structure. It can be shown [14,118] that the equilibrium molecular density represents a global minimum in $\Omega[\rho]$ with respect to variations in $\rho(\mathbf{r})$ at fixed $\phi(\mathbf{r})$ and μ. Implementation of DFT consists of locating the minimum in $\Omega[\rho]$ based upon a knowledge of $A[\rho]$.

A central problem in the development of the theory is the formulation of a suitable expression for $A[\rho]$ for the nonuniform system of interest. In treatments of SFE, two kinds of approach have been pursued. The first approach is to make a functional Taylor expansion of $A_{ex}[\rho]$ the excess

Helmholtz energy over that of an inhomogeneous ideal gas with the same one-body density, about that of the liquid in powers of $\Delta\rho(\mathbf{r}) = \rho(\mathbf{r}) - \rho_L$, where ρ_L is the density of the liquid. To second order this expansion has the form

$$A_{ex}[\rho] = A_{ex}[\rho_L] + \mu_{ex}(\rho_L)\int \Delta\rho(\mathbf{r})d\mathbf{r}$$
$$-\frac{kT}{2}\int\int c(|\mathbf{r} - \mathbf{r}'|; \rho_L)\Delta\rho(\mathbf{r})\Delta\rho(\mathbf{r}')d\mathbf{r}d\mathbf{r}' + \cdots \qquad (2.30)$$

where $\mu_{ex}(\rho_L)$ is the excess chemical potential of the liquid with density ρ_L. Thus, to second order this expansion requires knowledge of the two-body direct correlation function of the uniform liquid, $c(|\mathbf{r} - \mathbf{r}'|; \rho_L)$, which can be obtained from an established liquid state theory. For example, in the case of hard spheres the direct correlation function is usually approximated by the analytic result from the Percus–Yevick theory (PY) [119–121]. The third-order term in the expansion involves the three-body direct correlation function for the liquid, about which little reliable information is available, so that computations are usually limited to the second-order approximation. Since little is known about the convergence of the functional Taylor expansion for $A_{ex}[\rho]$, it is not possible to justify a priori the truncation at second order, and this is frequently cited as a weakness of this approach [14]. Nevertheless, the implementations of this approach by Ramakrishnan and Yussouf and by Haymet and Oxtoby served as an important stimulus to research in SFE over the last 20 years using DFT and other approaches.

2. Weighted Density Approximations

Another kind of approach to expressing $A[\rho]$ leads to what are referred to as weighted density approximations (WDAs). In the earliest of these approximations to be applied to SFE [122,123], $A_{ex}[\rho]$ is written in the form

$$A_{ex}[\rho] = \int \rho(\mathbf{r})\psi_{ex}(\bar{\rho}(\mathbf{r}))d\mathbf{r} \qquad (2.31)$$

where ψ_{ex} is the excess free energy per molecule of the uniform system and $\bar{\rho}(\mathbf{r})$ is a weighted density given by

$$\bar{\rho}(\mathbf{r}) = \int \rho(\mathbf{r}')w(|\mathbf{r} - \mathbf{r}'|; \bar{\rho}(\mathbf{r}))d\mathbf{r}' \qquad (2.32)$$

where $w(|\mathbf{r} - \mathbf{r}'|; \bar{\rho}(\mathbf{r}))$ is a weight function that is to be determined by imposing some auxiliary condition. For example, since the two-body direct correlation function is the second order functional derivative of Eq. (2.29) with respect to $\rho(\mathbf{r})$, a suitable auxiliary condition is to require that the correct bulk liquid direct correlation function be obtained in the uniform limit. The modified WDA (MWDA) of Denton and Ashcroft [124] and the generalized effective liquid approximation (GELA) of Lutsko and Baus [125,126] use a position independent weighted density. $A_{ex}[\rho]$ is given by

$$A_{ex}[\rho] = A_{ex}[\rho_{eff}] \qquad (2.33)$$

where ρ_{eff} is defined by

$$\rho_{eff} = \frac{1}{N} \int \int \rho(\mathbf{r}')w(|\mathbf{r} - \mathbf{r}'|; \rho_{eff})d\mathbf{r}d\mathbf{r}' \qquad (2.35)$$

In the MWDA, the weight function is determined by requirement that the direct correlation function be reproduced in the uniform limit. In the GELA, the effective liquid state is required to yield both the direct correlation function and the free energy of the uniform liquid. These WDAs, especially the GELA, give among the most accurate SFE predictions for hard spheres. A further modification of the MWDA has been recently proposed that incorporates information about the static lattice into the functional [127].

A third type of WDA approximation has been extensively developed by Rosenfeld and his co-workers [128–130] and by Kierlik and Rosinberg [131] based on the concept of "fundamental measures." In this approach, $A_{ex}[\rho]$ is expressed as

$$A_{ex}[\rho] = kT \int \phi_{ex}(\{\bar{\rho}_\alpha(\mathbf{r})\})d\mathbf{r} \qquad (2.36)$$

where ϕ_{ex} is an excess free energy density evaluated in terms of a set of weighted densities, $\{\bar{\rho}_\alpha(\mathbf{r})\}$, given by

$$\bar{\rho}_\alpha(\mathbf{r}) = \int \rho(\mathbf{r}')\omega^{(\alpha)}(\mathbf{r} - \mathbf{r}')d\mathbf{r}' \qquad (2.37)$$

where the set of weight functions, $\{\omega^{(\alpha)}(\mathbf{r} - \mathbf{r}')\}$, is independent of density and is related to fundamental geometrical measures of the molecules. The weight functions and ϕ_{ex} were originally constructed so that for the uniform fluid ϕ_{ex} is the excess free energy density obtained from the scaled particle theory [132] or Percus–Yevick compressibility equation of state [120]. However, it was shown that this fundamental measures functional (FMF) did not predict a freezing transition [131], although it gives very accurate results for other inhomogeneous fluid applications. More recently, Rosenfeld

[129,130] has developed adaptations of his approach that do predict a freezing transition. Moreover, they provide results for the solid-phase properties at high density that are more accurate than those given by other functionals.

Prior to applications to SFE, WDAs were originally proposed to give a more realistic description of molecular packing effects in interfacial systems, and Evans [14] has given an interesting discussion of their relative suitability in the context of interfaces and SFE. WDAs are nonperturbative in character and so are not subject to the criticism leveled at the functional Taylor expansion approach.

An additional issue in the development of the density functional theory is the parameterization of the trial function for the one-body density. Early applications followed the Kirkwood–Monroe [17,18] idea of using a Fourier expansion [115–117,133]. More recent work has used a Gaussian distribution centered about each lattice site [122]. It is believed that the latter approach removes questions about the influence of truncating the Fourier expansion upon the DFT results, although departures from Gaussian shape in the one-body density can also be important as has been demonstrated in computer simulations [134,135].

Many of the papers on DFT have focused primarily on the hard-sphere system, and it is for this system that most success has been achieved. However, DFT has also been applied to the Lennard-Jones 12-6 system, binary mixtures, nonspherical molecules, and coulombic systems. We will discuss some of these applications later in the chapter as we review what is known about the phase diagrams of various models systems.

3. Physical Insights on the Hard-Sphere Phase Diagram from a DFT Perspective

Baus [10] has presented an interesting argument about how the hard-sphere phase transition can be understood from the perspective of DFT. The argument is based on an analysis of the difference between the Helmholtz energies of the solid and fluid phases as a function of the density, ρ. This can be written exactly within the context of DFT as

$$\frac{\Delta A[\rho]}{kT} = \int \rho(\mathbf{r}) \ln \frac{\rho(\mathbf{r})}{\rho} d\mathbf{r} - \int_0^1 \int \int \Delta\rho(\mathbf{r})\Delta\rho(\mathbf{r}')$$

$$\times c^{(2)}(\mathbf{r}, \mathbf{r}'; [\rho(\mathbf{r}; \lambda)]) d\mathbf{r} d\mathbf{r}' d\lambda \qquad (2.38)$$

where

$$\rho(\mathbf{r}; \lambda) = \rho + \lambda[\rho(\mathbf{r}) - \rho]. \qquad (2.39)$$

For hard spheres the configurational internal energy is zero so that Eq. (2.38) is also the negative of the entropy difference between the two phases. Baus calls the contribution to the entropy difference from the first term on the right-hand side of Eq. (2.38) the configurational entropy difference and that from the second term the correlational entropy difference. (This choice of terminology is somewhat unfortunate since there is an established convention in classical thermodynamics [136] under which a configurational quantity is that part of the quantity that depends upon the density.) Baus argues that the relative stability of the two phases is determined by a competition between the two terms. Actually, we can pursue the same line of argument without making the identification between free energy and entropy for the hard-sphere system. The first ("configurational") term in Eq. (2.38) is positive at all densities so this always stabilizes the fluid relative to the solid. Thus, the stability of the solid at high density is due to its having a substantially lower value of the "correlational" free energy than the fluid. This argument is similar to the kind of arguments that explain vapor-liquid coexistence in the context of mean field theory. Ashcroft and his co-workers [137] have used a similar argument in discussing the equation of state of the hard sphere solid in the MWDA approximation. They use the fact that the Helmholtz energy functional can be written as a sum of an inhomogeneous ideal gas term and the excess term $A_{ex}[\rho]$ to make a similar decomposition of the pressure into ideal and excess contributions via

$$P = -\left(\frac{\partial A}{\partial V}\right)_{T,N} = P_{id} + P_{ex} \qquad (2.40)$$

They find that the P_{id} term is large and positive for all solid states increasing with the density, while the P_{ex} term is negative for low solid densities and positive for high solid densities. They argue that negative values of P_{ex} indicate an effective attraction between hard spheres in the solid.

3. A Two-Phase Approach—SFE Calculations Using a Theory Appropriate to Each Phase

Up to this point we have focused on the use of a single theory to describe both phases in a calculation of SFE. A somewhat less ambitious approach is to use a theory appropriate to each phase in the calculations. In particular, the cell theory can be used for the solid phase and a liquid state theory (e.g., a hard-sphere equation of state or thermodynamic perturbation theory) used for the fluid phase. This approach turns out to be at least as accurate as either the two-phase cell theory or DFT approaches described above and is often more accurate. Moreover, it has been more successful in the treatment of systems more complex than hard spheres.

The application of this approach to the hard-sphere system was presented by Ree and Hoover in a footnote to their paper on the hard-sphere phase diagram. They made a calculation where they used Eq. (2.27) for the solid phase and an accurate equation of state for the fluid phase to obtain results that are in very close agreement with their results from MC simulations. The LJD theory in combination with perturbation theory for the liquid state free energy has been applied to the calculation of solid-fluid equilibrium for the Lennard–Jones 12-6 potential by Henderson and Barker [138] and by Mansoori and Canfield [139]. Ross has applied a similar approch to the exp-6 potential. A similar approach was used for square well potentials by Young [140]. More recent applications have been made to nonspherical molecules [100,141] and mixtures [101,108,109,142].

4. Thermodynamic Perturbation Theory and Other Variations on the van der Waals Approach

Following the success of thermodynamic perturbation theories for liquid state properties [30]. It was natural to apply them to solid phase properties. As we described in our introduction, the van der Waals theory of Longuet-Higgins and Widom [31] is the simplest version of such a theory. The most successful of the liquid state perturbation theories is the theory of Weeks, Chandler, and Andersen (WCA) [143], and this was applied to the Lennard-Jones 12-6 solid by Weis [144]. He found that the theory was not as accurate in the solid phase as for the liquid phase and attributed this to the inaccuracy of the treatment of the reference system properties in the WCA theory. Subsequently, the WCA theory has been generalized by Kang et al. [145,146] to circumvent this problem. They pointed out that the choice of reference potential in the WCA theory is less suitable at very high liquid densities near the freezing point and or in the solid phase. They modified the WCA division of the intermolecular potential to create a reference system whose range decreases with increasing density. The resulting theory gives very good results for the Lennard-Jones 12-6, exp-6, and inverse power potentials. Jackson and van Swol [147] applied perturbation theory to a truncated and shifted Lennard-Jones model, and predicted that significant regions of the pressure-temperature phase diagram are stable to the hcp phase; however, they also found that fcc remained the stable phase at melting. In order to apply thermodynamic perturbation theory to the solid phase it is necessary to have accurate results for the hard sphere radial distribution function in the solid phase. Empirical fits to MC simulation results for this quantity have been given for both the fcc [148] and hcp [147] solids. Recently, a version of the WCA theory has been developed that uses a theoretical rather than simulation approach to the hard sphere distribution function [149].

The van der Waals approximation discussed in Section I.A applies the mean field approximation in the solid phase in the same way as in the fluid phase. Baus and co-workers [150,151] have recently presented an alternative formulation in which the localization of the molecules in the solid phase is taken into account. They have applied this to the understanding of trends in the phase diagrams of systems of hard spheres with attractive tails as the range of attractions is changed. For the mean field term in the solid phase they use the static lattice energy for the given interaction potential and crystal lattice. A similar approach was used earlier to incorporate quad-rupole-quadrupole interactions into a van der Waals theory calculation of the phase diagram of carbon dioxide [152].

5. Melting and Freezing Rules

Over the years several semi-empirical melting or freezing rules have emerged in an attempt to correlate SFE with other features of the solid or fluid phases [153]. A distinguishing characteristic of these rules is that they are all formulated in terms of properties of just one of the phases. The best known of these is the Lindemann rule [154], which states that the melting point of a solid correlates with the mean-squared atomic displacement. When this quantity exceeds a particular value, the solid is observed to melt. This rule is actually a loose statement about the mechanical stability of the solid, rather than a statement of its thermodynamic stability relative to the fluid.

The Ross melting rule [155] is expressed instead in terms of a threshold value of an excess free energy. It is based on a postulate of invariance of a scaled form of the partition function. A useful formulation of the statement requires the further assumption of a cell model for the solid. In terms of the free energy, the Ross melting rule has

$$\beta A - \ln \rho + 1 - \beta \Phi_{\text{lat}} = \text{constant} \tag{2.41}$$

that is, the Helmholtz free energy, minus its ideal-gas value, minus the energy of the perfect lattice, is a constant on the melting curve. Ross' rule has an advantage over Lindemann's in that it can be applied in two dimensions (Lindemann's cannot because the mean-square displacement is infinite in two dimensions).

Other melting rules are formulated in terms of properties of the saturated liquid phase, focusing in particular on some feature of the structure as quantified by the radial distribution function $g(r)$. Ideally, these rules could be applied to predict the freezing transition of a molecular model using a correlation function given by (say) an integral equation theory for $g(r)$.

Hansen and Verlet [156] observed an invariance of the intermediate-range (at and beyond two molecular diameters) form of the radial distribution function at freezing, and from this postulated that the first peak in the structure factor of the liquid is a constant on the freezing curve, and approximately equal to the hard-sphere value of 2.85. They demonstrated the rule by application to the Lennard-Jones system. Hansen and Schiff [157] subsequently examined $g(r)$ of soft spheres in some detail. They found that, although the location and magnitude of first peak of $g(r)$ at crystallization is quite sensitive to the intermolecular potential, beyond the first peak the form of $g(r)$ is nearly invariant with softness. This observation is consistent with the Hansen–Verlet rule, and indeed Hansen and Schiff find that the first peak in the structure factor $S(k)$ at melting varies only between 2.85 ($n = 8$) to 2.57 (at $n = 1$), with a maximum of 3.05 at $n = 12$.

The melting law of Raveché, Mountain, and Streett [158] postulates that the ratio $g(r_{min})/g(r_{max})$ is a constant on the melting curve. Here $g(r_{max})$ is the value of the radial distribution function at its first maximum, and $g(r_{min})$ is the value at its first (non zero) minimum. Raveché et al. base this rule on a consideration of the number of nearest neighbors about a particle in the saturated fluid.

More recently Giaquinta and Giunta [159] proposed another rule to portend the onset of a phase transition. The key quantity is a type of multiparticle-excess entropy, defined

$$\Delta s \sim s - s_{id} - s_2 \tag{2.42}$$

where s_{id} is the entropy of the ideal gas at the density and temperature of the fluid, and s_2 is a type of pairwise contribution to the entropy

$$s_2 = -\frac{1}{2}\rho \int [g(r)\ln g(r) - g(r) + 1]\,dr \tag{2.43}$$

where $g(r)$ is the radial distribution function. Fluid conditions at which Δs approaches zero seem to correlate well with the onset of a phase transition, either fluid-fluid or fluid-solid.

We remain skeptical of the value of any theory or rule of SFE that considers only one of the phases. No one questions that the condition of thermodynamic equilibrium between the solid and fluid phases depends on features of both. Treatments that consider only one of the phases must therefore be limited in their ability to yield quantitative predictions. In most cases the rules are invoked after the fact, upon completion of an SFE calculation or measurement. The rules are then tested against the actual outcome, primarily as a matter of curiosity. In this regard we note that any test of a rule is incomplete without some characterization of how the rule metric behaves as one moves off of the solid-fluid coexistence curve. A melting-line invariant that is also nearly invariant away from melting does not serve well as a melting-point marker.

IV. PHASE DIAGRAMS FOR MOLECULAR MODELS

A. Hard Spheres

As described in Section I, the first rigorous calculation of a phase diagram for a model potential was completed for the hard-sphere model by Hoover and Ree [24]. Hoover and Ree evaluated the solid-phase free energy via integration of the equation of state from the single-occupancy gas, while the fluid free-energy was evaluated similarly, integrating from the unconstrained ideal gas. Frenkel and Ladd [25] examined the hard-sphere freezing transition as a prototype application of their free-energy technique. Using a completely different methodology, and paying careful attention to error analysis and finite-size effects, they obtained a result in complete accord with the Hoover–Ree calculation. Clearly, the hard-sphere freezing transition is firmly established quantitatively.

In Table I we show a comparison of phase coexistence properties for the hard sphere solid calculated from various theories. The three most accurate density functional theories are shown together with two sets of predictions based on cell theory. The first cell theory prediction shown is from Barker's

TABLE I
A Comparison of Solid-Fluid Coexistence Properties for Hard Spheres Calculated via Various Theories with Results from Computer Simulation. Values of the Lindemann Parameter, L, for the Melting Solid are also Given

	$\rho_f \sigma^3$	$\rho_s \sigma^3$	$\Delta\rho\sigma^3$	$P\sigma^3/kT$	L
Simulation	0.943	1.041	0.098	11.7	0.129
Cell Theory (SCF)	0.919	1.056	0.137	12.5	0.113
Cell theory (LJD)	0.933	1.048	0.115	11.2	0.084
DFT (MWDA)	0.909	1.044	0.126	10.9	0.097
DFT (GELA)	0.945	1.041	0.096	11.9	0.100
DFT (FMF)	0.937	1.031	0.094	12.3	0.101

The simulation results for coexistence properties are from Hoover and Ree [24] and the value of L is from the work of Ohnesorge et al. [135]. The result from Barker's SCF theory [94] is from the leading term (order α^2) in an expansion of the mean square displacement in powers of $\alpha = (\rho_{cp}/\rho) - 1$ and may be an underestimate of the true value from that theory.

SCF, self-consistent field; LJD, Lennard-Jones and Devonshine; MWDA, modified weighted-density approximation; GELA, generalized effective liquid approximation; FMF, fundamental measures functional.

SCF approximation, which is described in Section III.A. The second set of cell theory results were obtained by using the approach described in Section III.C using the LJD theory for the solid phase and an accurate equation of state for the fluid phase. All the predictions of coexistence properties are in good agreement with the MC simulation data, the results from the FMF and GELA being particularly impressive. The theoretical estimates of the Lindemann parameter (defined as the root mean square displacement per molecule divided by the nearest neighbor distance in the static lattice) are generally too low, especially for the last four theories listed. The value marked cell theory is the LJD result. The low value in this case reflects the neglect of correlation effects in that theory, which should tend to increase the mean square displacement.

Much of the interest in hard-sphere phase transitions now centers on the question of the relative stability of the fcc versus the hcp phases. Frenkel and Ladd [25] examined this question too, and concluded that, although the fcc structure was more stable, the free-energy difference between these phases was too small to resolve within the precision of their calculations. More recently, Woodcock [27,160] returned to the single-occupancy concept to determine the hcp-fcc free-energy difference. The single-occupancy equations of state for these two crystals differ significantly over only a small range of densities, around the limit of mechanical stability. By focusing his simulation efforts in this region, Woodcock was able to ascertain that fcc is the more stable phase, by about 0.005 RT. However, Bolhuis *et al.* [276] pointed out that this result lies well outside the confidence limits of the Frenkel-Ladd calculation, and they presented new data confirming the Frenkel-Ladd result. The most precise calculation has been reported by Bruce et al. [26] using the technique reviewed in Section II.A.3. They too find that the fcc phase is stable at melting, but by only 0.00083 RT. Pronk and Frenkel [28], using the same method, attempted to reproduce this value, but instead obtained a slightly higher result of 0.00112 RT. However, all these calculations indicate that the fcc structure is the thermodynamically stable one. Further confirmation is provided by the lattice-switch calculations of Mau and Huse [275]. They looked at a broader range of stacking sequences, with an interesting perspective on how the local sequences in different planes "interact" as a function of their separation—they cast their analysis in the language of a one-dimensional Ising model. They conclude that the fcc structure is again the most stable of all.

The fcc/hcp stability question is not easily addressed experimentally either. However, studies of colloidal hard spheres [161], including zero-gravity crystallization experiments performed on the U.S. space shuttle [162], indicate too that fcc is the more stable phase, although stacking faults are pervasive [163]. Pronk and Frenkel [28] have applied molecular simulation to estimate that these faults will anneal out spontaneously, but only over a period of months.

B. The Square-Well Model

The simplest attractive hard-sphere model is the square-well potential, for which the energy is constant (and negative) over some range extending beyond the hard repulsive core; outside of this range the energy is zero, that is,

$$
\begin{aligned}
u(r) &= 0 & r &< \sigma \\
u(r) &= -\varepsilon & \sigma &< r < \sigma + \lambda \\
u(r) &= 0 & \sigma &+ \lambda < r
\end{aligned}
\tag{2.44}
$$

Phase transitions in the square-well model were studied extensively by Young [140,164] using a cell model. Young and Alder [165] and Kofke [48] have also reported some less complete molecular simulation results. In regard to its freezing behavior, the square-well model exhibits an exceedingly rich phase diagram. Depending on the range of the attractive well, the pressure-temperature diagram may exhibit multiple regions in which hcp, fcc, or bcc are stable (in addition to vapor and liquid regions). In a few cases, the shapes of these regions, and their relation to each other in the pressure-temperature plane, coincides very well (albeit qualitatively) with experimental phase diagrams of the elements. However, the model is in some ways too rich. The interplay between the range of the potential on one hand, and the distances to successive neighbor shells on the other, can lead to some very abrupt changes in the shape of the phase diagram, with entire regions vanishing as the range of the potential is increased by a differential amount. This is a consequence of the lack of any gradual attenuation of the pair energy with distance, so for a suitable range of the potential, fifth nearest neighbors (for example) interact with an atom just as strongly as nearest neighbors. As the range of attraction is varied below the value that permits some neighbor shell to interact when at close packing, the phase diagram undergoes a qualitative shift. This behavior is an artifact of the square-well model, and does not reflect the nature of any real system.

Thus, on the whole, the square-well potential is not a suitable model for the solid phase. An exception arises in the case of very short-ranged potentials, in which case the detailed structure of the attractive region of the potential is not important. By "short range" we refer to potentials in which the attractive interactions extend no more than about 20% beyond the size of the repulsive core. These models are suitable for describing certain macromolecules, such as the fullerenes, and colloidal systems. As the range of the attractive portion of the potential becomes smaller, the difference between the critical temperature and the triple-point temperature narrows. Ultimately a point is reached in which this difference vanishes, and for potentials of shorter attractive range there exists no (stable) liquid-vapor coexistence behavior; the transition is pre-empted by the freezing transition [166]. The same qualitative result has been observed in more complex short-range model potentials. These include sphericalized models for C_{60} [167,168], as well as a hard-core attractive Yukawa model of colloidal particles [169,170].

Ten Wolde and Frenkel [171] have made the very interesting observation that this hidden transition can nonetheless profoundly influence the crystallization behavior of the system. Fluids that are in the vicinity of this submerged critical point display substantial density fluctuations, just as they do when near a usual critical point. The crystallization mechanism in these instances proceeds by a route in which the fluid fluctuates to a solid-like density before arranging itself into a crystal form. This is in contrast to a mechanism in which the crystal first nucleates into a very small crystal, which then grows as it encounters additional fluid molecules. This understanding can contribute to the difficult art of crystallizing proteins. In fact, successful crystallizations have been known to be associated with a fluid opalescence that previously was not considered to be in any way related to the same effect seen in critical fluids. Density-functional approaches have since been applied and found to support the ten Wolde–Frenkel hypothesis [172].

For extremely narrow ranges of attraction (less than 7% of the hard core), the square-well model (or any comparable model with the same range of attraction) exhibits an isostructural solid-solid transition. In this transition the solid undergoes a condensed-to-expanded transition between two crystals of the same symmetry, much like a fluid liquid-to-vapor transition, complete with a true critical point [173,174] (such transitions are found also in longer-ranged square-well models, but again this is an unrealistic artifact of this model). The locations of the coexistence curves (the expanded- and condensed-solid densities) depend strongly on the range of attraction, but the critical temperature is insensitive to this quantity. In fact, through some very clever transformations, Bolhuis et al. [174] were able to demonstrate that the transition persists—the critical temperature remains finite (nonzero)—even in the limit of an infinitely narrow range of attraction. In a separate study [175], Bolhuis and Frenkel found that this type of transition exists also in repulsive short-ranged "shoulder" potentials.

C. Soft Spheres and the Lennard-Jones Model

As an alternative to adding attraction, the hard-sphere model may be made more realistic by making the repulsion softer. The soft-sphere or inverse-power model is the prototype for introducing softness to the potential. This potential is of the form

$$u(r) = \varepsilon(\sigma/r)^n \qquad (2.45)$$

where n is a parameter that governs the softness of the repulsion. As n grows, the potential hardens and becomes shorter ranged, reaching the hard-sphere model in the limit $n \to \infty$; for $n \leq 3$ range of the potential is infinite and its thermodynamics is undefined unless a neutralizing background potential is introduced. In particular, the case $n = 1$ defines the so-called one-component plasma [176,177]. The parameters ε and σ are redundant, and it is easy to see that the only relevant potential parameter (other than n) is the group $\varepsilon\sigma^n$. This feature results in several simplifications to the model [70].

The stable form of the one-component plasma at freezing is bcc, while that for hard spheres is fcc. For large n, the bcc phase is not only thermody-namically unstable but is mechanically unstable with respect to shear. For intermediate n a bcc-fcc-fluid triple point arises at about $n = 6$ [178,179], providing the demarcation of the regions of fcc and bcc freezing. The first molecular simulation study of the effect of the parameter n on the freezing diagram was performed by Hoover and co-workers [69,70,180]; a few subsequent studies [57,181] examined particular values of n. The complete n-dependence (from hard spheres to n approaching 3), including the emergence of the bcc phase, was evaluated by the Gibbs-Duhem integration method [178,179], with n taken as the thermodynamic field variable of integration; the integration was begun with knowledge of the hard-sphere solid-fluid transition pressure.

The $n = 12$ soft sphere model is the high-temperature limit of the 12-6 Lennard-Jones (LJ) potential. Agrawal and Kofke [182] used this limit as the starting point for another Gibbs–Duhem integration, which proceeded to lower temperatures until reaching the solid-liquid-vapor triple point. The complete solid-fluid coexistence line, from infinite temperature to the triple point, can be conveniently represented by the empirical formula [182]

$$p = \beta^{-5/4}\exp(-0.4759\beta^{1/2})[16.89 - 7.19\beta - 3.028\beta^2] \qquad (2.46)$$

where the pressure p and reciprocal temperature β are given in units of the LJ size and energy parameters.

In comparison to the "simpler" models just discussed, the Lennard-Jones phase diagram is relatively uninteresting. However, the model has been used as a test of various melting rules reviewed in Section III.E. This examination was performed by Agrawal and Kofke in their study of the freezing behavior

[182] (the subsequently developed freezing rule of Giaquinta and Giunta was of course not examined). The simulations show that the rule of Raveche et al. provides the best characterization of the solid-fluid transition in the Lennard–Jones model; its application would yield a freezing density within 1% of the correct value, over the entire range of temperatures from the triple point to infinite temperature (i.e., pure soft-sphere repulsion).

Several theoretical investigations of SFE for the Lennard-Jones 12-6 potential have been made. DFT has been applied by several groups. Two approaches have been taken to the treatment of the liquid phase properties for input to the DFT. Haymet and his co-workers [183,184] have used the mean spherical approximation applied directly to the 12-6 liquid and also parameterized computer simulation results. Other workers have used first-order perturbation theory with a hard-sphere reference system or mean field theory [122,123,185]. Generally the coexistence densities obtained from these approaches are in quite good agreement with computer simulation results. Henderson and Barker [138] presented predictions for the solid-fluid equilibrium in the 12-6 system in which perturbation theory was used for the fluid phase and the LJD cell theory for the solid phase. They found good agreement with experimental data for argon for the solid-fluid coexistence pressure versus temperature. Cottin and Monson repeated these calculations using an accurate equation of state for the 12-6 fluid [109]. They found good agreement with the simulation results of Hansen and Verlet [156] and with those of Agrawal and Kofke [182], however systematic errors were apparent in the coexistence densities. They found that significant improvement could be achieved by including an approximate resummation of the cell cluster expansion corrections to the LJD theory.

DFT has been much less successful for the soft repulsive sphere models. The definitive study of DFT for such potentials is that of Laird and Kroll [186] who considered both the inverse power potentials and the Yukawa potential. They showed that none of the theories existing at that time could describe the fluid to bcc transitions correctly. As yet, there is no satisfactory explanation for the failure of the DFTs considered by Laird and Kroll for soft potentials. However, it appears that some progress with such systems can be made within the context of Rosenfeld's fundamental measures functionals [130].

D. Yukawa Model

The Yukawa model is described by a screened-Coulomb pair potential

$$u(r) = \frac{\varepsilon}{r/\sigma}\exp\left[-\kappa(r/\sigma - 1)\right] \qquad (2.47)$$

Here κ is an inverse screening length, and ε describes the strength of the interaction. The Yukawa model is used to describe charge-stabilized colloidal systems, and it is also appropriate to characterize the behavior of small particles in plasma environments. In the limit of zero screening the Yukawa model reduces to the one-component plasma, which is also a limiting case of the inverse-power (soft-sphere) model; for large κ the model becomes more like the hard-sphere potential. To stiffen up the short-range repulsion, the Yukawa model can be modified by the addition of a hard core. Usually this is done while changing the sign of the Yukawa potential, thereby turning it into an attractive tail. The model so formed then exhibits vapor-liquid coexistence.

Several molecular simulation studies have been published in the past decade regarding the solid-phase coexistence behavior of purely repulsive [187–195] and attractive [169,170,196,197] Yukawa models; binary mixtures have also been examined approximately [198].

The picture that emerges from these studies is qualitatively very much like that found for other simple models. As with the soft-sphere model, the purely repulsive, short-ranged Yukawa model (large κ) freezes into a fcc structure, and as the potential is made softer and longer ranged the bcc solid becomes stable at freezing. A fcc-bcc-fluid triple point arises at intermediate values of κ [194]; in fact, a plot of the Yukawa potential at this screening coincides almost perfectly with the soft-sphere potential at its fcc-bcc-fluid triple point [179]. The hard-sphere model with a short-ranged attractive Yukawa tail (large κ) exhibits a regime in which the vapor-liquid phase transition sinks beneath the sublimation curve; it also exhibits a condensed-to-expanded isostructural solid transition. Both of these behaviors have been observed in the square-well and other models.

Thus the qualitative features of the solid-fluid coexistence behavior of the Yukawa model are familiar from the models discussed above. Given the aim of the potential to model colloidal systems, one might hope for more interesting behaviors. For example, recent experiments [199] indicate that the Yukawa model inadequately describes qualitative features of the ordering transition of highly charged colloidal systems, missing in particular a re-entrant transition to the fluid phase with increasing charge density.

E. Gaussian Core Model

Stillinger and Webber [200,201] have studied the phase behavior of the Gaussian core model. In this model the pair potential is given by

$$u(r) = u_0 \exp[-(r/l)^2] \tag{2.48}$$

This model has the mathematical advantage that the coefficients in the asymptotic inverse temperature expansion of the free energy may be evaluated analytically. It also exhibits very interesting behavior in its solid-fluid phase diagram. In particular, it has a maximum in its melting temperature as a function of density so that the system can exhibit both positive and negative volumes of mixing. It is remarkable that such a simple model can exhibit behavior normally associated with more complex systems with molecular anisotropy or hydrogen bonding.

F. Solid-Fluid Equilibrium in Mixtures

Solid-fluid phase diagrams of binary hard sphere mixtures have been studied quite extensively using MC simulations. Kranendonk and Frenkel [202–205] and Kofke [206] have studied the solid-fluid equilibrium for binary hard sphere mixtures for the case of substitutionally disordered solid solutions. Several interesting features emerge from these studies. Azeotropy and solid-solid immiscibility appear very quickly in the phase diagram as the size ratio is changed from unity. This is primarily a consequence of the nonideality in the solid phase. Another aspect of these results concerns the empirical Hume–Rothery rule, developed in the context of metal alloy phase equilibrium, that mixtures of spherical molecules with diameter ratios below about 0.85 should exhibit only limited solubility in the solid phase [207]. The simulation results for hard sphere tend to be consistent with this rule. However, it should be noted that the Hume–Rothery rule was formulated in terms of the ratio of nearest neighbor distances in the pure metals rather than hard sphere diameters. Thus, this observation should be interpreted as an indication that molecular size effects are important in metal alloy equilibria rather than as a quantitative confirmation of the Hume–Rothery rule.

Eldridge et al. [208–212] have studied the SFE for hard sphere mixtures with size ratios where substitutionally ordered solid solutions (also known as compounds) can be formed. Three types of compound have been found to be stable: AB, AB_2, and AB_{13}. The AB compound has the NaCl structure and the AB_2 compound has the aluminium boride (AlB_2) structure. The AB_{13} compound unit cell consists of eight simple cubic cells of the large spheres with an icosahedral array of small spheres at the center of each cube. For much of the range of size ratios where they are stable, the AB and AB_2 compounds have higher maximum packing densities than their phase-separated components and on the basis of simple free volume arguments this would be expected to contribute to their stability. This is not the case for the AB_{13} compound, although its maximum packing density is not much lower than that of the phase-separated components. The idea that hard spheres could form such solid phase mixtures emerged in the work of Murray and Sanders [213,214], who discovered the AB_2 and AB_{13} structures in samples of natural opal. Several interesting features of the packing are discussed in that work. These compounds have also been seen in colloidal systems that closely mimic hard-sphere behavior [215–218] and more recently in non-hard-sphere systems including mixtures of C_{60} with carbon tetra-chloride [219,220].

DFT studies of binary hard-sphere mixtures predate the simulation studies by several years. The earliest work was that of Haymet and his co-workers [221,222] using the DFT based on the second-order functional Taylor expansion of the $A_{ex}[\rho]$. Although this work has to some extent been superceded, it was a significant stimulus to much of the work that followed both with theory and computer simulations. For example, it was Smithline and Haymet [221] who first analyzed the Hume–Rothery rule in the context of hard sphere mixture behavior and who first investigated the stability of substitutionally ordered solid solutions. The most accurate DFT results for hard-sphere mixtures have come from the WDA-based theories. In particular the results of Denton and Ashcroft [223] and those of Zeng and Oxtoby [224] give qualitatively correct behavior for hard spheres forming substitutionally disordered solid solutions.

The cell theory plus fluid phase equation of state has been extensively applied by Cottin and Monson [101,108] to all types of solid-fluid phase behavior in hard-sphere mixtures. This approach seems to give the best overall quantitative agreement with the computer simulation results. Cottin and Monson [225] have also used this approach to make an analysis of the relative importance of departures from ideal solution behavior in the solid and fluid phases of hard-sphere mixtures. They showed that for size ratios between 1.0 and 0.7 the solid phase nonideality is much more important and that using the ideal solution approximation in the fluid phase does not change the calculated phase diagrams significantly.

In reviewing the phase diagrams of hard-sphere mixtures, Cottin et al. [226] made note of the fact that binary hard-sphere mixtures exhibit many of the features of binary phase diagrams of organic mixtures as surveyed by Matsuoka [227]. In a survey of 1500 mixtures Matsuoka found that 95% of the mixtures exhibited behavior that could be described by one of six types of phase diagram. Only one of these phase diagram types, which exhibits a congruently melting compound, is not observed for hard-sphere mixtures. The point to be made here is not that hard-sphere mixtures are a quantitative model for binary organic systems but that molecular size differences should also play a significant role in determining the solid-fluid phase behavior of more complex mixtures.

Theoretical approaches have been applied to binary Lennard-Jones 12-6 mixtures by Rick and Haymet, who used DFT [222], and by Cottin and Monson [109], who used their cell theory plus fluid phase equation of state approach. The latter approach gave qualitatively good agreement with experimental results for binary mixtures involving argon, krypton, and methane. Quantitative agreement with experiment was found to be sensitive to the choice of parameters used in the Lennard-Jones potentials. Very recently, Hitchcock and Hall [228] have presented a Monte Carlo simulation study of solid-fluid phase equilibrium in binary Lennard-Jones mixtures focusing on solid phases that are substitutionally disordered solid solutions. Their results agree quite well with the theoretical predictions of Cottin and Monson, but the simulations give systematically higher melting points than those predicted by the theory. Hitchcock and Hall have also used their simulations to study how the phase diagram changes with the Lennard-Jones parameters. Consistent with the findings of studies of hard-sphere binaries, they observe the strong influence of size parameters on the qualitative nature of the phase diagram, but they also observe that the energy-parameter ratio of the two species can also strongly affect the type of phase behavior exhibited by the mixture.

In somewhat earlier work, Vlot et al. [229,230] made calculations of Lennard–Jones binary mixtures in which the pure components are identical but in which the unlike interactions have departures from the Lorentz–Berthelot combining rules. They use this as a model of mixtures of enantiomers. A variety of solid-fluid phase behavior can be obtained from the model. Both substitutionally ordered and substitutionally disordered solid solutions were found to occur.

G. Nonspherical Hard-Core Molecules

Nonspherical hard core particles are an important model of the effects of molecular shape in determining phase diagrams. Such effects range from the occurrence of liquid crystal phases to the stability of plastic crystal solid phases. Like the hard-sphere model, they can also serve as the basis of perturbation and generalized van der Waals theories for systems with attractive forces. Here we will focus on the behavior that occurs for models with anisotropies appropriate for the treatment of small linear molecules such as nitrogen, ethane, or carbon dioxide and where liquid crystallinity does not occur. The primary issue here is the extent of orientational order in the solid phase and the influence this has on the phase diagram. Three models of linear molecules have been extensively studied by MC simulation: the hard diatomic (dumbbell), the hard ellipsoid, and the hard spherocylinder. The qualitative behavior of these model systems turns out to be quite similar for the anisotropies of interest here [231]. We will focus on results for the hard dumbbell system inasmuch as since these have been studied most extensively in the context of SFE, and the model has also been the most frequently studied using theories of SFE. The ellipsoid [232–236] and spherocylinder [237,238] models have been studied primarily in relation to their liquid crystalline properties for larger anisotropies, a topic outside the scope of the present review. Moreover, the properties of the hard ellipsoid and spherocylinder models have recently been the subject of a review in this series [236].

The hard dumbbell model consists of two hard spheres of diameter σ held together by a rigid bond of length L. SFE in this model has been extensively studied by Singer and Mumaugh [239] and by Vega et al. [60,231,240]. Singer and Mumaugh showed that hard dumbbells freeze into an orientationally disordered or plastic crystal phase for mild anisotropies. The plastic crystal phase is similar to the β-phase of solid nitrogen. Vega et al. determined orientationally ordered crystal structures that allow hard dumbbells to reach their maximum packing density [60]. They showed that there are at least three different orientationally ordered structures that can reach the same maximum packing density. The plastic crystal phase becomes unstable with respect to these orientationally ordered solid phases for larger anisotropies and at high pressure. The structures of these orientationally ordered phases are related to the stable solid phases of the halogens. The coexistence densities for hard dumbbells are illustrated in Figure 2.7. The dependence of the freezing properties of hard dumbbells upon L is really quite remarkable. For example, the fluid phase of hard

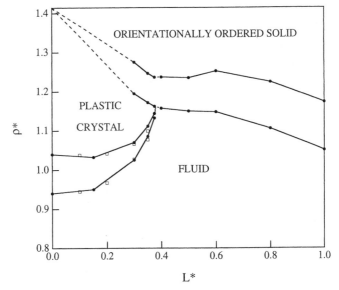

Figure 2.7. Reduced coexistence densities ($\rho^* = \rho d^3$, where d is the diameter of the sphere with the same volume as the dumbbell) for hard dumbbells versus reduced bondlength ($L^* = L/\sigma$). The filled circles and lines are the results of Vega et al. [60,240] and the open squares are the results of Singer and Mumaugh [239]. The dashed lines give estimates of the plastic crystal to orientationally ordered solid.

dumbbells may be stable at volume fractions up to 25% greater than the fluid phase of hard spheres. The strongest dependence upon the anisotropy occurs for values of L just less than where the plastic crystal becomes unstable $(L \sim 0.38 \, \sigma)$. In this region the coexistence properties show a sharp increase with increasing L and the density change on freezing becomes very small. Interestingly, nitrogen (which is often modeled as a dumbbell with $L = 0.33 \, \sigma$) exhibits a relatively small density change on freezing into its β-phase compared with that seen in the freezing of argon. On this basis, Vega et al. [60,240] argued that the fluid to plastic crystal transition in hard dumbbells provides a good structural model for the freezing of liquid nitrogen.

For the case of dumbbells formed from tangent spheres, $L = \sigma$, it is also possible to pack the dumbbells in orientationally disordered structures in which the spheres of the dumbbell lie on an fcc lattice but the centers of mass form an aperiodic structure. This idea has been explored more extensively for two-dimensional dumbbells by Wojciechowski et al. [241,242]. The configurational degeneracy of the aperiodic structure renders it more stable than the orientationally ordered structures at all densities, even though the effect upon the equation of state and the free energy without the contribution from the degeneracy is quite small [60]. The freezing of hard dumbbells into such structures has also been studied by Bowles and Speedy [243].

Haymet and co-workers [244] and Chandler and co-workers [245] have applied the DFT to hard dumbbells. The theories differ in their approach to the calculation of the fluid state pair correlations for nonspherical molecules. The theory developed by Chandler and co-workers [245–247] is based upon the interaction site formalism whereas that of Haymet and co-workers is based upon the full orientation dependent direct correlation function. Smithline et al. [244] found that the plastic crystal is always stable with respect to orientationally ordered structures and that within the accuracy of their method there appears to be no difference in the stabilities of the fcc and hcp plastic crystal structures. McCoy et al. [245] reached similar conclusions, although their study was restricted to orientationally disordered solid phases. Interestingly, both groups found that as the diatomic bond length was increased, a point was reached beyond which the fluid phase was stable with respect to all the solid phases considered. The maximum bond length for which the plastic crystal is stable is significantly less than that found in the simulations. It seems plausible that it is failure of approximations in the DFT that leads to the observed differences. Paras et al. [100] have applied the cell theory plus fluid phase equation of state approach to hard dumbbells and obtained significantly better agreement with the simulation results than was obtained from the DFT and the approach can describe the SFE for orientationally ordered solids. This approach has been extended to heteronuclear dumbbells by Rainwater and co-workers [248].

Paras et al. [152] have extended the Longuet–Higgins–Widom generalized van der Waals theory to the solid-fluid equilibria of diatomic molecules using the hard dumbbell model as the reference system. One of the principal results of this work was a prediction of the variation of the reduced triple point temperature (defined as the triple point temperature as a fraction of the critical temperature) with molecular anisotropy. When the stable solid phase is a plastic crystal the reduced triple point temperature as decreases with increasing molecular anisotropy. This is consistent with the observed behavior for nitrogen and oxygen, both of which freeze into plastic crystals. When the stable solid phase is orientationally ordered, the triple point temperatures show a slight increase with increasing molecular anisotropy. This is to some extent consistent with the observed behavior for the halogens. An adaptation of the approach to systems with quadrupolar interactions offers a plausible explanation for the especially high reduced triple point temperature and large density change on melting of carbon dioxide.

H. Models of Chain Molecules

In recent work some computer simulation studies have been presented dealing with the important problem of calculating solid-fluid phase equilibrium for chain molecules. Malanoski and Monson [62] and Polson and Frenkel [63] have studied SFE for chains of tangent spheres. Both groups have adapted the Einstein crystal method of calculating the free energy to chain molecules. In the work of Malanoski and Monson [62] freely jointed tangent hard sphere chains were considered. The goal was to investigate the feasibility of calculating the free energy in the solid phase. Polson and Frenkel [63] considered the case of tangent Lennard-Jones 12-6 spheres and investigated the effect of variable chain stiffness. They found that increasing the chain stiffness stabilized the solid phase.

The n-alkanes are an important prototypical system for studying the effect of molecular shape and chain length on solid-fluid phase diagrams of chain molecules. The solid-phase equilibrium properties of n-alkanes exhibit odd-even effects that are related to the way in which the chains are packed in the solid phase [249]. Such effects cannot be described with models that do not restrict the C–C–C bond angles. Malanoski and Monson [65] have considered united atom hard sphere chain models that give a geometrically realistic model of the n-alkanes. They found that if the hard sphere chain results were used as reference system data in a generalized van der Waals or mean field theory for the phase diagram, the predictions for the chain length dependence of the ratio of the triple point temperature to the critical temperature, T_t/T_c, were in qualitatively good agreement with the experimental results if, in addition to the chain structure, the torsional potential was modeled realistically. Polson and Frenkel [66] have presented a calculation of the melting line for a model of n-octane that includes both carbon and hydrogen forces centers. Good agreement with experiment was obtained. Polson and Frenkel emphasized the considerable computational investment required to investigate this more complex model. The approach of Malanoski and Monson [65] allows for a much wider ranging study of the effects of chain length and chain flexibility upon the phase behavior, although at the cost of being able to achieve quantitative agreement with experiment.

DFT has been extended to SFE in chain molecules by Yethiraj et al. [250] and applied to the freezing of polyethylene. Now that simulation results are available for molecular models of chain molecules, it will be interesting to investigate the performance of DFT for such model systems. The cell theory has recently been applied to the SFE in the hard chain models considered by Malanoski and Monson [62] with results comparable in accuracy to those achieved for hard spheres and hard dumbbells [251].

I. Polydisperse Systems

Polydispersity arises in systems composed of particles characterized by a property (e.g., particle diameter) that spans a continuum of values. Small molecules exhibit discrete properties, so they do not form polydisperse mixtures. Only at the level of macromolecules and colloidal aggregates does polydispersity become an issue. Here variations in particle size are known to influence the ordering into a solid phase. Experimentally it has been observed that colloidal systems will not form a solid phase if the size polydispersity (as measured by the standard deviation of the particle-size distribution) is greater than about 5% to 10% of the average diameter [252].

The hard-sphere model was originally formulated to understand the behavior of atomic and molecular systems, but in some ways it is a even more useful model for simple colloidal systems, where the range of attraction can be made very short relative to the particle size [162,253]. In this context it is quite natural to consider the effect of size polydispersity on the freezing behavior of hard spheres. Various theories [254,255], including an application of DFT by McRae and Haymet [256], and simulation results [257–259] have supported the notion that a sufficiently polydisperse hard-sphere system will not form a solid in the way that pure hard spheres do. The solid and fluid phases exhibit different symmetries, so the termination of the coexistence region with increasing polydispersity cannot represent a critical point as it is usually understood (Bolhuis and Kofke [260] have thus avoided the use of "critical polydispersity" in referring to this point, instead choosing to call it the "terminal polydispersity"). All of the earlier studies approximate the transition as one exhibiting no fractionation, meaning that the size distribution is the same in both phases. This is reasonable if the systems are sufficiently monodisperse, but if one is interested in examining and understanding the terminal polydispersity it is a poor approximation.

Bolhuis and Kofke [260] performed the first rigorous calculation of the freezing behavior of hard spheres as a function of size polydispersity. They employed the Gibbs–Duhem integration technique to trace the solid-fluid coexistence line as the variance of the (Gaussian) activity-ratio distribution was increased from zero, the monodisperse hard-sphere limit. Working with the activity ratio directly, rather than the composition distribution, has the advantage of making it easy to enforce equality of chemical potential distributions between the phases. It also permits for a natural termination of the coexistence line with increasing overall polydispersity. What is found is that the solid phase reaches a limit of about 5.5%, while the coexisting fluid, owing to fractionation, can adopt arbitrarily large polydispersities. Below this terminal polydispersity of the solid, reentrant melting behavior is observed: at low pressure and density the fluid is stable; upon compression, a crystalline solid forms; further compression leads again to a melting transition, with a more monodisperse solid in equilibrium with a more polydisperse dense amorphous phase (most likely a glass). Note that the polydispersity of each phase is measured relative to its average diameter, so by this definition a more polydisperse fluid can be obtained from a narrow parent distribution if it receives an abundance of the smaller particles. The stability region of the solid is presented in Figure 2.8. The reentrant behavior was emphasized in a subsequent study that applied an approximate mean-field treatment to the same system [261].

Away from the nearly monodisperse limit, and where fractionation is significant, the width of the diameter distribution in the solid is determined not by the width of the parent distribution, but instead by consideration of how well large and small particles can cohabit the same lattice structure [262]. Here the saturated solid phase can be characterized by a single dimensionless group, constructed from a ratio of the pressure to the slope of the chemical potential distribution across the range of diameters present in the solid. This is a great simplification in the characterization of this phase: it leads to a "universal" form (within the hard-sphere model) for the solid-phase "shadow curve," one that is independent of the composition of the fluid from which it is precipitating. Kofke and Bolhuis [262] used this observation as the basis for a cell theory treatment of the high-density saturated solid, and thereby obtained the upper stability limit presented in Figure 2.8.

Cuesta and co-workers [263,264] have looked at the issue of demixing in several types of polydisperse fluids, but the issue of compound formation

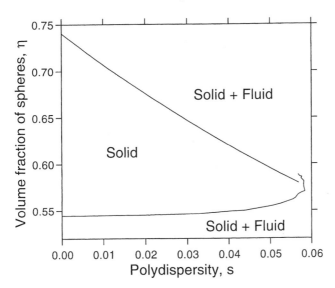

Figure 2.8. Region of thermodynamic stability of the solid phase of polydisperse hard spheres. Volume fraction η occupied by the spheres is plotted against the polydispersity s, defined as the standard deviation of the sphere diameter distribution, divided by the mean. Lower curve shows results [260] from semigrand Monte Carlo simulations of hard spheres with a Gaussian activity distribution. The upper curve is computed from a semiempirical cell model [262] for a linear chemical-potential distribution; it is $\eta = \frac{\pi}{6} 2^{1/2} (1 + 1.563\,s)^{-3}$. As described in Ref. [262], the envelope is a (nearly) invariant melting curve for unimodal polydisperse hard spheres. The freezing lines describing the coexisting fluid phases are off the diagram to the right and/or below, and (unlike the melting envelope shown here) have a shape and location that depend on the details of the overall composition distribution.

and demixing in solid-phase polydisperse systems has not yet been addressed in any molecular simulation study.

More recently, Lacks and Weinhoff [265] have examined the mechanical stability of r^{-12} soft spheres as a function of size polydispersity. In their study they minimize the potential energy of an fcc array of polydisperse spheres, and examine how the structure of the minimum-energy configuration changes with increasing polydispersity. An appropriate structural order parameter for the fcc solid is found to drop precipitously for polydispersities in the range of 10% to 15%. The absence of energy minima in the ordered phase indicates that such a system is not mechanically stable, and therefore cannot be thermodynamically stable.

J. Systems with Electrostatic Interactions

Classical electrostatic interactions play an important role in determining the structure and thermodynamics of condensed phase systems. Here we discuss three contexts where the effects of such interactions have been incorporated into simple molecular models of solid-fluid coexistence.

1. Quadrupolar and Dipolar Systems

An extensive study of the phase diagrams of the quadrupolar hard dumbbell molecular model has been made [266]. The goal of this work was to understand the interplay between the molecular shape and the quadrupole-quadrupole interactions in determining SFE for linear quadrupolar molecules. Quadrupole-quadrupole interactions tend to stabilize lower density solids than would be found on the basis of hard core packing. Vega and Monson [266] also used these results as reference system data in a refined version of the generalized van der Waals theory of solid-fluid equilibrium [152]. This theory provides a quite accurate picture of the phase diagrams of quadrupolar molecules like carbon dioxide and acetylene, which have very high triple point temperatures in relation to their critical temperatures. A key ingredient in this behavior is the occurrence of an expanded solid phase that is stabilized by the quadrupole-quadrupole interactions. However, the coupled effect of molecular shape on the thermodynamic properties is also important.

Rainwater and his co-workers have extended their work on heteronuclear hard dumbbells to the treatment of dipolar systems [267]. They have used a generalized van der Waals theory similar to that used for quadrupolar molecules [152,266] to calculate the phase diagram and have compared their results with experiment for methyl chloride. The theory correctly predicts the effect of the dipole moment on the stable crystal structure.

2. Hydrogen Bonding

A comprehensive program of understanding of the molecular basis of SFE must also address the role of short-range directional forces present in hydrogen-bonded systems such as water. The study of the liquid-solid transition for molecular models of water has received some attention recently. Some of these studies have focused on the use of realistic or semirealistic intermolecular potentials developed in the study of liquid water. For example, the crystallization from the liquid phase via molecular dynamics has been considered by Svishchev and Kusalik [268] and Baez and Clancy [269] have determined the fluid-solid transition for two water potentials. Another type of study focuses on trying to identify the essential features that are needed in a molecular model to reproduce qualitatively the thermodynamics and structure in the neighborhood of the solid-fluid transition. In this context Speedy [270] has studied the dense phases of a simple tetravalent network-forming model system using molecular dynamics. This model exhibits several interesting features including a transition from an ice-like phase to a more dense amorphous structure. This behavior resembles the melting of ice under pressure. A key feature of Speedy's model is the presence of repulsions between pairs of nonbonded molecules in the network. These repulsions are sufficiently long ranged to act between next nearest neighbors. Speedy argues that this is the key ingredient in the model that allows it to mimic the high-density behavior of water. A less attractive feature of the model is that the connectivity of the network is permanent. This feature is presumably acceptable for the properties of the solid but seems less appropriate for the liquid where fluctuations in the network connectivity are more important. Indeed there is evidence that the crystal-to-amorphous phase transition in Speedy's model terminates at a critical point.

Vega and Monson [64] have recently presented a study of SFE for a molecular model that was developed in somewhat the same spirit as that of Speedy, but that permits fluctuations in the connectivity of the hydrogen bond network. This is the "primitive model of water" (PMW) developed by Nezbeda and his co-workers [271] and consisting of hard spheres with tetrahedrally arranged square well association sites. The PMW exhibits short-ranged repulsion and short-ranged directional forces that are saturated when the molecules are tetrahedrally coordinated. Thus it is capable of describing the effects of association. Moreover, network formation is reversible so that changes in connectivity associated with first-order phase transitions can be modeled. Vega and Monson considered two solid phases and a fluid phase of the PMW. The idea was to investigate the relative stability of high- and low-density solid phases and the role this plays in the phase diagram. They showed that a low-density (ice-I-like) solid phase exhibits reentrant melting (i.e., the low-density solid could be isothermally compressed to give a liquid phase of higher density) but that this transition occurs at states where both fluid and low-density solid are metastable with respect to a high density (ice-VII-like) solid. In this version of the model the high-density phase is stable down to much lower pressures than might have been anticipated.

Although these simplified models of hydrogen-bonded systems give a far from complete picture of the solid-fluid phase behavior of water, this kind of approach to identifying the key features required in the molecular model is an instructive one. Indeed, the inability of the PMW to generate reentrant melting of the low-density solid at thermodynamically stable states is an important result. It shows us that more than just short-range directional forces are required for this to occur.

3.   Ionic Systems—The Restricted Primitive Model

One of the simplest realistic molecular models of an ionic solution is the restricted primitive model. This consists of an equimolar mixture of oppositely charged but equal-sized hard spheres in a dielectric continuum.

Although most of the studies of this model have focused on the fluid phase in connection with the theory of electrolyte solutions, its solid-fluid phase behavior has been the subject of two recent computer simulation studies in addition to theoretical studies. Smit et al. [272] and Vega et al. [142] have made MC simulation studies to determine the solid-fluid and solid-solid equilibria in this model. Two solid phases are encountered. At low temperature the substitutionally ordered CsCl structure is stable due to the influence of the coulombic interactions under these conditions. At high temperatures where packing of equal-sized hard spheres determines the stability a substitutionally disordered fcc structure is stable. There is a triple point where the fluid and two solid phases coexist in addition to a vapor-liquid-solid triple point. This behavior can be qualitatively described by using the cell theory for the solid phase and perturbation theory for the fluid phase [142]. Predictions from density functional theory [273] are less accurate for this system.

V. CONCLUSION

It has long been known from experiment that the thermodynamic phase behavior of solids can be very rich. Polymorphism, compound formation, demixing, and other effects are all very common, and even the interpretation of the temperature-composition phase diagram of solids can challenge the brightest newcomer to the field. Given all this richness, it is a surprise and a pleasure to learn that most, if not all, of these complexities can be explained qualitatively with the simplest molecular models. Hard spheres form solids; two-component hard-sphere mixtures form compounds and azeotropes, and demix; polydispersity bounds freezing of hard spheres just as it does in real colloidal spheres. Extensions of the hard-sphere model to nonspherical molecules can describe phenomena ranging from orientational order in diatomic solids through the melting points of chain molecules and can even begin to describe the anomalies in the SFE of hydrogen bonded systems. It is a true triumph of statistical mechanics and molecular modeling to be able to explain such complexity from such simple origins.

In solids, as in liquids, macroscopic behavior is a consequence of microscopic structure. Even more, in solids the structure defines the thermodynamic phase, and deviations from the nominal structure are true anomalies—defects. In contrast, defects are so prevalent in fluids that the term loses currency; fluids are described instead by fluctuations, reflecting the diminished (albeit consequential) role of structure in fluid-phase behavior. Structure in solids is a much more cooperative and large-scale phenomenon than in liquids. This means that changes in the structure of solids do not happen incrementally or in isolation. Changes in structure are actually changes in thermodynamic phase. The richness of the phase behavior of solids arises from the all-or-nothing nature of the response of structure to changes in state or in the Hamiltonian.

In both fluids and solids structure is determined by the parts of the intermolecular potential that vary most quickly with the intermolecular separation and relative orientation—we refer to these as the most forceful parts of the potential. Usually this statement is interpreted narrowly, referring only to the repulsive core, which leads one to the conventional wisdom that the shape of a molecule (its steric exclusion region) is the primary determinant of structure. However, in truth the more general statement applies, and thus gives a more comprehensive explanation of the origins of structure. The attractive part of the potential can be as forceful—varying rapidly over a short range—as the repulsive core. In such cases attraction plays a crucial role in determining structure. For fluids the effects are usually more quantitative than qualitative, but in solids the outcome can be much more dramatic. The effect of forceful attraction is demonstrated in several of the models reviewed here, including the short-ranged square well and the hydrogen-bonding prototypes.

It is axiomatic to say that the forceful pair configurations occupy only a small part of the potential-energy surface. Thus details of what happens in these narrow regions are much less important than the basic size, shape, and position of the regions themselves. Consequently, simple models of the type reviewed here suffice to characterize them. Given this sufficiency, and that these regions largely govern the macroscopic phase behavior, it follows that simple models should be capable of capturing the complexities of phase equilibria involving solids. This outcome is what the research in this field has amply demonstrated.

Perhaps more than any other tool, molecular simulation has played an indispensable role in the development of the insights into molecular behavior we have reviewed in this Chapter. Before simulation was possible, even the existence of a stable hard-sphere crystal was in doubt, despite many years of attention to the question. Much of the progress reviewed here has occurred in the past decade, coinciding with the widespread availability of very powerful, inexpensive computers; equally important have been advances in molecular simulation methodology as applied to solid phases.

This is not to say that the theoretical work on the problem of SFE has not been important. The search for theoretical methods for predicting the solid-fluid transition ranging from the early work of Kirkwood to the most recent studies with DFT have made an important contribution to our understanding, as well as serving as a stimulus to other efforts in experiment and computer simulation. Moreover, our general understanding of the statistical mechanics of condensed phase systems has benefited considerably from this work. On the other hand, theoretical approaches have been of limited utility for systems beyond spherical molecules. For example, except for the two-phase cell theory plus fluid phase equation of state approach discussed in Section III.C, little success has been achieved in dealing with the freezing of molecular liquids. Even for spherical molecules the problem of SFE for systems with soft-core repulsions remains largely unsolved. In a climate where very fast computers are increasingly widely available at low cost there is tendency for theoretical work of this kind to be supplanted by research programs entirely based on MC or MD simulation. Nevertheless, there is no substitute for the insights achieved through the careful formulation of approximate solutions to the statistical mechanics.

Acknowledgments

The authors acknowledge support from the U.S. Department of Energy for research in this area to PM at the University of Massachusetts (contract DE–FG02-90ER14150) and to DK at SUNY-Buffalo (contract DE-FG02-96ER14677). PM is grateful to Xavier Cottin, Anthony Malanoski, Eleanor Paras, and especially Carlos Vega for their contributions to research described here.

REFERENCES

1. D. Frenkel, in *Liquids, Freezing and the Glass Transition*, edited by J. P. Hansen, D. Levesque, and J. Zinn-Justin (North-Holland, Amsterdam, 1991), pp. 693.

2. A. K. Arora and B. V. R. Tata, *Curr. Opinion Coll. Interf. Sci.* **78**, 49 (1998).

3. A. D. Dinsmore, J. C. Crocker, and A. G. Yodh, *Curr. Opinion Coll. Interf. Sci.* **3**, 5 (1998).

4. T. L. Breen, J. Tien, S. R. J. Oliver, T. Hadzic, and G. M. Whitesides, *Science* **284**, 948 (1999).

5. Y. N. Xia, J. A. Rogers, K. E. Paul, and G. M. Whitesides, *Chem. Rev.* **99**, 1823 (1999).

6. B. Kuchta and R. D. Etters, *Computers Chemistry* **19**, 205 (1995).

7. S. L. Price, *Phil. Mag. B* **73**, 95 (1996).

8. S. L. Price, *J. Chem. Soc. Faraday* **92**, 2997 (1996).

9. B. Kuchta, K. Rohlender, D. Swanson, and R. D. Etters, *J. Chem. Phys.* **106**, 6771 (1997).

10. M. Baus, *J. Stat. Phys.* **48**, 1129 (1987).

11. A. D. J. Haymet, *Science* **236**, 1076 (1987).

12. A. D. J. Haymet, *Ann. Rev. Phys. Chem.* **38**, 89 (1987).

13. M. Baus and J. F. Lutsko, *Physica A* **176**, 28 (1991).

14. R. Evans, in *Fundamentals of Inhomogeneous Fluids*, edited by D. Henderson (Dekker, New York, 1992).

15. A. D. J. Haymet, in *Fundamentals of Inhomogeneous Fluids*, edited by D. Henderson (Marcel-Dekker, New York, 1992), pp. 363.

16. H. Lowen, *Phys. Rep.* **237**, 249 (1994).

17. J. G. Kirkwood and E. Monroe, *J. Chem. Phys.* **8**, 845 (1940).

18. J. G. Kirkwood and E. Monroe, *J. Chem. Phys.* **9**, 514 (1941).

19. J. G. Kirkwood, E. K. Maun, and B. J. Alder, *J. Chem. Phys.* **18**, 1040 (1950).

20. B. J. Alder and T. E. Wainwright, *J. Chem. Phys.* **27**, 1208 (1957).

21. W. W. Wood and J. D. Jacobson, *J. Chem. Phys.* **27**, 1207 (1957).

22. N. A. Metropolis, A. W. Rosenbluth, M. N. Rosenbluth, A. H. Teller, and E. Teller, *J. Chem. Phys.* **21**, 1087 (1953).

23. W. G. Hoover and F. H. Ree, *J. Chem. Phys.* **47**, 4873 (1967).

24. W. G. Hoover and F. H. Ree, *J. Chem. Phys.* **49**, 3609 (1968).

25. D. Frenkel and A. J. C. Ladd, *J. Chem. Phys.* **81**, 3188 (1984).

26. A. D. Bruce, N. B. Wilding, and G. J. Ackland, *Phys. Rev. Lett.* **79**, 3002 (1997).

27. L. V. Woodcock, *Nature* **385**, 141 (1997).

28. S. Pronk and D. Frenkel, *J. Chem. Phys.* **110**, 4589 (1999).

29. P. W. Bridgman, *Proc. Acad. Arts Sci.* **70**, 1 (1935).

30. J. P. Hansen and I. R. McDonald, *Theory of Simple Liquids*, 2nd ed. (Academic Press, New York, 1986).

31. H. C. Longuet-Higgins and B. Widom, *Mol. Phys.* **8**, 549 (1964).

32. M. P. Allen and D. J. Tildesley, *Computer Simulation of Liquids* (Clarendon Press, Oxford, 1987).

33. J. M. Haile, *Molecular Dynamics Simulation: Elementary Methods* (Wiley, New York, 1992).

34. D. Frenkel and B. Smit, *Understanding Molecular Simulation: From Algorithms to Applications* (Academic Press, New York, 1996).

35. A. P. Sutton, *Phil. Trans. R. Soc. Lond. A* **341**, 233 (1992).

36. D. Frenkel, in *Computer Simulation in Chemical Physics*, edited by M. P. Allen and D. J. Tildesley (Kluwer Academic, New York, 1993), pp. 93.

37. P. Kollman, *Chem. Rev.* **32**, 2395 (1993).

38. M. P. Allen, in *Proceedings of the Euroconference on Computer Simulation in Condensed Matter Physics and Chemistry*, Vol. 49, edited by K. Binder and G. Ciccotti (Como, Italy, 1996), pp. 255.

39. D. A. Kofke and P. T. Cummings, *Mol. Phys.* **92**, 973 (1997).

40. R. W. Zwanzig, *J. Chem. Phys.* **22**, 1420 (1954).

41. K. Denbigh, *Principles of Chemical Equilibrium*, 3rd ed. (Cambridge University, Cambridge, 1971).

42. B. Widom, *J. Chem. Phys.* **39**, 2808 (1963).

43. J. L. Jackson and L. S. Klein, *Phys. Flu.* **7**, 228 (1964).

44. J. P. Valleau and D. N. Card, *J. Chem. Phys.* **57**, 5457 (1972).

45. C. H. Bennett, *J. Comp. Phys.* **22**, 245 (1976).

46. G. M. Torrie and J. P. Valleau, *J. Comp. Phys.* **23**, 187 (1977).

47. D. A. Kofke and P. T. Cummings, *Fluid Phase Equil.* **150**, 41 (1998).

48. D. A. Kofke, in *Monte Carlo Methods in Chemistry*, Vol. 105, edited by D. M. Ferguson, J. I. Siepmann, and D. G. Truhlar (New York, John Wiley & Sons, 1998).

49. M. Watanabe and W. P. Reinhardt, *Phys. Rev. Lett.* **65**, 3301 (1990).

50. C. Jarzynski, *Phys. Rev. A* **46**, 7498 (1992).

51. L. W. Tsao, S. Y. Sheu, and C. Y. Mou, *J. Chem. Phys.* **101**, 2302 (1994).
52. M. de Koning and A. Antonelli, *Phys. Rev. E* **53**, 465 (1996).
53. S. R. Phillpot and J. M. Rickman, *J. Chem. Phys.* **94**, 1454 (1991).
54. A. M. Ferrenberg and R. H. Swendsen, *Phys. Rev. Lett.* **61**, 2635 (1988).
55. A. M. Ferrenberg and R. H. Swendsen, *Phys. Rev. Lett.* **63**, 1195 (1989).
56. A. M. Ferrenberg, D. P. Landau, and R. H. Swendsen, *Phys. Rev. E* **51**, 5092 (1995).
57. H. Ogura, H. Matsuda, T. Ogawa, N. Ogita, and A. Ueda, *Prog. Theoret. Phys.* **58**, 419 (1977).
58. J. M. Rickman and S. R. Phillpot, *J. Chem. Phys.* **95**, 7562 (1991).
59. S.-Y. Sheu, C.-Y. Mou, and R. Lovett, *Phys. Rev. E* **51**, R3795 (1995).
60. C. Vega, E. P. A. Paras, and P. A. Monson, *J. Chem. Phys.* **96**, 9060 (1992).
61. L. A. Baez and P. Clancy, *Mol. Phys.* **86**, 385 (1995).
62. A. P. Malanoski and P. A. Monson, *J. Chem. Phys.* **107**, 6899 (1997).
63. J. M. Polson and D. Frenkel, *J. Chem. Phys.* **109**, 318 (1998).
64. C. Vega and P. A. Monson, *J. Chem. Phys.* **109**, 9938 (1998).
65. A. P. Malanoski and P. A. Monson, *J. Chem. Phys.* **110**, 664 (1999).
66. J. M. Polson and D. Frenkel, *J. Chem. Phys.* **111**, 1501 (1999).
67. M. Born and K. Huang, *Dynamical Theory of Crystal Lattices* (Clarendon Press, Oxford, 1954).
68. W. G. Hoover, A. C. Hindmarsh, and B. L. Holian, *J. Chem. Phys.* **57**, 1980 (1972).
69. W. G. Hoover, M. Ross, K. W. Johnson, D. Henderson, J. A. Barker, and B. C. Brown, *J. Chem. Phys.* **52**, 4931 (1970).
70. W. G. Hoover, S. G. Gray, and K. W. Johnson, *J. Chem. Phys.* **55**, 1128 (1971).
71. E. Smargiassi and P. A. Madden, *Phys. Rev. B* **51**, 117 (1995).
72. D. A. McQuarrie, *Statistical Mechanics* (Harper & Row, New York, 1976).
73. J. R. Morris and K. M. Ho, *Phys. Rev. Lett.* **74**, 940 (1995).
74. M. Parrinello and A. Rahman, *Phys. Rev. Lett.* **45**, 1196 (1980).
75. P. Najfabadi and S. Yip, *Scripta Metallurgica* **17**, 1199 (1983).
76. S. Yashoneth and C. N. R. Rao, *Mol. Phys.* **54**, 245 (1985).
77. S. Nosé and F. Yonezawa, *J. Chem. Phys.* **84**, 1803 (1985).
78. A. Mulder, P. J. Michels, and J. A. Schouten, *J. Chem. Phys.* **106**, 8806 (1997).
79. J. R. Morris, C. Z. Wang, K. M. Ho, and C. T. Chan, *Phys. Rev. B* **49**, 3109 (1994).
80. D. Frenkel and J. P. McTague, *Ann. Rev. Phys. Chem.* **31**, 491 (1980).
81. K. Bagchi, H. C. Andersen, and W. Swope, *Phys. Rev. E* **53**, 3794 (1996).
82. K. Bagchi, H. C. Andersen, and W. Swope, *Phys. Rev. Lett.* **76**, 255 (1996).
83. D. A. Kofke and E. D. Glandt, *Fluid Phase Equil.* **29**, 327 (1986).
84. M. Mehta and D. A. Kofke, *Chem. Eng. Sci.* **49**, 2633 (1994).
85. D. A. Kofke, *Mol. Phys.* **78**, 1331 (1993).
86. D. A. Kofke, *J. Chem. Phys.* **98**, 4149 (1993).
87. R. LeSar, R. Najafabadi, and D. J. Srolovitz, *Phys. Rev. Lett.* **63**, 624 (1989).
88. R. Najafabadi and D. J. Srolovitz, *Phys. Rev. B* **52**, 9229 (1995).
89. S. M. Foiles, *Phys. Rev. B* **49**, 14930 (1994).

90. W. C. Swope and H. C. Andersen, *Phys. Rev. A* **46**, 4539 (1992).

91. D. R. Squire and W. G. Hoover, *J. Chem. Phys.* **50**, 701 (1969).

92. J. A. Barker, *Lattice Theories of the Liquid State* (Macmillan, New York, 1963).

93. J. A. Barker, *J. Chem. Phys.* **44**, 4212 (1966).

94. J. A. Barker, *J. Chem. Phys.* **63**, 632 (1975).

95. J. E. Lennard-Jones and A. F. Devonshire, *Proc. Roy. Soc.* (London) **A163**, 53 (1937).

96. J. E. Lennard-Jones and A. F. Devonshire, *Proc. Roy. Soc.* (London) **A165**, 1 (1938).

97. J. E. Lennard-Jones and A. F. Devonshire, *Proc. Roy. Soc.* (London) **A170**, 464 (1939).

98. J. E. Lennard-Jones and A. F. Devonshire, *Proc. Roy. Soc.* (London) **A169**, 317 (1939).

99. R. J. Buehler, R. H. Wentorf, J. O. Hirschfelder, and C. F. Curtiss, *J. Chem. Phys.* **19**, 61 (1951).

100. E. P. A. Paras, C. Vega, and P. A. Monson, *Mol. Phys.* **77**, 803 (1992).

101. X. Cottin and P. A. Monson, *J. Chem. Phys.* **99**, 8914 (1993).

102. W. L. Bragg and E. J. Williams, *Proc. Roy. Soc.* (London) **A145**, 699 (1934).

103. J. A. Pople and F. E. Karasz, *J. Phys. Chem. Solids* **18**, 28 (1961).

104. J. A. Pople and F. E. Karasz, *J. Phys. Chem. Solids* **20**, 294 (1961).

105. E. R. Cowley and J. A. Barker, *J. Chem. Phys.* **73**, 3452 (1980).

106. I. Prigogine and V. Mathot, *J. Chem. Phys.* **20**, 49 (1952).

107. Z. W. Salsburg and J. G. Kirkwood, *J. Chem. Phys.* **20**, 1538 (1952).

108. X. Cottin and P. A. Monson, *J. Chem. Phys.* **102**, 3354 (1995).

109. X. Cottin and P. A. Monson, *J. Chem. Phys.* **105**, 10022 (1996).

110. J. G. Kirkwood, *J. Chem. Phys.* **3**, 300 (1935).

111. R. Lovett, *J. Chem. Phys.* **66**, 1225 (1976).

112. J. J. Kozak, *Adv. Chem. Phys.* **40**, 229 (1979).

113. R. F. Kayser and H. J. Raveché, *Phys. Rev B* **22**, 424 (1980).

114. R. Lovett and F. P. Buff, *J. Chem. Phys.* **72**, 2425 (1980).

115. T. V. Ramakrishnan and M. Yussouff, *Solid State Commun.* **21**, 389 (1977).

116. T. V. Ramakrishnan and M. Yussouff, *Phys. Rev. B* **19**, 2775 (1979).

117. A. D. J. Haymet and O. D. W., *J. Chem. Phys.* **74**, 2559 (1981).

118. R. Evans, *Adv. Phys.* **28**, 143 (1979).

119. J. K. Percus and G. J. Yevick, *Phys. Rev.* **110**, 1 (1958).

120. M. S. Wertheim, *Phys. Rev. Lett.* **10**, 321 (1963).

121. M. S. Wertheim, *J. Math. Phys.* **5**, 643 (1964).

122. P. Tarazona, *Mol. Phys.* **52**, 81 (1984).

123. W. A. Curtin and N. W. Ashcroft, *Phys Rev A* **32**, 2909 (1985).

124. A. R. Denton and N. W. Ashcroft, *Phys. Rev. A* **39**, 470 (1989).

125. J. F. Lutsko and M. Baus, *Phys. Rev. Lett.* **64**, 761 (1990).

126. J. F. Lutsko and M. Baus, *Phys. Rev. A* **41**, 6647 (1990).

127. D. C. Wang and A. P. Gast, *J. Chem. Phys.* **110**, 2522 (1999).

128. Y. Rosenfeld, *Phys. Rev. Lett.* **63**, 980 (1989).

129. Y. Rosenfeld, M. Schmidt, H. Lowen, and P. Tarazona, *Phys. Rev. E* **55**, 4245 (1997).

130. Y. Rosenfeld and P. Tarazona, *Mol. Phys.* **95**, 141 (1998).

131. E. Kierlik and M. L. Rosinberg, *Phys. Rev. A* **42**, 3382 (1990).

132. H. Reiss, H. L. Frisch, and J. L. Lebowitz, *J. Chem. Phys.* **31**, 369 (1959).

133. A. D. J. Haymet, *J. Chem. Phys.* **78**, 4641 (1983).

134. D. A. Young and B. J. Alder, *J. Chem. Phys.* **60**, 1254 (1974).

135. R. Ohnesorge, H. Lowen, and H. Wagner, *Europhys. Lett.* **22**, 245 (1993).

136. J. S. Rowlinson and F. L. Swinton, *Liquids and Liquid Mixtures*, 3rd ed. (Butterworth, London, 1982).

137. A. R. Denton, N. W. Ashcroft, and W. A. Curtin, *Phys. Rev. E* **51**, 65 (1995).

138. D. Henderson and J. A. Barker, *Mol. Phys.* **14**, 587 (1968).

139. G. A. Mansoori and F. B. Canfield, *J. Chem. Phys.* **51**, 4967 (1969).

140. D. A. Young, *J. Chem. Phys.* **58**, 1647 (1973).

141. C. Vega and P. A. Monson, *Mol. Phys.* **85**, 413 (1995).

142. C. Vega, F. Bresme, and J. L. F. Abascal, *Phys. Rev. E* **54**, 2746 (1996).

143. J. D. Weeks, D. Chandler, and H. C. Andersen, *J. Chem. Phys.* **54**, 5237 (1970).

144. J.-J. Weis, *Mol. Phys.* **28**, 187 (1974).

145. H. S. Kang, C. S. Lee, T. Ree, and F. H. Ree, *J. Chem. Phys.* **82**, 414 (1985).

146. H. S. Kang, T. Ree, and F. H. Ree, *J. Chem. Phys.* **84**, 4547 (1986).

147. G. Jackson and F. van Swol, *Mol. Phys.* **65**, 161 (1988).

148. J. M. Kincaid and J.-J. Weis, *Mol. Phys.* **34**, 931 (1977).

149. C. Rascon, L. Mederos, and G. Navascues, *Phys. Rev. Lett.* **77**, 2249 (1996).

150. A. Daanoun, C. F. Tejero, and M. Baus, *Phys. Rev. E* **50**, 2913 (1994).

151. T. Coussaert and M. Baus, *Phys. Rev. E* **52**, 862 (1995).

152. E. P. A. Paras, C. Vega, and P. A. Monson, *Mol. Phys.* **79**, 1063 (1993).

153. W. G. Hoover and M. Ross, *Contemp. Phys.* **12**, 339 (1971).

154. F. A. Lindemann, *Phys. Z.* **11**, 609 (1910).

155. M. Ross, *Phys. Rev.* **184**, 233 (1969).

156. J.-P. Hansen and L. Verlet, *Phys. Rev.* **184**, 151 (1969).

157. J.-P. Hansen and D. Schiff, *Mol. Phys.* **25**, 1281 (1973).

158. H. J. Raveché, R. D. Mountain, and W. B. Streett, *J. Chem. Phys.* **61**, 1970 (1974).

159. P. V. Giaquinta and G. Giunta, *Physica A* **187**, 145 (1992).

160. L. V. Woodcock, *Faraday Discuss.* **106**, 325 (1997).

161. P. N. Pusey, W. van Megen, P. Bartlett, B. J. Ackerson, J. G. Rarity, and S. M. Underwood, *Phys. Rev. Lett.* **63**, 2753 (1989).

162. J. Zhu, M. Li, R. Rogers, W. Meyer, R. H. Ottewill, W. B. Russel, and P. M. Chaikin, *Nature* **387**, 883 (1997).

163. N. A. M. Verhaegh, J. S. van Duijneveldt, A. van Blaaderen, and H. N. W. Lekkerkerker, *J. Chem. Phys.* **102**, 1416 (1995).

164. D. A. Young, *Phase Diagrams of the Elements* (University of California Press, Los Angeles, 1991).

165. D. A. Young and D. J. Alder, *J. Chem. Phys.* **73**, 2430 (1980).

166. A. Gast, C. K. Hall, and W. B. Russel, *J. Coll. Inter. Sci.* **96**, 251 (1983).

167. A. Cheng, M. L. Klein, and C. Caccamo, *Phys. Rev. Lett.* **71**, 1200 (1993).

168. M. H. J. Hagen, E. J. Meijer, G. C. A. M. Mooij, D. Frenkel, and H. N. W. Lekkerkerker, *Nature* **365**, 425 (1993).

169. M. H. J. Hagen and D. Frenkel, *J. Chem. Phys.* **101**, 4093 (1994).

170. D. Frenkel, P. Bandon, P. Bolhuis, and M. Hagen, *Mol. Sim.* **16**, 127 (1996).

171. P. R. ten Wolde and D. Frenkel, *Science* **277**, 1975 (1997).

172. V. Talanquer and D. W. Oxtoby, *J. Chem. Phys.* **109**, 223 (1998).

173. P. Bolhuis and D. Frenkel, *Phys. Rev. Lett.* **72**, 2211 (1994).

174. P. Bolhuis, M. Hagen, and D. Frenkel, *Phys. Rev. E* **50**, 4880 (1994).

175. P. Bolhuis and D. Frenkel, *J. Phys.: Cond. Matt* **9**, 381 (1997).

176. S. G. Brush, H. L. Sahlin, and E. Teller, *J. Chem. Phys.* **45**, 2102 (1964).

177. G. S. Stringfellow, H. E. DeWitt, and W. L. Slattery, *Phys. Rev. A* **41**, 1105 (1990).

178. R. Agrawal and D. Kofke, *Mol. Phys.* **85**, 43 (1995).

179. R. Agrawal and D. A. Kofke, *Phys. Rev. Lett.* **74**, 122 (1995).

180. W. G. Hoover, D. A. Young, and R. G. Grover, *J. Chem. Phys.* **56**, 2207 (1972).

181. B. B. Laird and A. D. J. Haymet, *Mol. Phys.* **75**, 71 (1992).

182. R. Agrawal and D. Kofke, *Mol. Phys.* **85**, 23 (1995).

183. C. H. Marshall, B. B. Laird, and A. D. J. Haymet, *Chem. Phys. Lett.* **122**, 320 (1985).

184. B. B. Laird, M. J. D., and A. D. J. Haymet, *J. Chem. Phys.* **87**, 5449 (1987).

185. L. Mederos, G. Navascues, P. Tarazona, and E. Chacon, *Phys. Rev. E* **47**, 4284 (1993).

186. B. B. Laird and D. M. Kroll, *Phys. Rev. A* **42**, 4810 (1990).

187. K. Kremer, M. O. Robbins, and G. S. Grest, *Phys. Rev. Lett.* **57**, 2694 (1986).

188. M. O. Robbins, K. Kremer, and G. S. Grest, *J. Chem. Phys.* **88**, 3286 (1988).

189. E. J. Meijer and D. Frenkel, *J. Chem. Phys.* **94**, 2269 (1991).

190. G. duPont, S. Moulinasse, J. P. Ryckaert, and M. Baus, *Mol. Phys.* **79**, 453 (1993).

191. M. J. Stevens and M. O. Robbins, *J. Chem. Phys.* **98**, 2319 (1993).

192. R. T. Farouki and S. Hamaguchi, *J. Chem. Phys.* **101**, 9885 (1994).

193. S. Hamaguchi, R. T. Farouki, and D. H. E. Dubin, *J. Chem. Phys.* **105**, 7641 (1996).

194. S. Hamaguchi, R. T. Farouki, and D. H. E. Dubin, *Phys. Rev. E* **56**, 4671 (1997).

195. E. J. Meijer and F. El Azhar, *J. Chem. Phys.* **106**, 4678 (1997).

196. E. Lomba and N. G. Almarza, *J. Chem. Phys.* **100**, 8367 (1994).

197. M. Hasegawa, *J. Chem. Phys.* **108**, 208 (1998).

198. C. Caccamo, D. Costa, and G. Pellicane, *J. Chem. Phys.* **109**, 4498 (1998).

199. J. Yamanaka, H. Yoshida, T. Koga, N. Ise, and T. Hashimoto, *Phys. Rev. Lett.* **80**, 5806 (1998).

200. F. H. Stillinger, *J. Chem. Phys.* **65**, 3968 (1976).

201. F. H. Stillinger and T. A. Weber, *J. Chem. Phys.* **68**, 3837 (1978).

202. W. G. T. Kranendonk and D. Frenkel, *J. Phys.: Condens. Matter* **1**, 7735 (1989).

203. W. G. T. Kranendonk and D. Frenkel, *Mol. Phys.* **72**, 715 (1991).

204. W. G. T. Kranendonk and D. Frenkel, *Mol. Phys.* **72**, 679 (1991).

205. W. G. T. Kranendonk and D. Frenkel, *Mol. Phys.* **72**, 699 (1991).

206. D. A. Kofke, *Molec. Sim.* **7**, 285 (1991).

207. R. E. Smallman, W. Hume-Rothery, and C. W. Haworth, *The Structure of Metals and Alloys* (The Institute of Metals, London, 1988).

208. M. D. Eldridge, P. A. Madden, and D. Frenkel, *Mol. Phys.* **79**, 105 (1993).

209. M. D. Eldridge, P. A. Madden, and D. Frenkel, *Nature* **365**, 35 (1993).

210. M. D. Eldridge, P. A. Madden, and D. Frenkel, *Mol. Phys.* **80**, 987 (1993).

211. M. D. Eldridge, P. A. Madden, P. N. Pusey, and P. Bartlett, *Mol. Phys.* **84**, 395 (1995).

212. E. Trizac, M. D. Eldridge, and P. A. Madden, *Mol. Phys.* **90**, 675 (1997).

213. M. J. Murray and J. V. Sanders, *Phil. Mag.* **42**, 721 (1980).

214. J. V. Sanders, *Phil. Mag.* **42**, 705 (1980).

215. P. Bartlett, *J. Phys. Condens. Matter* **2**, 4979 (1990).

216. P. Bartlett, R. H. Ottewill, and P. N. Pusey, *J. Chem. Phys.* **92**, 1299 (1990).

217. P. Bartlett and R. H. Ottewill, *J. Chem. Phys.* **96**, 3306 (1991).

218. P. Bartlett and P. N. Pusey, *Physica a* **194**, 415 (1993).

219. Y. Nagano and T. Tamura, *Chem. Phys. Lett.* **252**, 362 (1996).

220. Y. Nagano and T. Nakamura, *Chem. Phys. Lett.* **265**, 358 (1997).

221. S. L. Smithline and A. D. J. Haymet, *J. Chem. Phys.* **86**, 6486 (1987).

222. S. W. Rick and A. D. J. Haymet, *J. Chem. Phys.* **90**, 1188 (1989).

223. A. R. Denton and N. W. Ashcroft, *Phys. Rev. A* **42**, 7312 (1990).

224. X. C. Zeng and D. W. Oxtoby, *J. Chem. Phys.* **93**, 4357 (1990).

225. X. Cottin and P. A. Monson, *J. Chem. Phys.* **107**, 6855 (1997).

226. X. Cottin, E. P. A. Paras, C. Vega, and P. A. Monson, *Fluid Phase Equil.* **117**, 114 (1996).

227. M. Matsuoka, *Bunri Gijutsu (Separation Process Engineering)* **7**, 245 (1977).

228. M. R. Hitchcock and C. K. Hall, *J. Chem. Phys.* **110**, 11433 (1999).

229. M. J. Vlot, H. E. A. Huitema, A. deVooys, and J. P. vanderEerden, *J. Chem. Phys.* **107**, 4345 (1997).

230. M. J. Vlot, J. C. vanMiltenburg, H. A. J. Oonk, and J. P. vanderEerden, *J. Chem. Phys.* **107**, 10102 (1997).

231. C. Vega and P. A. Monson, *J. Chem. Phys.* **107**, 2696 (1997).

232. D. Frenkel, B. M. Mulder, and J. P. McTague, *Phys. Rev. Lett.* **52**, 287 (1984).

233. D. Frenkel and Mulder, *Mol. Phys.* **55**, 1171 (1985).

234. D. Frenkel and B. M. Mulder, *Mol. Phys.* **55**, 1171 (1985).

235. D. Frenkel, B. M. Mulder, and J. P. McTague, *Mol. Cryst. Liq. Cryst.* **128**, 119 (1985).

236. M. P. Allen, G. T. Evans, D. Frenkel, and B. M. Mulder, in *Advances in Chemical Physics*, Vol. LXXXVI, edited by I. Prigogine and S. A. Rice (John Wiley & Sons, New York, 1993).

237. S. C. McGrother, D. C. Williamson, and G. Jackson, *J. Chem. Phys.* **104**, 6755 (1996).

238. P. Bolhuis and D. Frenkel, *J. Chem. Phys.* **106**, 667 (1997).

239. S. Singer and R. Mumaugh, *J. Chem. Phys.* **93**, 1278 (1990).

240. C. Vega, E. P. A. Paras, and P. A. Monson, *J. Chem. Phys.* **97**, 8543 (1992).

241. K. W. Wojciechowski, *Phys. Lett. A* **122**, 377 (1987).

242. K. W. Wojciechowski, D. Frenkel, and A. C. Branka, *Phys. Rev. Lett.* **66**, 3168 (1991).

243. R. K. Bowles and R. J. Speedy, *Mol. Phys.* **87**, 1349 (1996).

244. S. J. Smithline, S. W. Rick, and A. D. J. Haymet, *J. Chem. Phys.* **88**, 2004 (1988).

245. J. D. McCoy, S. J. Singer, and D. Chandler, *J. Chem. Phys.* **87**, 4853 (1987).

246. D. Chandler, J. D. McCoy, and S. J. Singer, *J. Chem. Phys.* **85**, 5971 (1986).

247. D. Chandler, J. D. McCoy, and S. J. Singer, *J. Chem. Phys.* **85**, 5976 (1986).

248. S. C. Gay, P. D. Beale, and J. C. Rainwater, *Int. J. Thermophys.* **19**, 1535 (1998).

249. D. L. Morgan and R. Kobayashi, *Fluid Phase Equil.* **63**, 317 (1991).

250. A. Yethiraj, J. G. Curro, K. S. Schweizer, and J. D. McCoy, *J. Chem. Phys.* **98**, 1635 (1993).

251. A. P. Malanoski, C. Vega, and P. Monson, *Mol Phys.* **98**, 363 (2000).

252. P. Pusey, in *Les Houches, Session LI, Liquids, Freezing and Glass Transitions, NATO Advanced Study Institute, Series B: Physics*, edited by J. P. Hansen, D. Levesque, and J. Zinn-Justin (North-Holland, Amsterdam, 1991), Chap. 10.

253. S.-E. Phan, W. B. Russel, Z. Cheng, J. Zhu, P. M. Chaikin, J. H. Dunsmuir, and R. H. Ottewill, *Phys. Rev. E* **54**, 6633 (1996).

254. J. L. Barrat and J.-P. Hansen, *J. Physique* **46**, 1547 (1986).

255. P. Pusey, *J. Physique* **48**, 709 (1987).

256. R. McRae and A. D. J. Haymet, *J. Chem. Phys.* **88**, 1114 (1988).

257. E. Dickinson, *Faraday Discuss.* **65**, 127 (1978).

258. E. Dickinson, R. Parker, and M. Lal, *Chem. Phys. Lett.* **79**, 578 (1981).

259. E. Dickinson and R. Parker, *J. Physique Lett.* **46**, L229 (1985).

260. P. Bolhuis and D. A. Kofke, *Phys. Rev. E* **54**, 634 (1996).

261. P. Bartlett and P. B. Warren, *Phys. Rev. Lett.* **82**, 1979 (1999).

262. D. A. Kofke and P. G. Bolhuis, *Phys. Rev. E* **59**, 618 (1999).

263. J. A. Cuesta, *Europhys. Lett.* **46**, 197 (1999).

264. J. Zhang, R. Blaak, E. Trizac, J. A. Cuesta, and D. Frenkel, *J. Chem. Phys.* **110**, 5318 (1999).

265. D. J. Lacks and J. R. Weinhoff, *J. Chem. Phys.* **111**, 398 (1999).

266. C. Vega and P. A. Monson, *J. Chem. Phys.* **102**, 1361 (1995).

267. S. C. Gay, P. D. Beale, and J. C. Rainwater, *J. Chem. Phys.* **109**, 6820 (1998).

268. I. M. Svishchev and P. G. Kusalik, *Phys. Rev. Lett.* **73**, 975 (1994).

269. L. A. Baez and P. Clancy, *J. Chem. Phys.* **103**, 9744 (1995).

270. R. J. Speedy, *J. Chem. Phys.* **107**, 3222 (1997).

271. J. Kolafa and I. Nezbeda, *Mol. Phys.* **61**, 161 (1987).

272. B. Smit, K. Esselink, and D. Frenkel, *Mol. Phys.* **87**, 159 (1996).

273. J. L. Barrat, *J. Phys. C* **20**, 1031 (1987).

274. J. M. Polson, E. Trizac, S. Pronk and D. Frenkel, *J. Chem. Phys.* **112**, 5339 (2000).

275. S.-C. Mau and D. A. Huse, *Phys. Rev. E* **59**, 4396 (1999).

276. P. G. Bolhuis, D. Frenkel, S.-C. Mau and D. A. Huse, *Nature* **388**, 236 (1997).

IRREVERSIBLE MOTION ON MACROSCOPIC AND MOLECULAR TIMESCALES AND CHEMICAL DYNAMICS IN LIQUIDS

S. A. ADELMAN

Department of Chemistry, Purdue University
West Lafayette, IN 47907-1393

R. RAVI

Department of Chemical Engineering, Indian Institute of Technology
Kanpur, Kanpur-208016, India

I. INTRODUCTION

In this review we discuss some topics in the theory of irreversible processes [1–5], as they relate to the developing field of liquid phase chemical dynamics [6].

Our reason for writing this review is as follows. Following the pioneering work of Kramers [7], most activity in liquid phase reaction dynamics theory has focused on the familar "*slow variable/fast bath*" diffusive or Brownian motion [2] timescale regime [8–10]. However, there has been a recent surge of interest in the opposite and nontraditional "*fast variable/slow bath*" or *short timescale* regime. This interest has been spurred by the results of ultra-fast laser experiments [6,11–13] and molecular dynamics (MD) simulations [14–17] commencing in the late 1980s, which demonstrate significant limitations in the ranges of applicability of both the traditional slow variable rate theories [7–10] and also of the continuum friction models [18] usually employed to implement these theories. Moreover, the experiments and simulations indicate the need for fast variable theories of reaction dynamics; and for corresponding *molecular* friction models [19] valid on timescales of hundreds or even tens of femtoseconds. (The slow and fast variable regimes

Advances in Chemical Physics, Volume 115, edited by I. Prigogine and Stuart A. Rice.
ISBN 0-471-39331-2 © 2000 John Wiley & Sons, Inc.

have familiar analogues in gas phase chemical kinetics. Namely, they are similar respectively to the near adiabatic limit and near sudden limit molecular collision regimes.)

It is within this context that we thought it worthwhile to write a chapter that does the following:

1. It reviews some statistical mechanical topics germane to the points just raised.

2. It reviews our fast variable theory of liquid phase dynamics [19–26]; placing it in its context within statistical mechanics, and also relating it to the recent literature [6,11–17,27,28]. This theory was motivated by and is in accord with the familiar Arrhenius principle that only *atypically fast* reactant molecules have sufficient energy to surmount chemical activation barriers.

3. It points out a disharmony between the Arrhenius principle and the traditional slow variable theories of reaction dynamics. In agreement with the recent experiments [6,11–13] and simulations [14–17], this disharmony is expected to significantly restrict the regimes of applicability of the traditional theories.

The plan of this chapter is as follows. In Section II and the Appendix we review some basic topics in statistical mechanics that underlie questions concerning the regimes of validity of slow and fast variable descriptions of irreversible motion.

Section II gives an overview of the central issues, restricted to the specific problem of atomic motion in fluids. In the Appendix, this overview is expanded into a discussion of the general theory of irreversible processes [1–5], focusing especially on Onsager's work [1,3,4], including its extension to finite memory due to Mori [5].

Among the points we bring out are the following:

1. In liquid phase chemical kinetics, the reactant potential of mean force $W(S;x)$ *does not* play the same universal role that the reactant potential energy function $U(x)$ plays in gas phase kinetics. Rather the form of the driving potential of a liquid phase reaction depends on its timescale regime, with $W(S;x)$ driving only slow variable regime reactions.

2. As a result, the traditional picture of liquid phase reaction dynamics —solute motion in the potential of mean force damped by friction— holds only for slow variable regime reactions. For fast variable regime reactions the nonclassical picture of Eq. (3.36) and Figure 3.5 applies.

3. Mori's [5] generalized Langevin equation is rooted in a specific *macroscopic* phenomenology; namely, the flux-force Eq. (A.10). Consequently, despite its formal exactness, Mori's equation has a limited

rather than a universal domain of applicability, a point pertinent to solution chemistry inasmuch as generalized Langevin equations are used in that field.

In Section III, we combine the analyses of Section II and the Appendix with the results of recent experiments [13] and simulations [14] to discuss the slow variable description [7–10] of liquid phase activated barrier crossing. In Section IV, using these same ideas we review our fast variable theory of liquid phase reaction dynamics [19–26] from the standpoint of the recent literature [6,11–17,27,28]. Finally, in Section V we summarize our main points.

We will make no attempt here to review the traditional slow variable theories in detail, since comprehensive articles devoted to these theories already exist. See, for example, the reviews by Hynes [8], by Berne, Borkevec, and Straub [9], and by Hänggi, Talkner, and Borkovec [10].

II. TIMESCALES AND PARTICLE MECHANICS

In Section II we compare particle mechanics in the slow and fast variable timescale regimes. We start the discussion by showing the following. For damped *macroscopic particles*, the potential energy function whose minima locate the particle's points of static equilibrium also produces the forces which drive its dynamics. For damped *microscopic particles*, in contrast, the potential that determines the particle's statics may or may not produce the forces that drive its dynamics.

A. Macroscopic and Microscopic Particle Mechanics

We first examine the statics and dynamics of a one-dimensional macroscopic particle of mass M. Specifically, we study a relaxation process $x_F \rightarrow x_{F=0}$ that takes the particle from an initial point of static equilibrium x_F to a final point $x_{F=0}$ (Fig. 3.1).

1. Macroscopic Particle Statics and Dynamics

Consider, as is illustrated in Figure 3.1, a macroscopic particle of mass M and coordinate x. Assume that the particle is bound to the origin $x = 0$ by an internal potential $U(x)$. Further, assume that, at times, the particle is also subject to a constant external constraint force F. Note that when F is acting the particle's total potential energy is

$$U_F(x) = U(x) - xF. \qquad (3.1)$$

We will also require that for each F the particle have a unique point x_F of stable static equilibrium. This condition is met if $U_F(x)$ has a single extremum that is a minimum occurring at x_F. This final requirement is

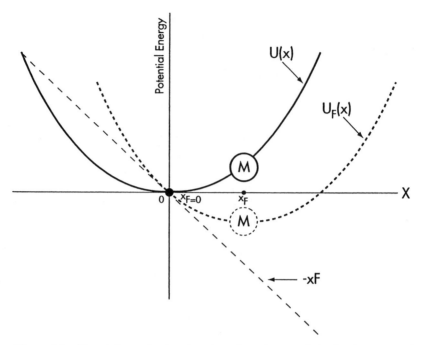

Figure 3.1. The statics and relaxation dynamics of a one-dimensional macroscopic particle. A particle of mass M bound to the origin by a potential $U(x)$ and also subject to a constraint force F, is initially in static equilibrium at the minimum x_F of $U_F(x) = U(x) - xF$. F is then removed so that $U_F(x) \to U(x)$, and the particle relaxes to a new point of static equilibrium $x_{F=0} = 0$, the minimum of $U(x)$. As discussed in the Appendix, Gibbs has noted that particle relaxation processes $x_{F=0} \to x_F$ are analogous to thermodynamic relaxation processes $\Gamma_n \to \Gamma_p$.

satisfied if $U(x)$ (1) has a convergent Maclaurin series expansion for all x, (2) is monotonically increasing for $|x| > 0$, and (3) has a positive second derivative at $x = 0$. Thus, henceforth, we will restrict ourselves to potentials $U(x)$ that conform to these three conditions. A simple example is the harmonic potential $U(x) = 1/2m\omega^2 x^2$.

For such $U(x)s$, moreover, $U_{F=0}(x) = U(x)$ has its minimum at $x = 0$. Thus, for the potentials under consideration, at $x = 0$

$$\frac{dU}{dx} = 0 \quad \text{and} \quad \frac{d^2U}{dx^2} > 0. \tag{3.2}$$

We next examine the particle's static equilibria, first assuming that the constraint force F is acting. To begin, we note that in the absence of friction the particle will never achieve static equilibrium. Rather, it will merely execute conservative motion in the potential $U_F(x)$. However, if a frictional force $-\zeta \dot{x}(t)$ arising from some internal or external source also acts, the particle will reach a point of static equilibrium x_F as $t \to \infty$. The condition that determines x_F may be derived from the particle's equation of motion in the presence of friction

$$M\ddot{x}(t) = -U'_F[x(t)] - \zeta \dot{x}(t). \tag{3.3}$$

Namely, setting $x(t) = x_F$ and $\dot{x}(t) = \ddot{x}(t) = 0$, the conditions for static equilibrium at x_F, in Eq. (3.3) yields the condition for x_F as

$$U'_F[x_F] = 0. \tag{3.4}$$

Finally, using Eq. (3.1), Eq. (3.4) for x_F may be recast as

$$\left[\frac{dU}{dx}\right]_{x=x_F} = F. \tag{3.5}$$

Next, assume that after the particle has achieved static equilibrium at x_F the constraint force F is removed so that $U_F(x) \to U(x)$. The particle will again begin to move according to Eq. (3.3) with ($U'_F[x]$ replaced by $U'[x]$), executing a damped motion that will eventually take it to a new position of static equilibrium $x_{F=0}$. The point $x_{F=0}$ is determined by the condition

$$\left[\frac{dU}{dx}\right]_{x=x_{F=0}} = 0 \tag{3.6}$$

which follows by setting $F = 0$ in Eq. (3.5). Eq. (3.6) shows that $x_{F=0}$ coincides with the minimum $x = 0$ of $U(x)$, and thus may be regarded as being determined by the minimization conditions of Eq. (3.2).

Note that in accord with our earlier comments, the same potential $U(x)$ that determines the particle's statics also drives its dynamics *during* the process $x_F \to x_{F=0}$, because $U(x)$ appears in both the condition for static equilibrium Eq. (3.2) and also the equation of motion Eq. (3.3) for $F = 0$.

We next turn to the statics and dynamics of a microscopic particle.

2.   Microscopic Particle Statics and Dynamics

Consider a particle of molecular size moving in one dimension in a pure fluid. We will denote the mass and coordinate of the particle by m and x; and will specify the fluid macroscopically by a set of parameters S (for example N, V, and E). We will denote the force exerted by the fluid on the particle at time t by $F[S; x(t)]$. Because of the particle's molecular size, its timescales are not necessarily longer than those of the fluid molecules. Consequently, $F[S; x(t)]$ in general depends on the particle's trajectory $x(\tau)$ for all times $\tau < t$ as well as on its instantaneous position $x(t)$. For the present qualitative purposes to account for this feature it is sufficient to assume that $F[S; x(t)]$ has the time-local but velocity-dependent form $F[S; x(t), \dot{x}(t)]$. We will further assume that the particle is also subject to a vacuum force $-U'_F(x)$ identical in form to that appearing in Eq. (3.3). That is, in analogy to Eq. (3.1) the vacuum potential energy of the particle is of the form $U_F(x) = U(x) - xF$. The total force acting on the particle is thus $F(S; x) = -U'_F + F(S; x)$. Hence, its equation of motion is

$$m\ddot{x}(t) = -U'_F[x(t)] + F[S; x(t)] \qquad (3.7)$$

where by assumption $F(S; x) = F(S; x, \dot{x})$. (Note for simplicity we use the same symbols $x, U[x]$, and so on for both the macroscopic and microscopic particle.)

In writing Eq. (3.7) we have implicitly assumed that for a given set of macroscopic parameters S, $F(S; x)$ represents a *single force*. However, $F(S; x)$ strictly should represent an *ensemble of forces*. This follows inasmuch as many microscopic fluid states are consistent with a given S; and because each microscopic state gives rise to a distinct $F(S; x)$. We will ignore this refinement here, and thus assume $F(S; x)$ is a single force.

To use Eq. (3.7) to study the behavior of the particle, we require an explicit form for $F(S; x)$. We will see that it is not straightforward to determine this form if the particle is moving. However, if the particle is stationary, $F(S; x)$ is just the fluid force $F_{eq}(S; x) \equiv F(S; x, \dot{x} = 0)$ conditional that the fluid + particle system is in thermodynamic equilibrium.

Let us assume the particle is indeed stationary, and thus take $F(S; x)$ as $F_{eq}(S; x)$. This choice, while not sufficient to determine the particle's trajectories $x(t)$ from Eq. (3.7), is sufficient to locate the its points of static equilibrium from that equation.

Thus we next turn to this latter problem, solving it by paralleling our earlier analysis of Eq. (3.3). To begin we define $W_F(S; x)$ by

$$W_F(S; x) = U_F(x) + w(S; x); \qquad (3.8)$$

where $U_F(x)$ is as in Eq. (3.1) and where $w(S; x)$ is defined to within a constant by

$$F_{eq}(S; x) = -w'(S; x). \qquad (3.9)$$

Eqs. (3.8) and (3.9) show that $W_F(S; x)$ is the particle's equilibrium fluid phase potential energy in the presence of the constraint force F.

Define $x_F(S)$ as the particle's fluid phase point of static equilibrium in the presence of F. The point $x_F(S)$ may be determined from the following fluid phase extension of Eq. (3.4)

$$W'_F[x_F(S)] = 0. \tag{3.10}$$

To derive Eq. (3.10) first set $x(t) = x_F(S)$, $F = F_{eq}$, and $\ddot{x}(t) = 0$ valid when the particle is at static equilibrium at $x_F(S)$ in Eq. (3.7) to yield the condition $-U'_F[x_F(S)] + F_{eq}[S; x_F(S)] = 0$. This condition and Eqs. (3.8) and (3.9) then yield Eq. (3.10).

Next, assume that after the particle has reached $x_F(S)$, the constraint force is suddenly removed. Then in analogy to the discussion after Eq. (3.5), removal of F will initiate the fluid phase relaxation process $x_F(S) \rightarrow x_{F=0}(S)$. The new position of static equilibrium $x_{F=0}(S)$ is determined by the following extension of Eq. (3.6)

$$\left[\frac{dW(S)}{dx} \right]_{x=x_{F=0}(S)} = 0 \tag{3.11}$$

where

$$W(S) = W(S; x) \equiv U(x) + w(S; x) \tag{3.12}$$

is the equilibrium fluid phase internal potential of the particle.

The above results may be summarized as follows. To obtain microscopic particle equations of static equilibrium like Eqs. (3.10) and (3.11), one merely replaces $U(x)$ with $W(S; x)$ in the corresponding macroscopic particle equations, in this case Eqs. (3.4) and (3.6). (Of course this simple correspondence holds only if $W[S; x]$ obeys the conditions required of $U[x]$, as stated after Eq. (3.1). This occurs, for example, for a spherical particle moving in a homogeneous fluid.)

Thus, microscopic particle statics is very similar to and hardly more complex than macroscopic particle statics. We will see, however, that the similarity of microscopic and macroscopic particle behavior will not persist when we move from statics to dynamics.

However, before turning to dynamics, we note that microscopic and macroscopic statics actually do differ in one significant way. This occurs because only a particle of molecular size is sensibly affected by microscopic fluid fluctuations like those noted after Eq. (3.7).

Because of these fluctuations, after the particle has reached static equilibrium at $x_{F=0}(S)$ it has a range of possible positions x' rather than the single position $x_{F=0}(S)$. The point $x_{F=0}(S)$ is merely the particle's most probable position.

The probability distribution $P(S; x')$, which gives the likelihood that the particle has the position x' after static equilibrium is reached, is the following Boltzmann distribution

$$P(S; x') = P(S; x^*)\exp\left[-\frac{\Delta W(S)}{kT}\right];\qquad (3.13)$$

where $x^* \equiv x_{F=0}(S)$ and where

$$\Delta W(S) = W(S; x') - W(S; x^*),\qquad (3.14)$$

with $W(S; x)$ as in Eq. (3.12).

We now leave statics and turn to dynamics.

We will deal with the following problem. Find the particle's trajectories $x(t)$ that occur during the fluid phase relaxation process $x_F(S) \to x_{F=0}(S)$. Solution of this problem involves more than a straightforward integration of Eq. (3.7) because, as indicated, for a moving particle the form of $F(S; x)$ is not *a priori* evident. To see this let us evaluate $F(S; x)$ in two limiting cases.

Case 1. The particle's velocity $\dot{x}(t) \to 0$ and thus its motion is infinitely slow. Then throughout the process $x_F(S) \to x_{F=0}(S), F(S; x)$ is the *thermodynamic equilibrium force* $F_{eq}(S; x) = F(S; x, \dot{x} = 0)$ that governs the statics.

Case 2. The particle's velocity $\dot{x}(t) \to \infty$ and thus its motion is infinitely fast. Then throughout the process $x_F(S) \to x_{F=0}(S), F(S; x)$ is the *instantaneous force* $F_{inst}(S; x) = F(S; x, \dot{x} = \infty)$; namely, the force the fluid would exert if it remained in its initial configuration of thermodynamic equilibrium with the particle at $x_F(S)$.

For intermediate $\dot{x}(t), F(S; x)$ falls between the two extreme values $F_{eq}(S; x)$ and $F_{inst}(S; x)$.

In summary, ambiguity clouds the form that $F(S; x)$ takes on during the relaxation process $x_F(S) \to x_{F=0}(S)$.

Thus is accord with our earlier comments, the potential $W(S; x)$ that determines the particle's statics from Eq. (3.11) or more generally its equilibrium behavior from Eq. (3.13), may or may not drive its dynamics during the process $x_F(S) \to x_{F=0}(S)$.

We next turn to the problem of determining the form of $F(S; x)$ for a moving microscopic particle; and, hence, the particle's trajectories $x(t)$.

While it is not easy to fully solve this problem, it is relatively simple to solve it in two limiting timescale regimes. These are the traditional slow variable regime, and the opposite and nontraditional fast variable regime noted in Section I.

For both regimes we must evalute the total force $F[S; x(t)]$ acting on the particle and then solve its equation of motion

$$m\ddot{x}(t) = F[S; x(t)] \tag{3.15}$$

where from Eq. (3.7)

$$F[S; x(t)] = -U'[x(t)] + F[S; x(t)]. \tag{3.16}$$

We will proceed as follows. For the slow variable regime we will evaluate $F[S; x(t)]$ as a corrected form of the force for the idealized infinity slow process of Case 1 after Eq. (3.14). Correspondingly, for the fast variable regime we will evaluate $F[S; x(t)]$ as a corrected form of the force for the idealized infinity fast process.

The particle's slow variable equations of motion turn out to be identical in form to the phenomenological equations of damped motion of a macroscopic particle immersed in, say, a viscous fluid. For example, compare the slow variable Eqs. (3.24) and (3.25) with the phenomenological eqs. (A.30) and (A.31) for macroscopic sedimentation processes.

Thus, recalling the argument of Section II.A.1, in the slow variable regime the potential $W(S; x)$ that determines the particle's statics also drives its dynamics.

In contrast, the particle's fast variable equation of motion is nonclassical; and a potential $V(S; x)$ that is very different from $W(S; x)$ drives the dynamics. (Similar problems occur in the general theory of irreversible processes. There they are resolved by assuming slow variable relaxation [1–5], see Appendix.)

We will first consider the slow variable regime.

B. Slow Variable Dynamics and the Traditional Concept of Dissipation

The slow variable regime applies to processes $x_F(S) \rightarrow x_{F=0}(S)$ throughout which the particle's velocity $\dot{x}(t)$ is much less than the mean thermal velocity of a fluid molecule.

1. The Slow Variable Equations of Motion

We obtain the particle's slow variable equations of motion as follows:

1. Using the model form $F(S; x, \dot{x})$ for $F(S; x)$ of Section II.A.2 and also using Eq. (3.16), Eq. (3.15) becomes

$$m\ddot{x}(t) = F[S; x(t), \dot{x}(t)] \tag{3.17}$$

where

$$F[S; x(t), \dot{x}(t)] \equiv -U'[x(t)] + F[S; x(t), \dot{x}(t)]. \qquad (3.18)$$

2. Making the slow variable assumption that throughout $x_F(S) \rightarrow x_{F=0}(S)$ $\dot{x}(t)$ is very small permits one to approximate F by the following linear order power series expansion

$$F[S; x(t), \dot{x}(t)] \doteq F[S; x(t), \dot{x}(t) = 0]$$
$$+ \left\{ \frac{\partial F[s; x(t), \dot{x}(t)]}{\partial \dot{x}(t)} \right\}_{\dot{x}(t)=0} \dot{x}(t). \qquad (3.19)$$

3. From Case 1 after Eq. (3.14) and Eq. (3.18) it follows that

$$F[S; x(t), \dot{x}(t) = 0] = -U'[x(t)] + F_{eq}(S; x) \equiv F_{eq}[S; x(t)] \qquad (3.20)$$

where $F_{eq}(S; x)$ is the total thermodynamic equilibrium force.

4. Using Eqs. (3.9), (3.12), and (3.20) $F_{eq}(S; x)$ may be written as

$$F_{eq}(S; x) = -\frac{dW(S; x)}{dx} \qquad (3.21)$$

where $W(S; x)$ defined in Eq. (3.12) is the particle's equilibrium potential or *potential of mean force*.

5. Defining the particle's *friction coefficient* $\zeta(S)$ by

$$\zeta(S) = -\left\{ \frac{\partial F[S; x(t), \dot{x}(t)]}{\partial \dot{x}(t)} \right\}_{\dot{x}(t)=0} \qquad (3.22)$$

it follows that

$$\left\{ \frac{\partial F[S; x(t), \dot{x}(t)]}{\partial \dot{x}(t)} \right\}_{\dot{x}(t)=0} \dot{x}(t) = -\zeta(S)\dot{x}(t). \qquad (3.23)$$

6. Finally comparing Eqs. (3.17)–(3.23) yields the particle's slow variable equation of motion as

$$m\ddot{x}(t) = -\frac{dW[S; x(t)]}{dx(t)} - \zeta(S)\dot{x}(t). \qquad (3.24)$$

Before going on, we will specialize Eq. (3.24) to a high friction and weak force or *strongly overdamped* system. Given the arguments of Appendix Section F.9, this may be done by letting $m\ddot{x}(t) \to 0$ in Eq. (3.24) to yield

$$\dot{x}(t) = -\zeta^{-1}(S)\frac{dW[S; x(t)]}{dx(t)}. \tag{3.25}$$

Note that in contrast to Eq. (3.7), Eqs. (3.24) and (3.25) are closed equations of motion and thus may be straightforwardly integrated. Therefore, they provide a solution to the problem of finding the particle's trajectories $x(t)$ during the process $x_F(S) \to x_{F=0}(S)$.

The validity of this solution, of course, hinges on the viability of the slow variable approximation to $F[S; x(t)]$ used to convert Eq. (3.15) into Eq. (3.24); namely,

$$F[S; x(t)] \doteq -\frac{dW[S; x(t)]}{dx(t)} - \zeta(S)\dot{x}(t). \tag{3.26}$$

To assess this viability we next point out the physics implied by the slow variable assumption.

2. Slow Variable Physics

From Eq. (3.26) the slow variable assumption amounts to approximating $F[S; x(t)]$ as a sum of two force terms; namely, (a) the *equilibrium force* $F_{eq}[S; x(t)] = -dW[S; x(t)]/dx(t)$, which governs the particle's statics; and (b) a *nonequilibrium correction* $-\zeta(S)\dot{x}(t)$, identical in form to the phenomenological frictional force acting on a macroscopic particle.

The interpretation of these force terms is as follows: (a) $-dW[S; x(t)]/dx(t)$ is the force for an idealized infinitely slow process ($\dot{x} \to 0$) throughout which the system stays in perfect thermodynamic equilibrium; and (b) $-\zeta(S)\dot{x}(t)$ corrects $-dW[S; x(t)]/dx(t)$ to lowest order in $\dot{x}$ for breakdowns of strict equilibrium due to fluid lag, which occur because of the finite rate ($\dot{x} \neq 0$) or real processes.

From this intepretation, it is evident that the slow variable assumption of very small $\dot{x}(t)$ requires that throughout $x_F(S) \to x_{F=0}(S)$ the fluid + particle system remain near to thermodynamic equililbrium.

We may now discuss the validity of the slow variable equations of motion.

3. Validity of the Slow Variable Solution

Referring to the previous comments we note the following concerning the validity of Eqs. (3.24) and (3.25).

1. Eq. (3.26) coincides in form with the phenomenological force law for a damped macroscopic particle. Thus for Eqs. (3.24) and (3.25) to be valid, an equation derived from macroscopic measurements must be sufficient to account for the microworld physics experienced by the particle.
2. For Eqs. (3.24) and (3.25) to be valid the fluid + particle system must stay in near thermodynamic equilibrium during $x_F(S) \rightarrow x_{F=0}(S)$.
3. For Eq. (3.25) to be valid: additionally (i) the particle must be strongly overdamped; and (ii) as shown in the Appendix Section F.9, its motion on timescales short compared to the relaxation time τ_A for particle accelerations must be observationally irrelevant.

We next extend Eqs. (3.24) and (3.25) to include the effects of fluctuations.

4.   Slow Variable Dynamics Including Fluctuations

We noted in Section II.A.2 that a microscopic particle has a range of possible positions x' of final static equilibrium rather than the single position $x_{F=0}(S)$. This range, recall, is governed by the Boltzmann distribution Eq. (3.13).

We next extend the treatment of static fluctuations of Section II.A.2 to dynamics. We do this by adding random forces $\tilde{f}(t)$ and $f(t) = \zeta^{-1}(S)\tilde{f}(t)$ of the standard Brownian motion type [2] to Eqs. (3.24) and (3.25). This yields the following equations of motion

$$m\ddot{x}(t) = -\frac{dW[S;x(t)]}{dx(t)} - \zeta(S)\dot{x}(t) + \tilde{f}(S;t) \qquad (3.27)$$

and

$$\dot{x}(t) = -\zeta^{-1}(S)\frac{dW[S;x(t)]}{dx(t)} + f(S;t). \qquad (3.28)$$

Eq. (3.27) is the particle's Langevin equation, and thus is equivalent to its *Fokker-Planck equation*. Eq. (3.28) applies only to the motion of a strongly overdamped particle course-grained over intervals $\sim \tau_A$; and then is equivalent to the particle's *diffusion equation*.

From the present standpoint, the most important aspects of Eqs. (3.27) and (3.28) are that they substantively macroscopic equations; and that this macroscopic nature limits their applicability to microscopic phenomena. The essentially macroscopic nature of Eqs. (3.27) and (3.28) may, for example, be seen as follows. Suppose $W(S; x)$ is harmonic. Then Eqs. (3.24) and (3.25) determine the most probable trajectories predicted by Eqs. (3.27) and (3.28). This ensures that the macroscopic character of the former equations implies that of that latter.

We next point out that the near equilibrium physics of the slow variable regime gives rise to a very simple picture of particle motion and underlies the traditional concept of frictional damping.

5. The Slow Variable Picture of Particle Motion

The slow variable Eq. (3.26) yields $F[S; x(t)]$ as the sum of the conservative force $-dW[S; x(t)]/dx(t)$ and the dissipative force $-\zeta(S)\dot{x}(t)$. The former drives the process $x_F(S) \rightarrow x_{F=0}(S)$, while the latter gives rise to its irreversible character.

Thus, in the slow variable regime the following picture of particle motion applies:

SLOW VARIABLE PICTURE. In the slow

variable limit, irreversible particle dynamics may

be viewed as *motion in the potential of mean*

force $W(S; x)$ subject to simple frictional damping. (3.29)

We may now discuss the physical basis of the traditional concept of damping.

6. Near Adiabatic Limit Physics and the Traditional Concept of Dissipation

The traditional concept of frictional damping as due to a retarding force linear in the particle's velocity $\dot{x}$ first emerged from macroscopic phenomenology. However it is now clear that its physical origin is as follows:

PHYSICS OF SLOW VARIABLE DISSIPATION. The

classical slow variable damping force $-\zeta(S)\dot{x}(t)$ reflects

slight deviations from perfect thermodynamic equilibrium

arising because the particle's velocity is finite not vanishing. (3.30)

This concludes our study of the slow variable limiting form for $F[S; x(t)]$. We next discuss its fast variable limiting form.

C.   First Variable Dynamics and a Nonclassical
Concept of Dissipation

The fast variable regime applies to processes $x_F(S) \rightarrow x_{F=0}(S)$ throughout which the particle's velocity $\dot{x}(t)$ is much greater than the mean thermal velocity of a fluid molecule.

To avoid complexity, we will limit the fast variable evaluation of $F[S; x(t)]$ to times t that are short relative to the fluid's motional response times $\sim \tau$. (An outline description of fast particle motion for times $t > \tau$ is given in Section IV.A and Figure 3.5.)

This restriction will limit us to the study of an initial segment of the particle's trajectory $x(t)$ of duration $\sim \tau$. We impose it, however, since it greatly simplifies the evaluation of $F[S; x(t)]$; the simplification arising because the fluid remains nearly motionless for times $t \leq \tau$. (The initial segment of $x[t]$ that we can study is nontrivial, because a hyperthermal particle moves appreciably in time τ even though the fluid moves only slightly.)

We next turn to the short time evaluation of $F[S; x(t)]$.

1.   The Short Time Equation of Motion

As indicated earlier, the slow variable $F[S; x(t)]$ of Eq. (3.26) may be viewed as a corrected form of the $F[S; x(t)]$ for the idealized infinitely slow process (Case 1 after Eq. [3.14] for which $F[S; x(t)] = F[S; x(t), \dot{x}(t) = 0]$); the correction accounting for the fact that $\dot{x}(t)$ is nonvanishing for real processes.

We will similarly develop the short time $F[S; x(t)]$ as a corrected form of the $F[S; x(t)]$ for the idealized infinitely fast process (Case 2 after Eq. [3.14] for which $F[S; x(t)] = F[S; x(t), \dot{x}(t) = \infty]$); the correction accounting for the fact that $\dot{x}(t)$ is finite for real processes.

To do this, we first define the short time equivalent $V(S; x)$ of the slow variable $W(S; x)$. Namely, in analogy to Eqs. (3.9) and (3.10) we define $V(S; x)$ by

$$F_{\text{inst}}(S; x) = -v'(S; x) \tag{3.31}$$

and

$$V(S; x) \equiv U(x) + v(S; x) \tag{3.32}$$

where $F_{\text{inst}}(S; x) = F(S; x, \dot{x} = \infty)$ is the instantaneous fluid force described earlier (see Case 2 after Eq. [3.14]).

We will, henceforth, refer to $V(S; x)$ as the *instantaneous potential*.

Let us denote the total force acting on the particle in the infinitely fast process by $F_{inst}(S;x)$. This force is given by

$$F_{inst}(S;x) = -U'(x) + F_{inst}(S;x). \tag{3.33}$$

From Eqs. (3.31)–(3.33) it follows that

$$F_{inst}(S;x) = -\frac{dV(S;x)}{dx}. \tag{3.34}$$

Eq. (3.34) is the short time analogue of the slow variable Eq. (3.21). It thus is a basic equation of the short time regime.

We next give the short time equivalent of the slow variable equation of motion Eq. (3.24).

This equivalent is the following equation of motion whose full form including fluctuations is derived in Refs. 20 and 21,

$$m\ddot{x}(t) = -\frac{dV[S;x(t)]}{dx(t)} + \int_0^t \Theta(t-t')\Delta x(t')dt' \tag{3.35}$$

where in Eq. (3.35) $\Delta x(t') \equiv x(t') - x_F(S)$ is the particle's displacement at time t' from its initial position $x(t' = 0) = x_F(S)$, and where $\Theta(t - t')$ is a function defined in Refs. 20 and 21 that describes the relaxation of the fluid in response to the displacement $\Delta x(t')$.

We next discuss the physics in Eq. (3.35).

2. Fast Variables: Near Sudden Limit Physics

This physics is as follows: (a) $-dV[S;x(t)]/dx(t)$ is the force for an idealized infinitely fast process ($\dot{x} \to \infty$) throughout which the system stays in perfect thermodynamic *dis*equilibrium (as described in Case 2 after Eq. [3.14]); and (b) $\int_0^t \Theta(t - t')\Delta x(t')dt'$ corrects $-dV[S;x(t)]/dx(t)$ to lowest order in Δx for breakdowns of strict disequilibrium due to fluid relaxation, which occur because of the finite rate ($\dot{x} \neq \infty$) of real processes.

We next describe the simple picture of particle motion that emerges in the short time regime as a consequence of this physics.

3. The Short Time Picture of Particle Motion

The short time Eq. (3.35) yields $F[S;x(t)]$ as the sum of the conservative force $-dV[S;x(t)]/dx(t)$ and the dissipative force $\int_0^t \Theta(t - t')\Delta x(t')dt'$. The former drives the process $x_F(S) \to x_{F=0}(S)$, while the latter gives rise to its irreversible character.

Thus the short time picture of irreversible particle motion is as follows:

> **SHORT TIME PICTURE.** In the short time
> limit, irreversible particle dynamics may be
> viewed as *motion in the instantaneous potential*
> *V(S; x) subject to weak relaxational damping.*   (3.36)

Finally we discuss the nonclassical concept of damping that emerges from the physics of the short time regime.

4.   Near Sudden Limit Physics and a Nonclassical Concept of Dissipation

This nonclassical concept of damping is as follows:

PHYSICS OF SHORT TIMESCALE DISSIPATION.

The nonclassical short timescale damping force $\int_0^t \Theta(t - t')\Delta x(t')dt'$

reflects *slight deviations from perfect thermodynamic* dis*equilibrium*
arising because the particle's velocity is finite not infinite.   (3.37)

D.   Summary

In summary, we have found two distinct limiting solutions to the problem of determining the trajectories $x(t)$ of a microscopic particle during the fluid phase relaxation process $x_F(S) \rightarrow x_{F=0}(S)$. The first solution is valid only for subthermal particles while the second has physical content, as opposed to formal exactness [20a], only for hyperthermal particles.

We have also found that very different pictures of the motion and very different concepts of dissipation emerge in the two limiting regimes. These are summarized in Eqs. (3.29) and (3.30) and Eqs. (3.36) and (3.37). In particular, the potential energy function that drives the process $x_F(S) \rightarrow x_{F=0}(S)$ is different in the two regimes; namely, it is the potential of mean force $W(S; x)$ in the slow variable regime, and the instantaneous potential $V(S; x)$ in the short time regime.

This concludes our comparison of particle motion in the slow and fast variable timescale regimes.

We next turn to the slow variable description of liquid phase activated barrier crossings.

III. THE SLOW VARIABLE DESCRIPTION OF ACTIVATED BARRIER CROSSING

The slow variable treatment of the particle relaxation process $x_F(S) \to x_{F=0}(S)$ of section II.B may be adapted to barrier crossing by making the following reinterpretations: (a) x is taken as the reaction coordinate of a solute system undergoing barrier passage; (b) $W(S;x)$ is taken as the system's reaction coordinate potential of mean force; (c) m is taken as the reduced mass factor for reaction coordinate motion; and (d) $x_F(S)$ and $x_{F=0}(S)$ are taken, respectively, as the transition state $x^\dagger$ and product well x_P values of x. Given these transcriptions, the process $x_F(S) \to x_{F=0}(S)$ is reinterpreted as the reactive barrier passage $x^\dagger \to x_P$.

A. The Kramers Model

Thus, the slow variable rate constant $k(T)$ may be found by: (a) solving either Eq. (3.27) or (3.28) to obtain the reaction coordinate trajectories $x(t)$; and then (b) performing a Boltzmann average over the ensemble of $x(t)$s to yield $k(T)$ [14a].

For realistic $W(S;x)$s like the one computed by Chandrasekhar, Smith, and Jorgensen [29] and reproduced in Figure 3.2, this computation must be performed numerically. However, as first shown by Kramers [7], $k(T)$ may be found analytically if one adopts the parabolic barrier approximation to $W(S;x)$; namely,

$$W(S;x) \doteq W(S;x^\dagger) - \tfrac{1}{2}m\omega_{\mathrm{PMF}}^2 y^2 \qquad (3.38)$$

where $y \equiv x - x^\dagger$ and where

$$\omega_{\mathrm{PMF}} \equiv \left[-\frac{W''(S_i x^\dagger)}{m} \right]^{1/2} \qquad (3.39)$$

is the parabolic barrier frequency of $W(S;x)$.

We will express Kramers' result for $k(T)$ in terms of the *transmission coefficient* $\kappa(T)$ defined by

$$\kappa(T) = k_{\mathrm{TST}}^{-1}(T)k(T) \qquad (3.40)$$

where $k_{\mathrm{TST}}(T)$ is the transition state theory estimate for $k(T)$. By solving equations equivalent to Eqs. (3.27) and (3.28) for the $W(S;x)$ of Eq. (3.38),

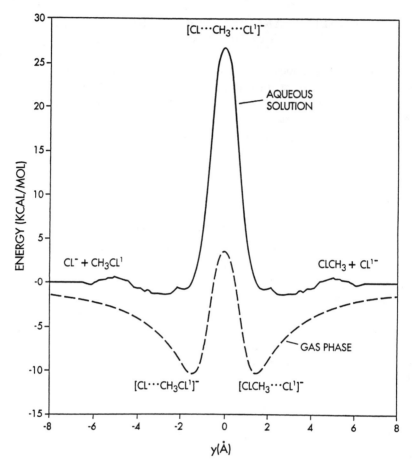

Figure 3.2. Reaction coordinate potentials for the aqueous Cl^-+CH_3Cl reaction near standard conditions computed by Chandrasekhar et al. [29]. The gas phase potential $U(x)$ and the potential of mean force $W(S;x)$ are plotted versus $y = x - x^\dagger$.

Kramers found respectively the following results for $\kappa(T)$

$$\kappa_{KR}(T) = \omega_{PMF}^{-1}\left\{\left[\frac{\zeta^2(S)}{4} + \omega_{PMF}^2\right]^{1/2} - \frac{\zeta(S)}{2}\right\} \tag{3.41}$$

and

$$\kappa_{KRD}(T) = \omega_{PMF}^{-1}\left[\frac{\omega_{PMF}^2}{\zeta(S)}\right]. \tag{3.42}$$

In accord with our earlier discussion, and the discussion in the Appendix, of limits like $m\ddot{x}(t) \to 0$ (needed to obtain Eq. [3.28] from Eq. [3.27]) Eq. (3.41) reduces to Eq. (3.42) in the limit of strong overdamping $\zeta(S) \gg 2\omega_{PMF}$.

The Kramers result for $\kappa(T)$ of Eq. (3.41) has been tested by Wilson and co-workers [14] in their MD simulations of model aqueous nucleophilic substitution reactions. Specifically, these authors determined by MD simulation both the "exact" $\kappa_{MD}(T)$ and Kramers $\kappa_{KR}(T)$ transmission coefficients for 12 $S_N 2$ systems [14a] and for one $S_N 1$ system [14b]. The coefficients $\kappa_{MD}(T)$ were determined from ensembles of reactive and nonreactive MD trajectories. The coefficients $\kappa_{KR}(T)$ were found from Eq. (3.41), with the parameters ω_{PMF} and $\zeta(S)$ being computed via an MD implementation of our partial clamping model [21]. Namely, ω_{PMF} and $\zeta(S)$ are computed via constrained MD simulations in which the reaction coordinate x is held fixed at its transition state value $x^\dagger$ while the remaining degrees of freedom of the solution are allowed to move freely subject to this single constraint.

The partial clamping model accurately describes the forces on x if it makes only *small amplitude excursions* $y = x - x^\dagger$ from $x^\dagger$ [21,23]. Thus, the model provides a valid computation of $\omega_{PMF}, \zeta(S)$, and other quantities that depend on the forces; for reactions in which the decision to react is made before y achieves large values.

In Table III we reproduce some of the MD results of Wilson and co-workers [14]. Especially in columns 4 and 5 we compare their results for $\kappa_{MD}(T)$ and $\kappa_{KR}(T)$ for the 13 model systems. The Kramers predictions for the transmission coefficients $\kappa_{KR}(T)$ are extremely poor, always being ~ 1–2 orders of magnitude smaller than the exact transmission coefficients $\kappa_{MD}(T)$.

This failure of the Kramers model is hardly surprising. Given the Arrhenius principle, one expects typical liquid phase reactions to occur in the fast variable rather than the slow variable timescale regime; thus invalidating the Langevin Eq. (3.27) and, hence, the Kramers Eq. (3.41).

The fast variable nature of liquid phase reactions is illustrated in Figure 3.3. There we reproduce a plot of Wilson and co-workers [14b] derived from MD studies of their model aqueous $S_N 1$ reaction. In Figure 3.3 the average reaction coordinate kinetic energy $\frac{1}{2}m\dot{x}^2$ computed from an ensemble of reactive trajectories is plotted as a function of time t. Notice that as early as -1 ps before $x^\dagger$ is reached, $\frac{1}{2}m\dot{x}^2$ is ~ 25 times the mean thermal value $\frac{1}{2}kT$; while in the last 0.1 ps before barrier passage $\frac{1}{2}m\dot{x}^2$ becomes as large as ~ 45 times $\frac{1}{2}kT$. Thus, the average reaction coordinate speed $\dot{x}$ can be ~ 5–$6\frac{1}{2}$ times its mean thermal value $(kT/m)^{1/2}$, and the reaction occurs in the fast variable timescale regime.

TABLE III

Transmission Coefficients κ Computed by Wilson and Co-workers [14] for Their Model Aqueous Nucleophilic Substitution Reactions

Model system	Gas phase barrier height (kcal/mole)	Rate of charge switching	κ_{MD}	κ_{KR}	κ_{GH}
1	4.9	0.5	0.49 ± 0.04	0.06 ± 0.03	0.56 ± 0.04
2	4.9	1.0	0.43 ± 0.05	0.03 ± 0.01	0.43 ± 0.04
3	4.9	2.0	0.32 ± 0.05	0.01 ± 0.006	0.33 ± 0.04
4	13.9	0.5	0.71 ± 0.04	0.03 ± 0.02	0.74 ± 0.07
5	13.9	1.0	0.54 ± 0.04	0.02 ± 0.01	0.57 ± 0.08
6	13.9	2.0	0.40 ± 0.05	0.004 ± 0.004	0.37 ± 0.10
7	31.9	0.5	0.90 ± 0.02	0.03 ± 0.03	0.88 ± 0.08
8	31.9	1.0	0.75 ± 0.03	0.05 ± 0.04	0.73 ± 0.05
9	31.9	2.0	0.50 ± 0.04	0.03 ± 0.02	0.57 ± 0.05
10	4.9	0.5	0.39 ± 0.05	0.019 ± 0.01	0.34 ± 0.07
11	4.9	1.0	0.37 ± 0.05	0.007 ± 0.005	0.32 ± 0.09
12	4.9	2.0	0.31 ± 0.05	0.01 ± 0.006	0.30 ± 0.04
$S_N 1$	–	–	0.53 ± 0.04	0.019 ± 0.001	0.58 ± 0.03

We next discuss the implications of fast variable physics for Kramers' model in a bit more detail. The main point is that for fast variable processes the near equilibrium assumption of the slow variable models is strongly violated. This implies especially that the driving force for typical liquid phase reactions is not supplied by the potential of mean force, as is required for the validity of Eq. (3.41).

That is, while $W(S; x)$ determines the force on a stationary or very slowly moving reaction coordinate, it has little to do with the force on a typical reaction coordinate moving with kinetic energies of thousands of Kelvins.

To expose the problematic nature of $W(S; x)$ in a graphic way, we next adopt a slightly different viewpoint. Namely, we note that typical *gas phase* reaction coordinate potentials $U(x)$ drop very steeply as x goes from $x^\dagger$ to x_P (Figs. 3.2 and 3.4). As a result, during $x^\dagger \to x_P$ the reaction coordinate x experiences extremely strong accelerating forces from $U(x)$. In the gas phase, these forces would cause an explosive departure of x from $x^\dagger$ toward x_p (Fig. 3.4). In the liquid phase for $W(S; x)$, to drive the reaction the solvent must synchronously explosively depart from its $x = x^\dagger$ equilibrium configuration toward its $x = x_p$ equilibrium configuration; because only then can the system stay in near equilibrium. Such a solvent departure is of course physically absurd. This is illustrated in Figure 3.4.

Thus, $W(S; x)$ cannot drive typical liquid phase chemical reactions, as is required for the validity of the Kramers model.

We have not yet commented on the diffusional Kramers model that yields Eq. (3.42) for κ. Recalling the discussion of Sections II.B.3 and II.B.4 and Appendix Section F.9, we merely note here that diffusional models require for their validity, in addition to near equilibrium physics, very strong overdamping and the irrelevance of short-time motion. However, both of these extra conditions are radically violated for typical reactions. First, such processes, as we will emphasize in Section IV are weakly damped, not strongly overdamped. This follows because, as mentioned, the relevant forces are extremely strong; and because the relevant friction is high frequency rather than zero frequency and, hence, is relatively small. Also, as shown by Wilson and co-workers [14], short-time motion is critical for typical reactions. For example, for the systems of Table I, $\kappa(T)$ is determined in the timescale regime for which the picture of motion of Eq. (3.36) is valid. Summarizing, we expect for typical reactions that Kramers' diffusion limit Eq. (3.42) is at least as unsatisfactory as his full Eq. (3.41).

Many extensions of the Kramers model have been described in the literature. Perhaps the most widely used is that of Grote and Hynes [30]. For this reason, we next briefly discuss the model of Grote and Hynes.

B. The Grote–Hynes Model

This model is based on the following generalized Langevin equation for the reaction coordinate

$$m\ddot{x}(t) = -\frac{dW[S; x(t)]}{dx(t)} - \int_0^t \zeta(S; t - \tau)\dot{x}(\tau)d\tau + \tilde{f}(S; t) \tag{3.43}$$

which is the finite memory extension of Eq. (3.27).

Grote and Hynes have solved Eq. (3.43) for the $W(S; x)$ of Eq. (3.38). Their result for the transmission coefficient $\kappa_{GH}(T)$ involves the Laplace domain friction kernel

$$\hat{\zeta}(S; z) \equiv \int_0^\infty \zeta(S; t)\exp(-zt)dt. \tag{3.44}$$

It takes the following form:

$$\kappa_{GH}(T) = \omega_{PMF}^{-1}\left\{\left[\frac{\hat{\zeta}^2(S; x_+)}{4} + \omega_{PMF}^2\right]^{1/2} - \frac{\hat{\zeta}(S; x_+)}{2}\right\}. \tag{3.45}$$

Notice that the Kramers and Grote–Hynes expressions for κ, Eqs. (3.41) and (3.45), are identical; except that the ordinary friction coefficient $\zeta(S)$ in

Eq. (3.41) is replaced by the z-dependent friction coefficient $\hat{\zeta}(S; x_+)$ in Eq. (3.45). The value of x_+ may be determined from the relation [30]

$$\kappa_{GH}(T) = \omega_{PMF}^{-1} x_+. \tag{3.46}$$

Namely, comparing Eqs. (3.45) and (3.46) yields the following self-consistent equation for x_+

$$x_+ = \left[\frac{\hat{\zeta}^2(S; x_+)}{4} + \omega_{PMF}^2\right]^{1/2} - \frac{\hat{\zeta}(S; x_+)}{2}. \tag{3.47}$$

If ω_{PMF} and $\zeta(S; t)$ are known Eqs. (3.44)–(3.47) permit one to evaluate $\kappa_{GH}(T)$.

To assess the Grote–Hynes model, we parallel the discussion of the Mori [5] identity given in Appendix Section F.15. Namely, noting that the generalized Langevin Eq. (3.43) stands in relation to the Langevin Eq. (3.27) in much the same way that Mori's Eq. (A.54) stands in relation to Onsager's Eq. (A.53); it follows from the discussion of Eq. (A.54) that the use of Eq. (3.43) is problematic for processes like typical liquid phase reactions, whose physics differs qualitatively from the macroworld slow variable physics that sets the form of Eq. (3.27). Thus, for example, from Eqs. (3.45) and (3.47) the Grote–Hynes transmission coefficient $\kappa_{GH}(T)$ contains as a main parameter the barrier frequency ω_{PMF} of the potential of mean force $W(S; x)$; a quantity that is not connected with the physics of barrier crossing dynamics in typical liquid phase reactions (Fig. 3.4).

The dependence of $\kappa_{GH}(T)$ on ω_{PMF} has led to serious misconceptions. For example, it is often stated [31] that for large ω_{PMF} or "sharp barriers," transition state theory provides a good approximation to $k(T)$; or equivalently from Eq. (3.40) that for sharp barriers $\kappa(T) \to 1$. This false statement apparently derives from the following argument. Eq. (3.47) seems to predict that as $\omega_{PMF} \to \infty, x_+ \to \omega_{PMF}$ and hence $\zeta(S; x_+) \to \zeta(S; \omega_{PMF}) \to 0$; implying from either Eqs. (3.45) or (3.46) that $\kappa_{GH}(T) \to 1$. Actually, as we will see in Sec IV, in the sharp barrier limit $k(T)$ approaches the "frozen solvent" result of Eq. (3.54) not $k_{TST}(T)$.

We next note that Wilson and co-workers have computed $\kappa_{GH}(T)$ for their model reactions via the MD method they used to determine $\kappa_{KR}(T)$. Their results for $\kappa_{GH}(T)$ are listed in column 6 of Table I. Comparing these results with the values for the exact transmission coefficient $\kappa_{MD}(T)$ in column 4 shows that, in contrast to the sharp disagreement between $\kappa_{KR}(T)$ and $\kappa_{MD}(T), \kappa_{GH}(T)$ and $\kappa_{MD}(T)$ are in good agreement.

These and similar results have been interpreted as confirmations of the Grote–Hynes model as a valid physical theory of liquid phase reactions, and, relatedly, of frequency-dependent friction as a valid concept for liquid phase chemistry. This interpretation is spurious. Actually the result $\kappa_{GH}(T) \approx \kappa_{MD}(T)$ shows only that for the systems of Table III the decision to react is made near the barrier top.

That is, while numerical accuracy of Kramers' Eq. (3.41) would validate the slow variable picture of Eq. (3.29), that of Grote and Hynes' Eq. (3.45) does not validate any physical model for the reaction dynamics. Rather, it validates only the accuracy of (a) the partial clamping model [21], used to compute the quantities in Eq. (3.45); and (b) the parabolic barrier approximation $U(x) \doteq U(x^{\dagger}) - 1/2 \, m\omega_g^2 y^2$ to the *gas phase* reaction coordinate potential. The accuracy of these, however, requires only that $y = x - x^{\dagger}$ remains small prior to the decision to react.

These comments may be rephrased as follows. Just as Mori's Eq. (A.54) is formally exact, Eq. (3.43) is nearly formally exact close to the barrier top. However, as noted in Appendix Section F.15, formal exactness does not equate to physical content. Thus, just as Eq. (A.54) can be merely a nugatory definition of the memory kernel $\beta^{-1}(t)$ rather than a real solution of the problem of finding an equation of thermodynamic relaxation; Eq. (3.43) is not necessarily a satisfactory solution of the central problem of determining the physical nature of the forces acting on the solute reaction coordinate. Rather far from the slow variable regime, Eq. (3.43) can be merely a formal definition of $\zeta(S;t)$ and $\tilde{f}(S;t)$ as those quantities that convert the slow variable force $-dW[S; x(t)]/dx(t)$ into the unrelated actual forces acting on the reaction coordinate.

These comments are closely related to questions of the validity of the frequency-dependent friction concept [13]. Namely, the use of frequency-dependent friction to correct Eq. (3.27) for finite solvent response times has physical content only for *low-frequency processes*, because if the solute's frequencies are too high the near equilibrium assumption is violated and $W(S; x)$ no longer drives the solute's motion.

Equivalently, for high-frequency or short time processes two effects must be accounted for; namely, the frequency dependence of the solvent response and the change $W \to V$ in the solute's driving potential. (See Eqs. (3.53) and (3.58).) The usual treatment [30] explicitly deals with only the first of these effects and, thus, for high-frequency processes does not identify all of the essential physics.

We will return to these points in Section IV.B.

To close Section III, we note some experimental results that may relate to the issues just discussed; namely, Sumi and Asano [13] have recently attempted to fit the viscosity dependence of isomerization rate constants for

N-benzylideneanalines in viscous organic solvents using Eq. (3.45). Making only very mild assumptions, they found that this fit required either an unphysically large and unphysically viscosity dependent decay time $\tau_{SC} \approx 10^5-10^{10}$ ps for $\zeta(S;t)$ or an unphysically large barrier frequency $\omega_{PMF} \approx 10^{15}-10^{18}s^{-1}$.

The reactions in question are typical with activation barriers of $\sim45-55$ kJ-mol^{-1}, and thus from Figure 3.4 cannot be driven by $W(S;x)$. The difficulties may well stem from the dependence of $\kappa_{GH}(T)$ on ω_{PMF}, and in particular on the false sharp barrier limit $k(T) \rightarrow k_{TST}(T)$.

We next briefly overview our fast variable description of liquid chemical dynamics.

IV. THE FAST VARIABLE DESCRIPTION OF LIQUID PHASE CHEMICAL DYNAMICS

We next quickly review, from the standpoint of the recent literature [6,11–17], our fast variable theory of liquid phase chemical processes [19–26]. This theory is designed to deal relatively realistically with the fast variable physics that is the signature of liquid phase chemical dynamics.

The plan of Section IV is as follows: In section IV.A, we qualitatively outline the general picture of reaction dynamics that emerges from fast variable physics. Next, in section IV.B, we examine liquid phase-activated barrier crossing in the short time regime of Section II.C. In Section IV.C we note that the "fast variable/slow bath" timescale separation also applies to liquid phase vibrational energy relaxation and then discuss that process from the fast variable standpoint. Finally, in Section IV.D, we discuss some related work of others.

A. The Fast Variable Picture of the Dynamics of Typical Liquid Phase Reactions

As indicated earlier, the basis of the fast variable physics of typical liquid phase reactions may be stated in two different but nearly equivalent forms.

Form 1. In liquids as in the gas phase reacting molecules are often moving with hyperthermal speeds (Fig. 3.3).

Form 2. The interactions in the liquid solution are asymmetric in that only the solute is subject to the extremely powerful accelerating forces produced by a typical gas phase potential energy surface $U(x)$ (see Fig. 3.4).

Because of these factors, typical reactions occur in a limit of strong thermodynamic disequilibrium exactly opposite to the near thermodynamic equilibrium limit of the slow variable models. Consequently, forces acting

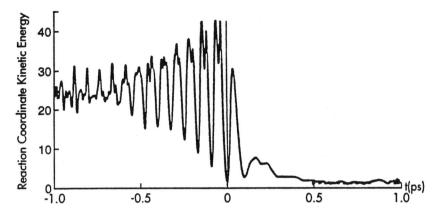

Figure 3.3. Reaction coordinate kinetic energy in units of $1/2\ kT$ computed by Wilson and co-workers for their model aqueous S_N1 reaction near standard conditions. Reproduced from fig. 4c of ref. 14b.

on both the solute and the solvent that act to restore thermodynamic equilibrium play an enhanced and essential role in typical solution reactions.

Our theory explains liquid phase reaction dynamics in terms of an interplay between the thermodynamic restoring forces and the forces deriving from the solute gas phase potential $U(x)$. The latter propel the

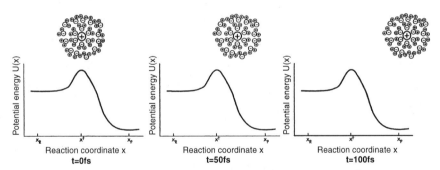

Figure 3.4. Breakdown of the slow variable picture of Eq. (3.29) for typical liquid phase reactions. In the gas phase, typical reaction coordinate potentials $U(x)$ cause an explosive departure of the reaction coordinate from $x^{\dagger}$ toward x_p. As shown, for the slow variable picture to be valid the solvent schematized as two solvation shells must make an unphysical synchronous explosive departure, because only then can near equilibrium between the solute schematized by a cation and the solvent be maintained.

To see the explosive nature of the gas phase solute departure, assume that $U(x)$ has a 1.0 eV activation barrier. For this typical value, a separable reaction coordinate leaving $x^{\dagger}$ with the mean thermal speed $(kT/m)^{1/2}$ will reach x_p with a speed of $\sim 8.4\ (kT/m)^{1/2}$. A car subject to a suitably scaled up forces and distances would accelerate from highway speed, ~ 65 miles per hour, to jet airplane speed, ~ 540 miles per hour, in a distance of $\sim 1-2$ yards.

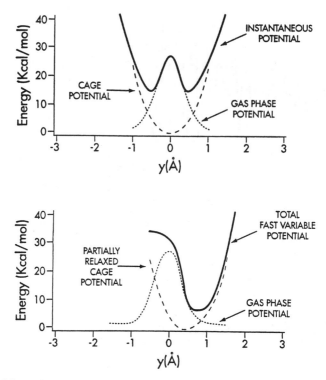

Figure 3.5. Schematic fast variable reaction coordinate potentials. (A) Short time "frozen solvent" [14,23] regime. At the shortest times typical reactions conform to Eq. (3.36), and thus are driven by the instantaneous potential $V(S;x)$ = gas phase potential $U(x)$ + cage potential $v(S;x)$. The double well form of $V(S;x)$ reflects the frozen solvent's capacity to transiently confine the solute. (B) Partially relaxed regime. At longer times, the short time picture of Eq. (3.36) breaks down due to a solvent relaxation [11,16] aimed at restoring thermodynamic equilibrium. This relaxation converts the instantaneous potential into the less confining total fast variable potential that can further drive the solute toward products.

solute out of equilibrium with the solvent while the former act to reinstate equilibrium.

The picture of reaction dynamics that emerges from this interplay is schematized in Figure 3.5, which we discuss next.

Assume that initially the solution is in thermodynamic equilibrium with the solute reaction coordinate fixed at $x^\dagger$ its transition state value. At $t = 0$ the reaction coordinate is released. The gas phase potential $U(x)$ then drives the solute toward products, initially leaving the solvent close to its original transition state configuration (Fig. 3.5a). In this initial "frozen solvent" regime [14,23], the short time picture of Eq. (3.26) applies.

Especially the solute moves on the instantaneous potential $V(S;x)$ defined in Eq. (3.32). In accord with Eq. (3.32), in Figure 3.5a we show $V(S;x)$ as a superposition of the gas phase potential $U(x)$ and a liquid phase potential $v(S;x)$. The liquid phase contribution $v(S;x)$, called the cage potential in Figure 3.5a, produces the thermodynamic restoring forces exerted on the reaction coordinate x by the frozen solvent. Thus, in the frozen solvent regime the reaction dynamics is dominated by the competition between $U(x)$, which "pushes" the solute toward products, and $v(S;x)$, which "pulls" the solute back to the transition state. This competition can transiently confine the solute, for example, by trapping it in the product well of $V(S;x)$. However, as shown in Figure 3.5b, this simple cage confinement picture of reaction dynamics soon breaks down due to a solvent relaxation aimed at re-equilibrating the solvent with the solute [11,16]. This solvent relaxation partially releases the frozen solvent restoring forces on the solute permitting $U(x)$ to further drive the reaction toward products.

In this manner, the solute dynamics is determined over the full course of the reaction by the gas phase forces and the liquid phase thermodynamic restoring forces.

We next discuss in a bit more detail activated barrier crossing processes for which the decision to react is made in the short time regime of Figure 3.5a.

B. Activated Barrier Crossing in the Short Time Regime

To do this we adapt the short time treatment of particle relaxation to barrier crossing by making transpositions like those given at the start of Section III. Namely, (a) x is reinterpreted as the solute's reaction coordinate; (b) $V(S;x)$ as its reaction coordinate instantaneous potential; (c) m as its reaction coordinate reduced mass factor; and (d) the process $x_F(S) \rightarrow x_{F=0}(S)$ as the reactive barrier passage $x^\dagger \rightarrow x_P$.

Given these transpositions, the fast variable rate constant may be found by solving Eq. (3.35) for the reaction coordinate trajectories $x(t)$ and then performing a Boltzmann average over the ensemble of $x(t)s$. For realistic $V(S;x)s$, for example, the one plotted in Figure 3.6, this evaluation must be performed numerically.

However, in analogy to the earlier closed form evaluations of the slow variable rate constant for $W(S;x)$ of Eq. (3.38), the short time rate constant may be found analytically if one adopts the following parabolic barrier approximation to $V(S;x)$:

$$V(S;x) \doteq V(S;x^\dagger) - \tfrac{1}{2}m\omega_{\mathrm{MIP}}^2 y^2 \qquad (3.48)$$

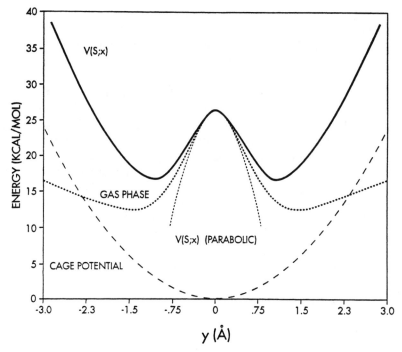

Figure 3.6. Frozen solvent reaction coordinate potentials for aqueous $Cl^- + CH_3Cl$ reaction near standard conditions versus $y = x - x^{\dagger}$. As in Figure 3.5A, the instantaneous potential $V(S;x)$ is the sum of the gas phase potential $U(x)$ and the cage potential $v(S;x) \cdot U(x)$ is that of Figure 3.2 and $v(S;x)$ is estimated [23] from the molecular dynamics results of ref. 14a. The parabolic approximation to $V(S;x)$ of Eq. (3.48) is also shown.

where $y = x - x^{\dagger}$ and where

$$\omega_{\mathrm{MIP}} \equiv \left[-\frac{V''(S;x^{\dagger})}{m} \right]^{1/2} \qquad (3.49)$$

is the parabolic barrier frequency of $V(S;x)$.

The short time transmission coefficient $\kappa_{\mathrm{ST}}(T)$ so obtained is [23]

$$\kappa_{\mathrm{ST}}(T) = \omega_{\mathrm{PMF}}^{-1} x_+ \qquad (3.50)$$

where x_+ solves the following self-consistent equation

$$x_+ = \omega_{\mathrm{MIP}}[1 + \omega_{\mathrm{MIP}}^{-2} \hat{\Theta}(x_+)]^{1/2}. \qquad (3.51)$$

TABLE IV

Transmission Coefficients and Other Parameters Computed by Wilson and co-workers [14a] for their Model Aqueous S_N2 Reactions. Model Systems are Those of Table III.

Model system	Gas phase barrier height (kcal/mole)	κ_{MD}	κ_{ST}	κ_{FS}	ω_{MIP} (cm^{-1})	ω_{PMF} (cm^{-1})	x_+ (cm^{-1})
1	4.9	0.49±0.04	0.56±0.04	0.37±0.03	190±9%	500±6%	280
2	4.9	0.43±0.05	0.43±0.04	0.22±0.02	190±9%	860±7%	370
3	4.9	0.32±0.05	0.33±0.04	0.14±0.01	230±11%	1730±9%	580
4	13.9	0.71±0.04	0.74±0.07	0.69±0.03	440±2%	640±6%	470
5	13.9	0.54±0.04	0.57±0.08	0.48±0.03	450±2%	930±9%	530
6	13.9	0.40±0.05	0.37±0.10	0.25±0.03	450±4%	1780±17%	650
7	31.9	0.90±0.02	0.88±0.08	0.87±0.02	910±0.4%	1040±5%	910
8	31.9	0.75±0.03	0.73±0.05	0.71±0.02	910±0.4%	1280±6%	930
9	31.9	0.50±0.04	0.57±0.05	0.52±0.02	910±0.8%	1740±7%	990

In Eq. 3.51

$$\hat{\Theta}(z) = \int_0^\infty \exp(-zt)\Theta(t)dt. \tag{3.52}$$

We next discuss the short time transmission coefficient $\kappa_{ST}(T)$ of Eq. (3.50), illustrating our points with the S_N2 simulation results of Wilson and co-workers [14a] reproduced in Table IV.

To start, we note that the short time $\kappa_{ST}(T)$ and Grote–Hynes $\kappa_{GH}(T)$ transmission coefficients are algebraically equivalent [23]. However, $\kappa_{ST}(T)$ and $\kappa_{GH}(T)$ are useful expressions in different physical regimes. Eqs. (3.50) and (3.51) for $\kappa_{ST}(T)$ provide a useful parameterization of $\kappa(T)$ only for reactions for which the rate constant $k(T)$ is determined by short time dynamics; while Eqs. (3.46) and (3.47) provide a useful parameterization only for reactions for which $k(T)$ is determined by slow variable dynamics. Nearly equivalently, Eqs. (3.50) and (3.51) apply to "sharp barrier" reactions, where the sharp barrier limit is defined as $\omega_{MIP} \to \infty$; while Eqs. (3.46) and (3.47) apply to "flat barrier" reactions, where the flat barrier limit is defined as $\omega_{PMF} \to 0$. (The sharp barrier limit is taken as $\omega_{MIP} \to \infty$, not as $\omega_{PMF} \to \infty$ as in Section III.B, isasmuch as sharp barrier reactions are short time, high-frequency processes for which ω_{MIP} is the physical barrier frequency. The converse argument yields the flat barrier limit as $\omega_{PMF} \to 0$.)

These comments have two implications:

1. Because of their algebraic equivalence, $\kappa_{GH}(T)$ and $\kappa_{ST}(T)$ are numerically identical and thus agree equally well with MD results. Com-

paring the $\kappa_{GH}(T)$ and $\kappa_{ST}(T)$ columns in Tables I and II illustrates this point for Wilson's S_N2 reactions. Thus, in accord with our comments in Section III.B, no physical model for reaction dynamics is validated by the agreement of $\kappa_{GH}(T)$ (or $\kappa_{ST}[T]$) with simulations. Rather, all that is validated by such agreement, as emphasized earlier, is that the decision to react is made near the barrier top.

Thus for the systems in Tables III and IV, the agreement of $\kappa_{GH}(T)$ and $\kappa_{ST}(T)$ with simulations alone is consistent with three possibilities: (1) the reaction occurs in the short time regime where $V(S;x)$ drives the reaction and thus $\kappa_{ST}(T)$ provides a useful parameterization of $\kappa(T)$; (2) the reaction occurs in the slow variable regime where $W(S;x)$ drives the reaction and thus $\kappa_{GH}(T)$ provides a useful parameterization; and (3) the reaction occurs in an intermediate regime and neither expression has much physical content.

2. Given the fast variable nature of typical reactions, Eqs. (3.50) and (3.51) are likely to often provide a more useful parameterization of $\kappa(T)$ than Eqs. (3.46) and (3.47) provide.

We next illustrate point 2 using MD results of Wilson and co-workers. To start, we specialize Eqs. (3.50) and (3.51) to the limit for which they have the most physical content; namely, the weak response limit $\omega_{MIP}^2 \gg \hat{\Theta}(\omega_{MIP})$. So specializing yields $\kappa_{ST}(T)$ as

$$\kappa_{ST}(T) \doteq \kappa_{FS}(T) + \frac{1}{2} \frac{\hat{\Theta}(\omega_{MIP})}{\omega_{MIP}\omega_{PMF}} \tag{3.53}$$

where in Eq. (3.53)

$$\kappa_{FS}(T) \equiv \omega_{PMF}^{-1} \omega_{MIP} \tag{3.54}$$

is the transmission coefficient for reaction coordinate motion in a hypothetical nonrelaxing or perfectly frozen solvent [32]. That is, $\kappa_{FS}(T)$ is the transmission coefficient for conservative motion in the potential of Eq. (3.48).

Eq. (3.53) gives $\kappa_{ST}(T)$ as a sum of the frozen solvent contribution $\kappa_{FS}(T)$ and a relaxation contribution $1/2 \ (\omega_{MIP}\omega_{PMF})^{-1}\hat{\Theta}(\omega_{MIP})$. The latter corrects $\kappa_{FS}(T)$ to first order for the effects of the solvent response to the zeroth order motion, which gives rise to $\kappa_{FS}(T)$. Because this zeroth order motion "probes" the solvent at the barrier frequency ω_{MIP}, the relaxation contribution is proportional to $\hat{\Theta}(z)$ evaluated for $z = \omega_{MIP}$. (Thus $\kappa_{ST}(T)$ of Eq. (3.53) is in accord with the short time picture of Eq. [3.36].)

Before comparing Eq. (3.53) with MD results, we note that the weak response limit $\omega_{MIP}^2 \gg \Theta(\omega_{MIP})$, is the converse of the strong overdamping limit $\zeta(S) \gg 2\omega_{PMF}$ of the diffusional Kramers model. Thus, Eq. (3.42) for

$\kappa_{KRD}(T)$ and Eq. (3.53) for $\kappa_{ST}(T)$ provide $\kappa(T)$ formulas that apply in exactly opposite limiting regimes; namely, the regimes of extreme overdamping for $\kappa_{KRD}(T)$ and extreme underdamping for $\kappa_{FS}(T)$.

We next compare with MD results.

For the reactions of Tables III and IV, $\kappa(T)$ is determined in the (nearly) frozen solvent regime of Figure 3.5a [14]. Thus, Eq. (3.53) should naturally map onto the $\kappa(T)s$ for these reactions.

Comparing the $\kappa_{FS}(T)$ and $\kappa_{ST}(T)$ columns in Table IV shows that this is indeed the case. For the low, medium, and high gas phase barrier reactions $\kappa_{FS}(T)$ accounts, respectively, for 42%–46%, 67%–93%, and 91%–99% of $\kappa_{ST}(T) \approx \kappa_{MD}(T)$. Moreover, in accord with Eq. (3.53), $\kappa_{FS}(T)$ becomes increasingly dominant as ω_{MIP} increases from its low barrier values of 190–230 cm^{-1} to its high barrier values of 910–990 cm^{-1}. Also, on physical grounds (Fig. 3.5b), the relaxation term in Eq. (3.53) is expected to increase $\kappa(T)$; as in needed to bridge the gap between $\kappa_{FS}(T)$ and $\kappa_{ST}(T)$.

Relatedly, from Eqs. (3.51)–(3.54), x_+ has the limiting form

$$x_+ = \omega_{MIP} + \frac{1}{2} \frac{\hat{\Theta}(\omega_{MIP})}{\omega_{MIP}}. \tag{3.55}$$

Thus, according to the short time Eq. (3.51), the barrier passage frequency x_+ is expected to "track" with ω_{MIP} and to approach the sharp barrier limiting value $x_+ \to \omega_{MIP}$ as $\omega_{MIP} \to \infty$. In contrast, (recall Section III.B) according to the Grote–Hynes Eq. (3.47), x_+ is expected to track with ω_{PMF} and apparently to approach the sharp barrier limiting value $x_+ \to \omega_{PMF}$ as $\omega_{PMF} \to \infty$. A comparison of the last three columns in Table IV shows that it is Eq. (3.51) not Eq. (3.47), which is in accord with the MD results.

Similarly, from either Eqs. (3.50), (3.51), and (3.54) or from eq. (3.53); it follows that in the sharp barrier limit $\omega_{MIP} \to \infty, \kappa_{ST}(T) \to \kappa_{FS}(T)$. Thus, from Eq. (3.40) the sharp barrier limit of the rate constant is

$$k(T) \to k_{TST}(T)\kappa_{FS}(T). \tag{3.56}$$

Thus, the false sharp barrier limit $k(T) \to k_{TST}(T)$ that emerges from either the Kramers Eq. (3.41) or the Grote–Hynes eqs. (3.45)–(3.47), is not predicted by the fast variable eqs. (3.50) and (3.51).

The false sharp barrier limit emerges from the slow variable eqs. (3.41) and (3.45), but not fast variable eqs. (3.50) and (3.51), for the following reason. Comparing Figures 3.2 and 3.6 shows that the slow variable potential $W(S; x)$ ignores barrier recrossings due to reflections of the reaction coordinate by the solvent cage that are properly accounted for by the fast variable potential $V(S;x)$. These reflections reduce $k(T)$ from $k_{TST}(T)$ by the factor $\kappa_{FS}(T)$.

This concludes our overview of liquid phase activated barrier crossing.

We next briefly discuss a second liquid phase chemical process, namely, the vibrational energy relaxation of high-frequency solute normal modes [33].

C. Vibrational Energy Relaxation of High-Frequency Solute Normal Modes

Fast variable physics underlies many processes other than typical liquid phase chemical reactions. In fact, this physics is expected to be important for most condensed phase molecular processes governed by the motion of internal solute degrees of freedom, that is, those other than molecular translational and rotational coordinates.

Perhaps the simplest process determined by fast variable physics is liquid phase vibrational energy relaxation (VER) [33]. We next outline a fast variable treatment of this process [24].

1. Fast Variable Physics and Liquid Phase VER

The "fast variable/slow bath" timescale separation arises in liquid phase VER because of a frequency mismatch between the solute and solvent molecules, rather than because of a speed difference, as in typical reactions. Despite this, the fast variable timescale separation yields a picture of solute VER very similar to the picture of short time-activated barrier crossing reflected in Eq. (3.53). We expect similar pictures to emerge for other processes.

Because relaxing solute normal mode frequencies are often much higher than typical solvent translational-rotational frequencies, the following short time picture of solute VER emerges [22,24]. In zeroth order the relaxing mode executes conservative harmonic notion in a hypothetical solvent that is nonresponsive to this motion. The velocity autocorrelation function $\dot{\chi}(t)$ for this zeroth order motion is

$$\dot{\chi}(t) = \cos\omega_\ell t \qquad (3.57)$$

where ω_ℓ, analogous to ω_{MIP} of Eq. (3.49), is the liquid phase frequency of the mode in the nonresponding solvent. In first order, the zeroth order motion "probes" the solvent at the frequency ω_ℓ inducing a weak response and an associated mode $\rightarrow$ solvent energy transfer. The velocity correlation function for the first order motion is [24]

$$\dot{\chi}(t) = \cos\omega_\ell t \exp\left[-\frac{1}{2}\beta(\omega_\ell)t\right] \qquad (3.58)$$

where $\beta(\omega_\ell)$ is the Fourier domain friction kernel of the mode evaluated at frequency ω_ℓ.

As depicted in Figure 3.7, Eq. (3.58) describes the motion of a harmonic oscillator in a regime of extreme underdamping exactly opposite to the regime of extreme overdamping discussed in Appendix Section F.11 and described by Eq. (A.47b).

Additionally, the damping term in Eq. (3.58) accounts for the mode to solvent energy transfer and leads to VER with relaxation time [24]

$$T_1 = \beta^{-1}(\omega_\ell). \tag{3.59}$$

Eq. (3.59) is compared with similar results of Oxtoby [33] and others in section 23 of reference 19.

2. Partial Clamping Implementation of Eq. (3.59)

T_1 may be found from Eq. (3.59) by determining ω_ℓ and $\beta(\omega)$ using our partial clamping model [21]. This method for T_1 is analogous to the one described in Section III for $\kappa(T)$. Namely, the partial clamping model is implemented by constraining the relaxing solute normal mode to have its equilibrium value while permitting the other solution degree of freedom to move freely subject to this constraint.

At analytic theory based on these ideas, which yields T_1 in terms of molecular parameters like masses and bondlengths and also equilibrium pair correlation functions of the liquid solution, has been described elsewhere [19,24].

We next discuss some MD implementations [15,27] of this method for T_1.

(Eq. [3.59], when implemented via the partial clamping model, has been referred to as an "extension of the Landau–Teller theory to the liquid phase." This description of our theory of liquid phase VER [24] is not accurate. There are three reasons for this: [1] The Landau–Teller theory [34,35] is a very primitive gas phase model that assumes a colinear atom-diatom system interacting via a crude one-dimensional repulsive exponential potential. Our theory treats vastly more complex systems at a much higher level of realism and thus is not properly regarded as a simple extension of the work of Landau and Teller. [2] The Landau–Teller model and our VER theory are based on very different assumptions and physical pictures [36]. For example, the Landau–Teller theory was developed to treat high temperature $[T \gtrsim 1000\,K]$ vibrational energy transfer processes and thus is based on assumptions [34,35] intended for very-high-energy collisions; while our theory in contrast derives from assumptions [19] tailored to treat the low-energy collisions that mediate liquid phase VER. Thus it is not clear, and has not been shown, that a systematic application of Landau and Teller's methods to liquid phase VER would yield results equivalent to eqs. [3.58] and [3.59]. [3] The Landau–Teller model applies only to vibrational energy transfer processes. Our method for VER in contrast is a special case of a general procedure that is applicable to many types of liquid phase chemical processes. This is evidenced by the similarities between eqs. [3.53] and [3.58].)

3.   MD Evaluations of T_1

MD implementations of the partial clamping model to yield T_1 from Eq. (3.59) have been reported by a number of workers [15,27]. Some of this work has recently been reviewed by Owrutsky, Raftery, and Hochstrasser [33]. Thus, we will restrict ourselves here to a few brief comments focusing on the work of Whitnell, Wilson, and Hynes [15a] and of Berne and co-workers [27].

To start, we note that validity of the partial clamping model as applied to the vibrational motion of a diatomic molecule dissolved in a Lennard-Jones fluid, has recently been examined by Berne et al. [27a]. Among the results obtained by these workers are the following:

1. The partial clamping approximation to the friction kernel $\beta(t)$ of the diatomic's vibrational coordinate was shown to become exact in the $\omega_\ell \to \infty$ limit. This result is in accord with our development of the partial clamping model from the assumption [21] of small amplitude solute vibrations.

2. For their systems, Berne et al. [27a] found that in practice the asymptotic $\omega_\ell \to \infty$ limit was achieved for remarkably small ω_ℓ. Specifically, they showed by MD simulation that the partial clamping and exact Mori [5] forms for $\beta(t)$ nearly coincided for ω_ℓ values only slightly larger than the most probable solvent translational frequency ω_{mp} (defined as the peak of the single solvent atom velocity autocorrelation function spectral density). This result suggests that for high-frequency solute normal modes for which $\omega_\ell \gg \omega_{mp}$, the partial clamping model should provide a useful approximation to $\beta(\omega)$ (the cosine transform of $\beta[t]$) needed to evaluate T_1 from Eq. (3.59).

3. Berne et al. also showed that the form of the solute normal mode velocity autocorrelation function $\dot{\chi}(t)$ approached, at least qualitatively, the fast variable form of Figure 3.7 for $\omega_\ell \gtrsim 3\omega_{mp}$ (while having the usual thermal form for $\omega_\ell \approx \omega_{mp}$; see Figure 1 of reference 27a), suggesting the validity of Eq. (3.59) for high-frequency modes.

In summary, the work of Berne et al. [27a] indicates that Eq. (3.59) implemented via the partial clamping model [21] is likely to provide a useful approximation to T_1 for high-frequency solute normal modes.

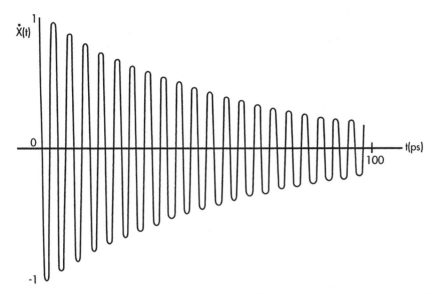

Figure 3.7. The velocity autocorrelation function $\dot{\chi}(t)$ of a high-frequency solute normal mode. The vibrational energy relaxation of a high-frequency solute normal mode occurs in a limit of extreme underdamping $\beta(\omega_\ell) \ll 2\omega_\ell$, yielding for the mode's velocity autocorrelation function $\dot{\chi}(t)$ the weakly damped cosine decay of Eq. (3.58).

This expectation was validated by the simulations of Whitnell, Wilson, and Hynes [15a] and Tuckerman and Berne [27b] (see their Table 2).

We next very briefly discuss the work of Whitnell, Wilson, and Hynes [15a]. These authors studied the VER of a model diatomic methyl chloride molecule in a model vibrating water solvent. In order to assess the effects of the CH_3Cl dipole moment, simulations were performed for three sets of charges on the CH_3Cl molecule. The three VER times T_1 were computed via an MD partial clamping evaluation of Eq. (3.59). In order to test this model, the three T_1 values were also computed by a full MD simulation of the relaxation process. The model and exact MD T_1's were found to be in good agreement for all three charge cases. For example, full MD simulation of the $v = 1 \rightarrow v = 0$ relaxation process yielded T_1 values of 106 ps, 4.95 ps, and 1.41 ps. The corresponding values found from Eq. (3.59) were 100 ps, 4.5 ps, and 1.4 ps (see Table 2 in the first of references 15a); showing that not only can the model yield good absolute T_1 values, but more importantly from the experimental standpoint [33,35] it can reproduce trends.

To conclude, we next touch on some of the additional recent work of others [11,16,17,28] that relates to questions of short timescale dynamics.

D. Some Additional Recent Discussions
of Short Timescale Dynamics

In Section IVD we first note some discussions of reaction dynamics [11,17] related to those of sections IVA and IVB; then we comment on some recent results obtained for time-dependent solvation dynamics [6,11,12,16]; and finally we make contact with Stratt's [28] instantaneous normal mode (INM) approach.

1. Fast Variable Physics and Reaction Dynamics

We first note that some of the physics described in Sections IIIA and IVA was discussed by Benjamin and Wilson [17] in connection with their MD studies of ICN photodissociation in Xe solution. For example these authors noted in their section VA that the general picture that emerges from their ICN simulations is "in many ways similar to that seen earlier in simulations of thermally activated reactions with appreciable barriers. ... In all these cases, the strong forces and accelerations produced by descending step barriers are such that the solvent environment cannot keep up in an adiabatic sense (i.e., cannot equilibrate) during the reactant motion." They also note that in rare gas solvents "in falling off a steep barrier, either in a photo-dissociation or in a thermal reaction, the initial acceleration can be so high..." that "the solvent mainly affects the products after they are 'up to speed'. ..." Also in their section 4A.1, Benjamin and Wilson [17] note that these features produce enhanced cage effects.

Rosenthal et al. [11] in connection with their laser studies of time-dependent solvation have noted that "the existence of a large amplitude, rapid inertial Gaussian component in the solvent response has substantial implication for theoretical descriptions of chemical reaction dynamics in solution. Theories involving continuum or overdamped approximations are likely to be inappropriate in many applications." In the same vein, Jimenez et al. [11] have noted in their observations of a similar inertial Gaussian component in the response of water that "the prominent ultrafast component observed in both atomic and molecular solutes should have a profound effect on aqueous reaction dynamics." For example, they note at the close of their paper (compare eqs. [3.36] and [3.53] and also Figure 3.5 of the present chapter) that "recrossing of the free energy barrier is influenced by the dynamics of the solvent response: if the solvent is able to respond to the change in charge distribution as the system crosses the transition state, it may be able to *stabilize the product* and *increase the overall reaction rate relative to a 'static' solvent* (italics ours). Our results suggest that solvents are able to respond to changes in reactant configuration on timescales relevant to reactive barrier crossing."

2. Time-Dependent Solvation Dynamics

Experiments [6,11,12] and simulations [6,16] have recently established that a large part of the relaxation of the nonequilibrium solvation around a solute immediately after its photoexcitation is due to the shortest timescale motions of the first shell solvent molecules. This result is in accord with the theoretical notions we have been developing since 1979 [20a], which after refinement [22] has led to our use of the Gaussian friction model as the basis of our molecular friction methods [19,24].

We expect that the inertial first shell motions will be even more important for the solvent response to fast variable probes than they are for time-dependent solvation; thus permitting, for example, a relatively simple evaluation of the relaxation contribution to $\kappa(T)$ in Eq. (3.53) from models that retain only the inertial part of the solvent response (e.g., those of reference 19 and 24).

3. Stratt's INM Theory

Stratt [28] has emphasized that on very short timescales liquids display solid-like motions, and has used this idea as the basis of his INM method for computing dynamical properties of liquids.

Since Stratt's work is well explained in the literature, we only note here that his ideas are closely related to ours. Especially our equivalent chain equations [20] provide an *exact* one-dimensional harmonic solid representation of the solvent effect on solute motions. We developed this representation in order to bring out the solid-like physics of the solvent effect evident on the shortest timescales.

V. SUMMARY

We have emphasized in this chapter that Arrhenius' principle implies that liquid phase chemical reactions occur in a nonclassical "fast variable" near sudden limit timescale regime rather than in the "slow variable" near adiabatic regime of standard irreversible statistical mechanics. Despite this, the traditional theories of liquid phase reaction dynamics [7–10] are of the slow variable type.

For this reason, in Section II and the Appendix we have presented a critical review of the standard Langevin, Onsager, Mori slow variable model of irreversible dynamics [1–5] that identifies the physical assumptions that underlie the model. Especially emphasized in this discussion is that the standard model emerged from attempts to explain purely macroscopic phenomena and is ultimately founded on fully macroscopic measurements.

In Section III, we discussed the implications of the assumptions of the standard model for the validity of the usual Kramers-type slow variable theories of liquid phase reaction rates [7–10]. Using the results of recent MD simulations and ultrafast laser experiments [6,11–17] we showed that the macroscopic roots of the standard model significantly restricts its applicability to chemical reactions in liquids and other fast variable processes. This is because such processes are mediated by the subpicosecond timescale components of the solvent response.

Finally, in Section IV we outlined our fast variable approach to short timescale and high frequency processes [19–26] and illustrated its utility using results from recent literature [6,11–17,27–29].

APPENDIX

This appendix is an expansion of the discussion of Section IIB into a short critical review of the standard Langevin, Onsager, Mori [1–5] slow variable model of irreversible processes. Specifically, we will outline here some basic topics in statistical and thermal physics [37], focusing on Onsager's slow variable theory of irreversible thermodynamic processes [1,3,4]. While a number of useful discussions of Onsager's theory exist in the literature [37,38], ours differs from most others in that it scrutinizes some of the theory's less frequently examined assumptions and thus identifies some of its important but not commonly discussed limitations.

At first glance a review of Onsager's theory might appear to have little or no relevance to the topics of this chapter. This, however, is not the case. This is because Onsager's theory, as extended to include memory by Mori [5] is the most general slow variable theory of irreversible motion. Thus, our examination of Onsager's work exposes limitations inherent in all slow variable models. Especially, it shows that the limitations of the Kramers-type models for reactions merely reflect the macroscopic scope of the general theory of irreversible processes [1,3–5].

In our discussion, for simplicity we will assume an *isolated one-component* macroscopic system comprised of N particles confined to a volume V with total energy E. Other types of macroscopic systems, however, may be treated similarly.

We begin with a synopsis of some familiar notions, pertinent to both classical equilibrium thermodynamics [39] and to its nonequilibrium extension due to Onsager [1,3,4].

A. Some Thermodynamic Concepts

We first summarize the thermodynamic specification of system states.

1. Thermodynamic Specification of System States

A *nonequilibrium state* $\Gamma_n(t)$ of the isolated system is specified at time t by N, V, and E; and by the instantaneous and prior values of any other n

extensive quantities $A_1, A_2, \ldots, A_n$, needed to provide a macroscopically complete description of the system. For an *equilibrium state* Γ_n the A's are maintained at time-independent values $A_1, A_2, \ldots, A_n$ by imposed constraints. An equilibrium state is thus specified by N, V, and E and by the *fixed values* $A_1, A_2, \ldots, A_n$. As an example of the A's, for a homogeneous but nonuniform system they might include the Fourier components of the system's mass and energy density each multiplied by V.

We next turn to thermodynamic processes.

2. *Thermodynamic Processes*

Thermodynamic processes are the changes $\Gamma_n \rightarrow \Gamma_p$ which take the system from an initial equilibrium state Γ_n to a final equilibrium state Γ_p.

3. *The Initiation and Outcome of a Thermodynamic Process*

A thermodynamic process may be initiated, for example, by removing the constraints on some of the A's, say, $A_{p+1}, \ldots, A_n$, while keeping the other macroscopic parameters at their initial values N, V, E, $A_1, \ldots, A_p$.

For a process so initiated, equilibrium thermodynamics predicts that the following will occur:

1. The unconstrained A's will relax spontaneously from their initial values $A_{p+1}, \ldots, A_n$ to final values $A^*_{p+1}, \ldots, A^*_n$ which are independent of both the initial values and of the kinetics of the process $\Gamma_n \rightarrow \Gamma_p$.

2. The values $A^*_{p+1}, \ldots, A^*_n$ are those that maximize the system's equilibrium entropy function $S(N, V, E, A_1, \ldots, A_n)$ with respect to $A_{p+1}, \ldots, A_n$ keeping $N, V, E, A_1, \ldots, A_p$ fixed. That is, the conditions that determine $A^*_{p+1}, \ldots, A^*_n$ are

$$\frac{\partial S}{\partial A_{p+i}} = 0 \quad \text{and} \quad \frac{\partial^2 S}{\partial A_{p+i} \partial A_{p+j}} < 0 \qquad (A.1)$$

where $i, j = 1, 2, \ldots, n - p$.

We next note that Gibbs [40] has emphasized that an analogy exists between thermodynamic equilibria and the much simpler static equilibria of mechanical systems (Fig. 3.1). We will briefly describe this analogy; since it helps to both motivate and explain Onsager's theory.

B. The Gibbs Analogy

We will illustrate the Gibbs analogy for a very simple mechanical system, namely, the one-dimensional macroscopic particle system illustrated in Figure 3.1 and discussed in Section II.A.1.

For this system the Gibbs analogy may be stated as follows. The thermo-dynamic equilibrium values of the macroscopic parameters $A_{p+1} \ldots, A_n$ are analogous to the static equilibrium values of the particle coordinate x. The removal of the constraints on $A_{p+1}, \ldots, A_n$ is analogous to the release of the constraint force F on x. The thermodynamic process $\Gamma_n \to \Gamma_p$ is analogous to the particle relaxation process $x_F \to x_{F=0} = 0$. Finally, the entropy maximization condition of Eq. (A.1) for $A_{p+1}^*, \ldots, A_n^*$ is analogous to the potential energy minimization condition of Eq. (3.2) for $x_{F=0} = 0$.

To motivate Onsager's theory, we next briefly describe both what ther-modynamics accomplishes and also what questions it leaves open.

C. The Content and Limitations of Equilibrium Thermodynamics

Classical thermodynamics may be viewed as a set of rules that predict the final equilibrium states of initially nonequilibrium macroscopic systems. These rules, while very powerful [39], are, of course, also limited inasmuch as they say nothing about the kinetics of thermodynamic change. Thus, while thermodynamics predicts the final equilibrium values $A_{p+1}^*, \ldots, A_n^*$ of the unconstrained A's from the rule of Eq. (A.1), it says nothing about the time-dependent values $A_{p+1}(t), \ldots, A_n(t)$ that these parameters take on during the relaxation process $\Gamma_n \to \Gamma_p$.

An analogous limitation holds for particle statics. For example, for the process $x_F \to x_{F=0}$ of Figure 3.1, statics predicts the particle's final equilibrium position $x_{F=0}$ from the rule of Eq. (3.2); but says nothing about the trajectories $x(t)$ which take the particle from x_F to $x_{F=0}$.

We next begin our development of Onsager's theory, which transcends some of the limitations of equilibrium thermodynamics.

D. Irreversible Processes: First Considerations

The rest of this appendix will focus on the following basic problem: find the time-dependent values $A_{P+1}(t), \ldots, A_n(t)$ that the unconstrained A's take during the irreversible process $\Gamma_n \to \Gamma_p$.

Working in the spirit of the Gibbs analogy, the solution of this problem apparently requires an extension of equilibrium thermodynamics that parallels the extension of particle statics needed to predict trajectories $x(t)$, namely, its extension to *particle dynamics*.

To make such an extension, however, is far from easy. We next point out some of difficulties that arise.

1. The Problem of the Equation of Motion

The most obvious problem is that no equation of motion is known for macroscopic parameters that is comparable in scope to Newton's equation for particle trajectories. Rather, only phenomenological thermodynamic equations of motion of limited validity exist.

2. The Problem of the Thermodynamic Forces

Even if a broadly valid thermodynamic equation of motion were available, no general prescription for choosing the thermodynamic forces occurring in it yet exists.

We next describe one of the problems involved in choosing thermodynamic forces.

3. The Timescale Problem and Thermodynamic Forces

Namely, a timescale problem similar to the one identified in Section II.A.2 for finite $\dot{x}(t)$ particle relaxation processes also occurs for irreversible (that is, finite $\dot{A}_{p+1}[t]\ldots\dot{A}_n[t]$) thermodynamic processes. Thus ambiguity clouds the form that the thermodynamic forces take on during irreversible relaxation processes $\Gamma_n \to \Gamma_p$. A familiar reflection of this ambiguity is that while for an ordinary conservative mechanical system a change in potential energy always equates to the negative work done by the system, for a macroscopic system a change in thermodynamic potential equates to the negative reversible but not irreversible work done by the system.

4. Onsager's Extension of Thermodynamics

Onsager has partially overcome the dual problem of formulating a thermodynamic equation of motion and of choosing the thermodynamic forces, thus enabling him to develop a nonequilibrium extension of classical thermodynamics in the spirit of the Gibbs analogy.

Before outlining Onsager's theory, we collect and quickly summarize some statistical mechanical principles needed to formulate the theory.

E. Some Statistical Mechanical Principles

We begin by recalling a few basic statistical mechanical notions [37]. These are as follows.

1. An isolated macroscopic system is specified microscopically by the *phase point* q, p of the atoms that comprise the system.

2. Many phase points are consistent with a single set of macroscopic parameter values. Thus associated with every macroscopic state $\Gamma_n(t)$ is an *ensemble* of microscopically distinct systems each with its own q, p.

3. As a consequence, within statistical mechanics a macroscopic state $\Gamma_n(t)$ is characterized by its *phase space probability distribution* $P_{\Gamma_n(t)}(q,p)$; defined so that $P_{\Gamma_n(t)}(q,p)\, dqdp$ is the fraction of the ensemble with phase points near q, p.

4. For an equilibrium state Γ_n, the form of $P_{\Gamma_n(t)}(q,p) = P_{\Gamma_n}(q,p)$ is given by the microcanonical ensemble equal *a priori* probability hypothesis (namely, all q, p consistent with state Γ_n occur with equal frequency in its ensemble). Thus, for an equilibrium state $P_{\Gamma_n(t)}(q,p)$ has the form

$$P_{\Gamma_n}(q,p) = \begin{cases} 0 & \text{if } q, \ p \text{ is outside } V_{\Gamma_n} \\ V_{\Gamma_n}^{-1} & \text{if } q, \ p \text{ is inside } V_{\Gamma_n} \end{cases} \tag{A.2}$$

where V_{Γ_n} is the phase space volume accessible to the ensemble associated with state Γ_n.

Next, let us consider from the standpoint of statistical mechanics the effects, earlier discussed thermodynamically, of the removal of the constraints on the macroscopic parameters $A_{p+1}, \ldots, A_n$ of the isolated system.

1.   Removal of the Constraints on $A_{p+1}, \ldots, A_n$.

According to statistical mechanics, the phase space volume accessible to an ensemble is restricted only in that it must be consistent with the values of constrained macroscopic parameters.

Thus, the removal of the constraints on $A_{p+1}, \ldots, A_n$ has the following two effects:

1. It permits the accessible phase space volume to greatly expand from the value V_{Γ_n} for state Γ_n to the value V_{Γ_p} for state Γ_p. The expansion occurs because—whereas V_{Γ_n} must be consistent with the $3 + n$ parameter values $N, V, E, A_1, \ldots, A_p, A_{p+1}, \ldots, A_n$—$V_{\Gamma_p}$ is required to be consistent with only the $3 + p$ parameter values $N, V, E, A_1, \ldots, A_p$. The weaker restrictions on V_{Γ_p} render $V_{\Gamma_p} \gg V_{\Gamma_n}$.

2. In state Γ_p removal of constraints permits the parameters $A_{p+1}, \ldots, A_n$ to fluctuate. This follows because the phase points in V_{Γ_p} are not restricted to be consistent with any particular set of values for these parameters. Hence for many points in V_{Γ_p} the values for $A_{p+1}, \ldots, A_n$ differ from their thermodynamic equilibrium values $A_{p+1}^*, \ldots, A_n^*$.

Based on the previous discussion, we next make some connections and comparisons of statistical mechanics and thermodynamics that are important for Onsager's theory.

2.   Some Links Between Statistical Mechanics and Thermodynamics

The connections and comparisons are as follows.

1. For the isolated system we have noted that removal of the constraints on $A_{p+1}, \ldots, A_n$ results in an expansion $V_{\Gamma_n} \rightarrow V_{\Gamma_p}$ of the system's

accessible phase space volume. However, according to statistical mechanics, it is this expansion that provides the driving force for the irreversible process $\Gamma_n \rightarrow \Gamma_p$. In contrast, within thermodynamics it is the increase in the system's entropy that drives the process $\Gamma_n \rightarrow \Gamma_p$. Thus one expects a close connection between a macroscopic system's accessible phase space volume and its thermodynamic entropy. We will return to this point shortly.

2. We have noted that according to statistical mechanics the parameters $A_{p+1}, \ldots, A_n$ fluctuate in the final equilibrium state Γ_p. Thus in statistical mechanics the unconstrained A's are *random variables* described by a probability distribution $P(\Gamma_p; A'_{p+1}, \ldots, A'_n)$, which gives the probability that $A_{p+1} = A'_{p+1}, A_{p+2} = A'_{p+2}$, and so on. In contrast, in thermodynamics the unconstrained A's are deterministic quantities with the *fixed values* $A^*_{p+1}, \ldots, A^*_n$. However, as we will shortly discuss, the most probable A's determined from $P(\Gamma_p; A'_{p+1}, \ldots, A'_n)$ coincide with thermodynamic equilibrium A's.

We next develop the form of $P(\Gamma_p; A'_{p+1}, \ldots, A'_n)$.

3.   The Form of $P(\Gamma_p; A'_{p+1}, \ldots, A'_n)$.

To start, we note that in analogy to Eq. (A.2), the phase space probability distribution for state Γ_p is given by

$$P_{\Gamma_p}(q, p) = \begin{cases} 0 & \text{if } q, \ p \text{ is outside } V_{\Gamma_p} \\ V_{\Gamma_n}^{-1} & \text{if } q, \ p \text{ is inside } V_{\Gamma_p} \end{cases} \qquad \text{(A.3)}$$

where V_{Γ_p} is the accessible phase space volume for state Γ_p.

Next let $V(\Gamma_p; A'_{p+1}, \ldots, A'_n)$ denote the accessible phase space volume when the system is in state Γ_p and the unconstrained A's have the values $A'_{p+1}, \ldots, A'_n$. (Thus, for example, $V[\Gamma_p; A_{p+1}, \ldots, A_n] = V_{\Gamma_n}$.)

Then $P(\Gamma_p; A'_{p+1}, \ldots, A'_n)$ may be expressed as

$$P[\Gamma_p; A'_{p+1}, \ldots, A'_n] = \int \cdots \int P_{\Gamma_p}[q, p] dq dp, \qquad \text{(A.4)}$$

where the phase space integral is over $V(\Gamma_p; A'_{p+1}, \ldots, A'_n)$. Comparing Eqs. (A.3) and (A.4) then yields the form of $P(\Gamma_p; A'_{p+1}, \ldots, A'_n)$ as

$$P(\Gamma_p; A'_{p+1}, \ldots, A'_n) = \frac{V[\Gamma_p; A'_{p+1}, \ldots, A'_n]}{V_{\Gamma_p}}. \qquad \text{(A.5)}$$

Eq. (A.5) may be recast as

$$P(\Gamma_p; A'_{p+1}, \ldots, A'_n) = P(\Gamma_p; A^*_{p+1}, \ldots, A^*_n) \frac{V(\Gamma_p; A'_{p+1}, \ldots, A'_n)}{V(\Gamma_p; A^*_{p+1}, \ldots, A^*_n)} \quad (A.6)$$

(which shows that the probability of a fluctuation $A^*_{p+1}, \ldots, A^*_n \to A'_{p+1}, \ldots, A'_n$ is determined solely by considerations of accessible phase space volume).

We have noted the close connection between thermodynamic entropy and accessible phase space volume. Because of this connection Eq. (A.6) may be recast as

$$P[\Gamma_p; A'_{p+1}, \ldots, A'_n] = P[\Gamma_p; A^*_{p+1}, \ldots, A^*_n] \exp\left[\frac{\Delta S(\Gamma_p)}{k}\right]. \quad (A.7)$$

In Eq. (A.7)

$$\Delta S(\Gamma_p) = S(\Gamma_p; A'_{p+1}, \ldots, A'_n) - S(\Gamma_p; A^*_{p+1}, \ldots, A^*_n) \quad (A.8)$$

where $S(\Gamma_p; A'_{p+1}, \ldots, A'_n)$ and $S(\Gamma_p; A^*_{p+1}, \ldots, A^*_n)$ are the entropies of the system for, respectively, the thermodynamic equilibrium states with macroscopic parameter values $N, V, E, A_1, \ldots, A_p, A'_{p+1}, \ldots, A'_n$ and $N, V, E, A_1, \ldots, A_p, A^*_{p+1}, \ldots, A^*_n$. Eqs. (A.7) and (A.8) may be derived from Eq. (A.6) using Boltzmann's entropy formula $S(\Gamma_p; A'_{p+1}, \ldots, A'_n) = k\ln P(\Gamma_p; A'_{p+1}, \ldots, A'_n)+$ constant.

Eqs. (A.7) and (A.8) are the main statistical mechanical results needed to develop Onsager's theory.

The relevance of Eqs. (A.7) asd (A.8) to Onsager's theory arises since these results, in the spirit of the Gibbs analogy, are the thermodynamic equivalents of the particle mechanics eqs. (3.13) and (3.14). For example, the most probable value of the particle coordinate x found by maximizing Eq. (3.13) coincides with its static equilibrium value $x_{F=0}(S)$ determined from Eq. (3.11). Similarly, the most probable values of the unconstrained A's found by maximizing Eq. (A.7) coincide with their thermodynamic equilibrium values $A^*_{p+1}, \ldots, A^*_n$ determined from Eq. (A.1). (Note, however, that this analogy between particle mechanics and thermodynamics is not perfect. This is because while for microscopic particles, typical fluctuations $x' - x^*$ of the particle's coordinate x can be comparable in magnitude to its initial displacement from final equilibrium $x_F(S) - x^*$; the macroscopic validity of thermodynamics requires that typical fluctuations $A'_{p+1} - A^*_{p+1}, \ldots, A'_n - A^*_n$ of $A_{p+1}, \ldots, A_n$ be negligible compared with their initial displacements from final equilibrium $A_{p+1} - A^*_{p+1}, \ldots, A_n - A^*_n$.)

As a consequence of this parallelism, Eqs. (A.7) and (A.8) permit the definition of a *thermodynamic potential* $W(\Gamma_p; A_{p+1}, \ldots, A_n)$ analogous to the particle potential $W(S; x)$ of Eqs. (3.13) and (3.14). The thermodynamic forces needed for Onsager's theory may then be derived from this potential via relations analogous to Eq. (3.9).

Given this background we may now turn to Onsager's theory [1,3,4], including its entension to finite memory due to Mori [5]. We both develop some of the basic equations of the Onsager–Mori theory, and also critically evaluate these equations in light of the physical assumptions which underlie them.

F.   The Onsager–Mori Relaxation Equations

We next outline the formulation of the Onsager–Mori relaxation equations.

Onsager's thermodynamic equation of motion rests on the *linear flux-force relations* of transport theory [38]. Thus, to start, we beiefly review these phenomenological equations.

1.   *The Linear Flux-Force Relations*

One of the most familiar linear flux-force relations is Ohm's law of electrical conductivity

$$J = \sigma E \qquad\qquad (A.9)$$

where J, E, and σ are, respectively, the current induced in a substance, the inducing electric field, and the substance's electrical conductivity.

Flux-force relations similar to Eq. (A.9) have been established experimentally for many transport processes. The general form of these relations is

$$J_i = \sum_{j=1}^{n} L_{ij} F_j. \qquad\qquad (A.10)$$

Thus, the flux-force equations linearly related n thermodynamics fluxes $J_1, J_2, \ldots$ to n conjugate external thermodynamic forces $F_1, F_2, \ldots$; with the proportionality constants L_{ij} being the n^2 transport coefficients for the process. Note that Eq. (A.9) is an example of Eq. (A.10) for $n = 1$; with $J_1 = J, L_{11} = \sigma$, and $F_1 = E$. Thermoelectric conduction and thermal diffusion are examples of processes that obey Eq. (A.10) for $n > 1$.

We may now describe Onsager's resolution of the problem of formulating a thermodynamic equation of motion.

2.   *Onsager's Thermodynamic Equation of Motion*

Onsager's thermodynamic equation of motion was chosen to accord with the linear flux-force Eq. (A.10). For the process $\Gamma_n \rightarrow \Gamma_p$ it is the following set of relaxation equations for the unconstrained A's:

$$\dot{A}_{p+i}(t) = \sum_{j=1}^{n-p} L_{p+i,p+j} X_{p+j}(t) \tag{A.11a}$$

with $i = 1, 2, \ldots, n - p$.

Notice that Eq. (A.11a) is identical in form to Eq. (A.10) so long as one interprets the $\dot{A}(t)$'s, the $X(t)$'s, and the L's in Eq. (A.11a) as, respectively, the thermodynamic fluxes, forces, and transport coefficients for the process $\Gamma_n \rightarrow \Gamma_p$. However, while the F's in Eq. (A.10) are external applied forces, the X's in Eq. (A.11a) are internal thermodynamic driving forces. Thus, in adopting Eq. (A.11a), Onsager has boldly postulated that the linear flux-force relations experimentally established for external forces hold as well for internal forces.

Eq. (A.11a) may be rewritten in matrix notation as

$$\dot{\mathbf{A}}(t) = \mathbf{L}\mathbf{X}(t) \tag{A.11b}$$

where $[\mathbf{L}]_{ij} = L_{p+i,p+j}$ and so on. Inverting Eq. (A.11b) yields the following equivalent form for Onsager's equation:

$$\mathbf{X}(t) = \mathbf{R}\dot{\mathbf{A}}(t) \tag{A.12}$$

where

$$\mathbf{R} = \mathbf{L}^{-1}. \tag{A.13}$$

We next give Onsager's resolution of the problem of choosing the thermodynamic forces.

3. Onsager's Choice of the Thermodynamic Forces

Onsager chose the thermodynamic forces $X_{p+j}(t)$ in Eq. (A.11a) as

$$X_{p+j}(t) = \frac{\partial S[\Gamma_p; A_{p+1}(t), \ldots, A_n(t)]}{\partial A_{p+j}(t)} \tag{A.14a}$$

where $S[\Gamma_p; A_{p+1}(t), \ldots, A_n(t)]$ is the entropy that the system would have if it were in the *thermodynamic equilibrium state* with macroscopic parameter values $N, V, E, A_1, \ldots, A_p, A_{p+1}(t), \ldots, A_n(t)$. Eq. (A.14a) may be rewritten concisely in matrix notation as

$$\mathbf{X}(t) = \frac{\partial S[\Gamma_p; \mathbf{A}(t)]}{\partial \mathbf{A}(t)} \tag{A.14b}$$

The usual justification for adopting Eqs. (A.14) is the notion [41] that it is the Second Law requirement that $S(\Gamma_p; \mathbf{A})$ must increase in a spontaneous process, which "drives" the process $\Gamma_n \rightarrow \Gamma_p$.

Eqs. (A.14) permit one to convert eqs. (A.11) into the following *closed equation of thermodynamic relaxation* for $\mathbf{A}(t)$:

$$\dot{\mathbf{A}}(t) = \mathbf{L} \frac{\partial S[\Gamma_p; \mathbf{A}(t)]}{\partial \mathbf{A}(t)}. \tag{A.15}$$

Eq. (A.15) is a basic equation of Onsager's theory. Notice that so far it rests solely on macroscopic phenomenology, since both the flux-force relations and the Second Law derive from purely macroscopic experiments.

Before going on, we show that Eq. (A.15) properly reduces to equilibrium thermodynamics in the limit $t \rightarrow \infty$. This result is the thermodynamic equivalent of the reduction of particle dynamics to particle statics, as exemplified by the derivation of Eq. (3.2) from Eq. (3.3).

4. The Reduction of Onsager's Theory to Equilibrium Thermodynamics

To recover equilibrium thermodynamics from Onsager's theory, first premultiply Eq. (A.15) by $\mathbf{R} = \mathbf{L}^{-1}$ and then set $\mathbf{A}(t) = \mathbf{A}^*$ and $\dot{\mathbf{A}}(t) = 0$ (the conditions for thermodynamic equilibrium in state Γ_p) in the result. This yields

$$\frac{\partial S(\Gamma_p; \mathbf{A}^*)}{\partial \mathbf{A}^*} = 0. \tag{A.16}$$

Eq. (A.16), however, is just the matrix form of the maximum entropy condition of Eq. (A.1) for $A_{p+1}^*, \ldots, A_n^*$. Thus, Onsager's theory reduces to classical thermodynamics in the "statics" limit.

This reduction, while necessary, is of course not sufficient for the validity of Onsager's theory. This follows since many types of relaxation dynamics are consistent with the "statics" of Eq. (A.16). Onsager's basic Eq. (A.15), as will become clear shortly, merely amounts to a selection of a particularly simple type of dynamics from the myriad of possible types.

For example, Eq. (A.19) yields relaxation dynamics consistent with Eq. (A.16) for all dissipative force functions F in accord with the condition $F(\Gamma_p; \mathbf{A}, \dot{\mathbf{A}} = 0) \propto \partial S(\Gamma_p; \mathbf{A})/\partial \mathbf{A}$. The situation is similar to that discussed after Eq. (3.14) where we found many types of particle dynamics are consistent with the statics of Eq. (3.11).

We next turn to a discussion of the physical content of Onsager's theory. In this discussion we will develop Onsager's Eq. (A.15) from slow variable assumptions, thus showing the close link between slow variable models and macroscopic phenomenology.

To start we describe an extension an extension of Eq. (A.15) due to Machlup and Onsager [4]. We begin by writing down a model equation of thermodynamic relaxation patterned after the particle mechanics eq. (2.7).

5. A Model Equation of Thermodynamic Relaxation

We assume that the kinetics of the process $\Gamma_n \to \Gamma_p$ is governed by the following model thermodynamic equation of motion

$$\tilde{\mathbf{M}}\ddot{\mathbf{A}}(t) = \mathbf{X}(t). \tag{A.17}$$

In Eq. (A.17), as before $\mathbf{X}(t)$ is the vector of thermodynamic forces while $\tilde{\mathbf{M}}$ is a symmetric matrix of phenomenological parameters introduced by Machlup and Onsager [4]. We adopt Eq. (A.17) inasmuch as it is the simplest equation of motion that is consistent with the Machlup–Onsager Eq. (A.24). Notice that Eq. (A.17) is similar in form to Newton's equation of motion for a particle system. Thus, we denote the matrix of phenomenological parameters by $\tilde{\mathbf{M}}$ in order to emphasize the analogy to particle masses. The analogy, however, is not perfect because $\tilde{\mathbf{M}}$ may be nondiagonal [4].

We choose the thermodynamic forces $\mathbf{X}(t)$ to have the following "velocity"-dependent form

$$\mathbf{X}(t) = \mathbf{F}[\Gamma_p; \mathbf{A}(t), \dot{\mathbf{A}}(t)]. \tag{A.18}$$

We adopt Eq. (A.18) since it is the simplest choice for $\mathbf{X}(t)$, which accounts for the nonconservative nature of the motion and for its timescale dependence, and which is also consistent with the results of Machlup and Onsager.

Our model equation of thermodynamic relaxation then follows from eqs. (A.17) and (A.18) as

$$\tilde{\mathbf{M}}\ddot{\mathbf{A}}(t) = \mathbf{F}[\Gamma_p; \mathbf{A}(t), \dot{\mathbf{A}}(t)]. \tag{A.19}$$

We next develop the Machlup–Onsager equation from Eq. (A.19) by making the familiar assumption that the macroscopic parameters $\mathbf{A}(t)$ are slow variables. This assumption is usually justified by the idea that a timescale separation exists between the parameters $\mathbf{A}(t)$ and their "bath variables," due to the macroscopic nature of the former and microscopic nature of the latter.

6. The Slow Variable Equation of Machlup and Onsager

We make the slow variable assumption that throughout the process $\Gamma_n \to \Gamma_p, \dot{\mathbf{A}}(t)$ is very small. Then we may proceed by paralleling our development in Section II.B.1. Thus, we approximate $\mathbf{F}$ in Eq. (A.19) by its linear order power series expansion in $\dot{\mathbf{A}}(t)$, namely,

$$\mathbf{F}[\Gamma_p; \mathbf{A}(t), \dot{\mathbf{A}}(t)] \doteq \mathbf{F}[\Gamma_p; \mathbf{A}(t), \dot{\mathbf{A}}(t) = 0]$$
$$+ \left\{ \frac{\partial \mathbf{F}^T[\Gamma_p; \mathbf{A}(t), \dot{\mathbf{A}}(t)]}{\partial \dot{\mathbf{A}}(t)} \right\}_{\dot{\mathbf{A}}(t)=0} \dot{\mathbf{A}}(t). \qquad (A.20)$$

In analogy to the discussion in Section II.B.1, the terms on the right-hand side of Eq. (A.20) have the following simple physical meanings:

1. The term $\mathbf{F}[\Gamma_p; \mathbf{A}(t), \dot{\mathbf{A}}(t) = 0]$ is an equilibrium thermodynamic force analogous to the particle force $F_{eq}(S; x)$ described in Case 1 after Eq. (3.14). We will thus denote this term by

$$\mathbf{F}_{eq}[\Gamma_p; \mathbf{A}(t)] \equiv \mathbf{F}[\Gamma_p; \mathbf{A}(t), \dot{\mathbf{A}}(t) = 0]. \qquad (A.21)$$

2. The term $\left\{ \frac{\partial \mathbf{F}^T[\Gamma_p; \mathbf{A}(t), \dot{\mathbf{A}}(t)]}{\partial \dot{\mathbf{A}}(t)} \right\}_{\dot{\mathbf{A}}(t)=0} \dot{\mathbf{A}}(t)$, because it is linear in the "velocities" $\dot{\mathbf{A}}(t)$, is a thermodynamic frictional force analogous to the particle force $-\zeta \dot{x}(t)$ in Eq. (3.3). Thus, we define the *thermodynamic friction matrix* $\tilde{\mathbf{R}}$ by

$$\tilde{\mathbf{R}} = -\left\{ \frac{\partial \mathbf{F}^T[\Gamma_p; \mathbf{A}^*, \dot{\mathbf{A}}(t)]}{\partial \dot{\mathbf{A}}(t)} \right\}_{\dot{\mathbf{A}}(t)=0}. \qquad (A.22)$$

Eqs. (A.21) and (A.22) yield Eq. (A.20) as

$$\mathbf{F}[\Gamma_p; \mathbf{A}(t), \dot{\mathbf{A}}(t)] = \mathbf{F}_{eq}[\Gamma_p; \mathbf{A}(t)] - \tilde{\mathbf{R}}\dot{\mathbf{A}}(t). \qquad (A.23)$$

Note that to obtain Eq. [A.23] we have made an additional assumption, namely, that $\partial \mathbf{F}^T[\Gamma_p; \mathbf{A}(t), \dot{\mathbf{A}}(t)]/\partial \dot{\mathbf{A}}(t) \doteq \partial \mathbf{F}^T[\Gamma_p; \mathbf{A}^*, \dot{\mathbf{A}}(t)]/\partial \dot{\mathbf{A}}(t)$. This assumption is valid because the corrections to Eq. (A.24) below obtained by dropping it are of the same order of smallness as the quadratic order $[\dot{\mathbf{A}}(t)]^2$ term omitted in the expansion of Eq. (A.20), so long as one invokes Onsager's slow variable assumption that $\dot{\mathbf{A}}(t)$ is very small and his additional assumption to be discussed shortly that $\mathbf{A}(t) - \mathbf{A}^*$ is also very small.

The Machlup–Onsager slow variable equation of thermodynamic relaxation then follows from eqs. (A.19) and (A.23) as

$$\tilde{\mathbf{M}}\ddot{\mathbf{A}}(t) = \mathbf{F}_{eq}[\Gamma_p; \mathbf{A}(t)] - \tilde{\mathbf{R}}\dot{\mathbf{A}}(t). \qquad (A.24)$$

Notice from eqs. (A.23) and (A.24) that the slow variable assumption amounts to approximating the total thermodynamic force as a sum of: (a) a zeroth-order term $\mathbf{F}_{eq}[\Gamma_p; \mathbf{A}]$ independent of $\dot{\mathbf{A}}$, which is the full force for an idealized infinitely slow ($\dot{\mathbf{A}} = 0$) process; and (b) a first-order term $-\tilde{\mathbf{R}}\dot{\mathbf{A}}$ linear in $\dot{\mathbf{A}}$, which corrects $\mathbf{F}_{eq}(\Gamma_p; \mathbf{A})$ for the finite rate ($\dot{\mathbf{A}} \neq \mathbf{0}$) of real processes.

To go further, we note that in analogy to Eq. (3.21) we expect an equilibrium thermodynamic potential $W(\Gamma_p; \mathbf{A})$ to exist such that

$$\mathbf{F}_{eq}(\Gamma_p; \mathbf{A}) = -\frac{\partial W(\Gamma_p; \mathbf{A})}{\partial \mathbf{A}}; \qquad (A.25)$$

where, as mentioned, the form of $W(\Gamma_p; \mathbf{A})$ follows from the correspondence between eqs. (3.13) and (A.7). Comparing these equations yields this form as

$$W(\Gamma_p; \mathbf{A}) = -TS(\Gamma_p; \mathbf{A}). \qquad (A.26)$$

We next define

$$\mathbf{M} = T^{-1}\tilde{\mathbf{M}} \quad \text{and} \quad \mathbf{R} = T^{-1}\tilde{\mathbf{R}} \qquad (A.27)$$

(where the $\mathbf{R}$'s in eqs. [A.13] and [A.27] will turn out to be identical). Then combining eqs. (A.24)–(A.27) yields the final form of the Machlup–Onsager equation as

$$\mathbf{M}\ddot{\mathbf{A}}(t) = \frac{\partial S[\Gamma_p; \mathbf{A}(t)]}{\partial \mathbf{A}(t)} - \mathbf{R}\dot{\mathbf{A}}(t). \qquad (A.28)$$

Eq. (A.28) is the thermodynamic equivalent of the particle mechanics Eq. (3.24).

Notice that Eq. (A.28) resembles the equation of motion of a macroscopic damped mechanical system comprised of $n - p$ particles moving according to the conservative force $\partial S/\partial \mathbf{A}$ and additionally subject to the frictional force $-\mathbf{R}\dot{\mathbf{A}}$.

Before going on, we point out some additional features of the Machlup–Onsager Eq. (A.28). Especially, we note that the slow variable assumption renders Eq. (A.28) valid only for a very limited and simple class of irreversible processes, namely, nearly reversible irreversible processes.

7. Some Features of the Machlup–Onsager Equation

We first comment on the physical meanings of the thermodynamic forces in Eq. (A.28).

Letting $\dot{\mathbf{A}}(t) \to \mathbf{0}$ in Eq. (A.28), the condition for a reversible process in which the system remains in strict thermodynamic equilibrium (compare

Case 1 after Eq. [3.14]) shows that for such processes the full thermo-dynamic force is the "conservative force" $\partial S/\partial \mathbf{A}$. Thus, the effects of departures from equilibrium or irreversibility arise solely from the "frictional force" $-\mathbf{R}\dot{\mathbf{A}}$.

We next explain what we mean by a "nearly reversible irreversible process."

While in a reversible process the system passes through a sequence of equilibrium thermodynamic states; in an irreversible process it passes through a series of nonequilibrium states. Denote by $\Gamma_{p'}(t)$ the system's actual nonequilibrium state at time t when it is undergoing the irreversible process $\Gamma_n \rightarrow \Gamma_p$. Denote by $\Gamma_{p'}$ the equilibrium system state with the same macroscopic parameter values as $\Gamma_{p'}(t)$. (That is, for state $\Gamma_{p'}$ the macro-scopic parameters are constrained to have the fixed values $N, V, E, A_1, \dots, A_p, A_{p+1}[t], \dots, A_n[t]$.)

If throughout the process $\Gamma_n \rightarrow \Gamma_p$, the states $\Gamma_{p'}(t)$ differ only slightly from the states $\Gamma_{p'}$, we will call $\Gamma_n \rightarrow \Gamma_p$ a nearly reversible irreversible process.

From these comments, it is evident that the Machlup–Onsager Eq. (A.28) only validly describes irreversible processes $\Gamma_n \rightarrow \Gamma_p$ of the nearly reversible type. The argument proceeds as follows. The deviations of $\Gamma_{p'}(t)$ from $\Gamma_{p'}$ vanish as $\dot{\mathbf{A}}(t) \rightarrow \mathbf{0}$. Thus, for the slow variable assumption ($\dot{\mathbf{A}}[t]$ is always small) to be valid for a process $\Gamma_n \rightarrow \Gamma_p$, throughout the process the departures of $\Gamma_{p'}(t)$ from $\Gamma_{p'}$ must be slight. Consequently, since Eq. (A.28) requires the slow variable assumption, it is limited to processes for which the departures are always slight, that is, to nearly reversible irreversible processes.

We may now obtain Onsager's original equation of thermodynamic relaxation, Eq. (A.15), as a limiting case of the Machlup–Onsager Eq. (A.28). (Compare the development of Eq. [3.25] from Eq. [3.24].)

8. Eq. (A.15) as a Limiting Case of Eq. (A.28)

Eq. (A.15) may be obtained from Eq. (A.28) by taking the formal limit

$$\mathbf{M}\ddot{\mathbf{A}}(t) \rightarrow \mathbf{0}, \tag{A.29}$$

premultiplying the resulting equation by $\mathbf{R}^{-1}$ and then using Eq. (A.13).

Before justifying the formal limiting process of Eq. (A.29) and hence the present "derivation" of Eq. (A.15), we note the following points.

1. We will see that to justify Eq. (A.29) we require a high-friction, weak-force (overdamped) system. This requirement is in complete harmony with, but is even more restrictive than, the assumption that, the unconstrained A's are slow variables.

2. Consequently, slow variable assumptions provide the physical basis of Onsager's Eq. (A.15).
3. Finally given point 2 it is evident that the Second Law does not guarantee that the actual driving force $\mathbf{X}$ for the process $\Gamma_n \rightarrow \Gamma_p$ is $\partial S/\partial \mathbf{A}$, where S is the system's equilibrium entropy function. This follows because the choice $\partial S/\partial \mathbf{A}$ requires the validity of not only the Second Law but also of the far more restrictive slow variable assumption.

These limitations of $S(\Gamma_p; \mathbf{A})$, parallel those noted in Section III for $W(S; x)$. Namely, just as $S(\Gamma_p; \mathbf{A})$ drives only nearly reversible processes, $W(S; x)$ drives only near equilibrium reactions $x^\dagger \rightarrow x_p$.

We next discuss the physics implied by the formal limit of Eq. (A.29). For simplicity we will make use of a one-dimensional particle dynamics model, similar to that of Figure 3.1, to illustrate the main points.

9. Particle Sedimentation: A Prototype Transport Process

We will consider an instructive prototype transport process. This process is the gravity (or ultracentrifuge) induced sedimentation of a collection of identical noninteracting macroscopic particles each of mass M. Assume that the particles are suspended in a long tube filled with a viscous fluid and that the tube is aligned parallel to the Z-axis. The particles thus descend in the negative Z-direction. Further assume that the fluid exerts only a simple frictional force $-\zeta \dot{z}(t)$ on each particle. Then the equation of motion for a typical particle is

$$M\ddot{z}(t) = -F_g - \zeta \dot{z}(t) \tag{A.30}$$

where $-F_g$ is the force of gravity.

We will regard Eq. (A.30) as a model for the Machlup–Onsager Eq. (A.28). Thus, we regard $M, z, -F_g$, and ζ in Eq. (A.30) as corresponding respectively to $\mathbf{M}, \mathbf{A}, \partial S/\partial \mathbf{A}$, and $\mathbf{R}$ in Eq. (A.28).

Using these correspondences and Eq. (A.13) yields the following model for Eq. (A.15):

$$\dot{z}(t) = -\zeta^{-1} F_g. \tag{A.31}$$

Similarly, the model for Eq. (A.29) is

$$M\ddot{z}(t) \rightarrow 0. \tag{A.32}$$

Note that in analogy to our "derivation" of Eq. (A.15) from eqs. (A.28) and (A.29), we may "derive" Eq. (A.31) from eqs. (A.30) and (A.32).

We next give a real derivation of Eq. (A.31) from Eq. (A.30). Our argument will make clear the physics implied by the formal limit of

Eq. (A.32). By analogy, we will then be able to infer the physics implied by Eq. (A.29).

We may derive Eq. (A.31) from Eq. (A.30) by solving Eq. (A.30) exactly for $\dot{z}(t)$ for the initial condition $\dot{z}(t = 0) = 0$. Solving Eq. (A.30) this way yields $\dot{z}(t)$ as

$$\dot{z}(t) = \dot{z}_T[1 - \exp(-t/\tau_A)] \tag{A.33}$$

where

$$\dot{z}_T = -\zeta^{-1}F_g \quad \text{and} \quad \tau_A = M\zeta^{-1}. \tag{A.34}$$

As expected, Eq. (A.33) describes (a) at short times, a linear growth $-\dot{z}(t) = M^{-1}F_g t$ of $-\dot{z}(t)$ due to gravitational acceleration; (b) at intermediate times, an exponential decay (with time constant τ_A) of the net acceleration, due to the increasing cancellation of the gravitational force $-F_g$ by the frictional force $-\zeta\dot{z}(t)$ arising from the growth of $-\dot{z}(t)$; and (c) at long times, complete cancellation of $-F_g$ by $-\zeta\dot{z}(t)$ yielding zero net acceleration and a constant terminal velocity $\dot{z}_T$.

It is now easy to assess the validity of eqs. (A.31) and (A.32). In brief, eqs. (A.33) and (A.34) yield

$$\lim_{t \gg \tau_A} \dot{z}(t) = \dot{z}_T = -\zeta^{-1}F_g. \tag{A.35}$$

From Eq. (A.35) it follows that Eq. (A.31) is valid only for times $t \gg \tau_A$, where τ_A is the relaxation time for particle accelerations. Thus, Eq. (A.32) has meaning only so long as one interprets it to say the following: sufficient time has elapsed for the gravitational and frictional forces to have come into balance, thus leaving the particle subject to zero net force and, hence, with zero acceleration.

Before going on we make a few more comments about the validity of eqs. (A.15) and (A.31), noting especially the following two points:

1. Eq. (A.31) is a limiting case of the slow variable Eq. (A.30), and thus requires that $\dot{z}(t)$ be very small. Given Eq. (A.31), this implies that the parameter $\zeta^{-1}F_g = (\zeta/M)^{-1}(F_g/M)$ must be small. (This restriction on $\zeta^{-1}F_g$ is similar to the condition that E must be small in order that Ohm's Eq. [A.9] be valid.)

2. Given Eq. (A.15), it follows that Eq. (A.31) is a useful equation of motion only if the distance traversed by the particle in time τ_A approximately $|\dot{z}_T\tau_A|$ is small relative to the total distance it travels ($\sim$length of the tube). From Eq. (A.34), this requires that the parameter $(F_g/M)(\zeta/M)^{-2}$ be small.

In summary, Eq. (A.31) validly describes the long time motions of particles with sufficiently large friction coefficients (per unit mass) ζ/M, subject to sufficiently weak forces (per unit mass) F_g/M.

A corresponding analysis of the Machlup–Onsager Eq. (A.28) yields similar conclusions. Namely, Eq. (A.15) validly describes the long time motions of macroscopic parameters with sufficiently large friction matrix elements $[\mathbf{R}]_{ij}$ and sufficiently small "mass" matrix elements $[\mathbf{M}]_{ij}$, subject to sufficiently weak thermodynamic forces $[\partial S(\Gamma_p; \mathbf{A})/\partial \mathbf{A}]_i$.

To obtain the standard form of Onsager's theory [37,38], we next linearize the thermodynamic forces in eqs. (A.15) and (A.28). This linearization reduces these equations to coupled damped harmonic oscillator equations of motion.

10. Onsager's Linear Transport Equations

Paralleling the harmonic oscillator expansion of the potential function of a mechanical system, we next approximate the equilibrium entropy function $S(\Gamma_p; \mathbf{A})$ by its quadratic order Taylor series expansion about $\mathbf{A}^*$ (the point at which S has its maximum). That is, we assume

$$S(\Gamma_p; \mathbf{A}) \doteq S(\Gamma_p; \mathbf{A}^*) - \frac{1}{2}\boldsymbol{\alpha}^T \left[\frac{\partial^2 S(\Gamma_p; \mathbf{A})}{\partial \mathbf{A}^2}\right]_{\mathbf{A}=\mathbf{A}^*} \boldsymbol{\alpha} \qquad (A.36)$$

where

$$\boldsymbol{\alpha} \equiv \mathbf{A} - \mathbf{A}^* \qquad (A.37)$$

is the displacement of $\mathbf{A}$ from its value $\mathbf{A}^*$ in the final equilibrium state Γ_p.

Because of Eq. (A.36), Onsager's theory is usually viewed as being limited to "close to equilibrium" irreversible processes. Actually, however, Onsager's theory is qualitatively valid for all nearly reversible irreversible processes for which throughout the process the true and parabolic forms for $S(\Gamma_p; \mathbf{A})$ never deviate qualitatively.

(The familiar notion [37,38] of a "close to equilibrium process" differs from our concept of a "nearly reversible process." For close to equilibrium processes, the initial and final equilibrium states Γ_n and Γ_p must have nearly identical A values. Thus, for such processes α is always small, and Eq. (A.36) is guaranteed to be valid. In contrast for nearly reversible processes, the $\mathbf{A}$ values for states Γ_n and Γ_p need not be close. Instead, for such processes the actual system states $\Gamma_{p'}(t)$ during the process must differ only slightly from the equilibrium states $\Gamma_{p'}$ with the same $\mathbf{A}$ values. Thus for nearly reversible processes α does not have to be small, and Eq. (A.36) may or may not be valid.)

Given Eq. (A.36), it is easy to obtain the linearized forms for Onsager's eqs. (A.15) and (A.28). Thus comparing eqs. (A.28), (A.36), and (A.37) yields

$$\mathbf{M}\ddot{\alpha}(t) = -\mathbf{M}^{1/2}\Omega^2\mathbf{M}^{1/2}\alpha(t) - \mathbf{R}\dot{\alpha}(t) \tag{A.38}$$

where the matrix Ω^2 is defined by

$$\Omega^2 = -\mathbf{M}^{-1/2}\left[\frac{\partial^2 S(\Gamma_p; \mathbf{A})}{\partial\mathbf{A}^2}\right]_{\mathbf{A}=\mathbf{A}^*}\mathbf{M}^{-1/2}. \tag{A.39}$$

Eq. (A.39) is the linearized form of the Machlup–Onsager Eq. (A.28). Similarly the linearized version of Eq. (A.15) is

$$\dot{\alpha}(t) = -\mathbf{L}\mathbf{M}^{1/2}\Omega^2\mathbf{M}^{1/2}\alpha(t). \tag{A.40}$$

We next define the quantities β and $\mathbf{y}(t)$ by

$$\beta = \mathbf{M}^{-1/2}\mathbf{R}\mathbf{M}^{-1/2} \tag{A.41}$$

and

$$\mathbf{y}(t) = \tilde{\mathbf{M}}^{1/2}\alpha(t) = \mathbf{M}^{1/2}T^{1/2}\alpha(t) \tag{A.42}$$

where we have used Eq. (A.27). Premultiplying Eq. (A.38) by $\mathbf{M}^{1/2}$ and using eqs. (A.41) and (A.42) then yields Eq. (A.38) as

$$\ddot{\mathbf{y}}(t) = -\Omega^2\mathbf{y}(t) - \beta\dot{\mathbf{y}}(t). \tag{A.43}$$

Using Eq. (A.13), Eq. (A.40) may similarly be exactly rewritten as

$$\dot{\mathbf{y}}(t) = -\beta^{-1}\Omega^2\mathbf{y}(t). \tag{A.44}$$

Finally, using eqs. (A.36), (A.37), (A.39), and (A.42) yields the following form for Eq. (A.7):

$$P(\Gamma_p; \mathbf{y}) = P(\Gamma_p; \mathbf{y} = \mathbf{0})\exp\left(-\frac{1}{2}\frac{\mathbf{y}^T\Omega^2\mathbf{y}}{kT}\right). \tag{A.45}$$

Next, note that from its definition $\partial^2 S/\mathbf{A}^2$ is a symmetric matrix. Also, according to Onsager's reciprocal relations [1,3,4], $\mathbf{R}$ is a symmetric matrix. Thus, from eqs. (A.39) and (A.31) and the symmetry of $\mathbf{M}$ it follows that both Ω^2 and β are symmetric matrices.

As a consequence:

1. Eq. (A.43) is identical to the equations of motion for a set of $n - p$ macroscopic coupled damped harmonic oscillators each of unit mass with the set having coordinate vector $\mathbf{y}(t)$, dynamical matrix Ω^2, and friction matrix β;
2. Eq. (A.44) is identical to equations of motion for the same system when Eq. (A.29) applies; and
3. Eq. (A.45) is identical to the configurational canonical ensemble distribution function for the same system as temperature T.

We next solve eqs. (A.43) and (A.44) and compare the results. This will help to further clarify the physical content of eq. (A.29).

11. Solutions of the Linear Transport Equations

For simplicity we will assume that there is only a single unconstrained A. Then eqs. (A.43) and (A.44) reduce to the following scalar equations

$$\ddot{y}(t) = -\Omega^2 y(t) - \beta \dot{y}(t) \tag{A.46a}$$

and

$$\dot{y}(t) = -\beta^{-1}\Omega^2 y(t). \tag{A.46b}$$

Similarly Eq. (A.45) becomes

$$P(\Gamma_p; y) = P(\Gamma_p; y = 0)\exp\left(-\frac{1}{2}\frac{\Omega^2 y^2}{kT}\right). \tag{A.46c}$$

The solution of Eq. (A.46a) is [42]

$$y(t) = y\exp\left(-\frac{1}{2}\beta t\right)\left[\cosh\left(\frac{1}{2}\beta_1 t\right) + \beta_1^{-1}\beta\sinh\left(\frac{1}{2}\beta_1 t\right)\right]$$
$$+ 2\beta_1^{-1}\dot{y}\exp\left(-\frac{1}{2}\beta t\right)\sinh\left(\frac{1}{2}\beta_1 t\right) \tag{A.47a}$$

while that of Eq. (A.46b) is

$$y(t) = y\exp(-\beta^{-1}\Omega^2 t). \tag{A.47b}$$

In eqs. (A.47), $y = y(t = 0)$ and $\dot{y} = \dot{y}(t = 0)$ while

$$\beta_1 = (\beta^2 - 4\Omega^2)^{1/2}. \tag{A.48}$$

Notice that eqs. (A.47a) and (A.48) describe the motion of (a) an *underdamped* harmonic oscillator for $\beta < 2\Omega$, (b) a *critically damped* oscillator for $\beta = 2\Omega$, and (c) and *overdamped* oscillator for $\beta > 2\Omega$.

We next specialize eq. (A.47a) to the limits of (a) *extreme overdamping* $\beta \gg 2\Omega$, and (b) *long time* $t \to \infty$. So specialized Eq. (A.47a) becomes

$$\lim_{\substack{t \to \infty \\ (\beta/2\Omega) \to \infty}} y(t) = (y + \beta^{-1}\dot{y})\exp(-\beta^{-1}\Omega^2 t). \tag{A.49}$$

To proceed further we note that Onsager and Machlup's [4] results imply the following Boltzmann form, similar to Eq. (A.46c), for the distribution of initial "velocities" $\dot{y}$

$$P(\Gamma_p; \dot{y}) = P(\Gamma_p; \dot{y} = 0)\exp\left(-\frac{1}{2}\frac{\dot{y}^2}{kT}\right). \tag{A.50}$$

Given Eq. (A.50) the average velocity $\langle \dot{y} \rangle = 0$, and hence

$$\lim_{\substack{t \to \infty \\ (\beta/2\Omega) \to \infty}} \langle y(t) \rangle = y \exp(-\beta^{-1}\Omega^2 t) \tag{A.51}$$

where $\langle y(t) \rangle$ is $y(t)$ of Eq. (A.49) averaged over the distribution of Eq. (A.50).

Note, however, that $\lim_{\substack{t \to \infty \\ (\beta/2\Omega) \to \infty}} \langle y(t) \rangle$ of Eq. (A.51) coincides with $y(t)$ of Eq. (A.47b). Thus, we have shown the following. The trajectory $y(t)$ found from the limiting Eq. (A.46b) coincides with the velocity averaged trajectory $\langle y(t) \rangle$ predicted by the "exact" Eq. (A.46a) so long as the system is strongly overdamped, and we require only the long time limit of the trajectory.

These results accord with our earlier conclusions concerning Eq. (A.29), namely, it is a high friction ($\beta \to \infty$), weak force ($\Omega^2 \to 0$), and long time ($t \to \infty$) limit.

We next augment Onsager's transport equations to include the effects of fluctuations.

12. Onsager's Linear Equations Including Fluctuations

We have seen how classical thermodynamics may be extended to include fluctuations; Boltzmann's eqs. (A.7) and (A.8) being key results of this extension. Onsager, within the context of his linear theory, has generalized Boltzmann's equilibrium fluctuation theory to dynamics. We next briefly discuss this generalization (compare Section II.B.4).

Onsager's treatment of nonequilibrium fluctuations rests on his *fluctuation-regression hypothesis* [1]. To explain this hypothesis, we first note the following.

As may be seen from arguments like those that led from Eqs. (A.15) to (A.16), Onsager's eqs. (A.38) and (A.40) properly reduce to equilibrium thermodynamics. Thus, Eqs. (A.38) and (A.40) predict that after the final equilibrium state Γ_p is reached, all motions of the unconstrained A's cease, and that these parameters remain forever fixed at their final state values $\mathbf{A}^*$.

However, in analogy to the breakdown of classical thermodynamics due to fluctuations (see Eq. [A.7]) Onsager's eqs. (A.38) and (A.40) must also break down. Namely, motions of the unconstrained A's do not cease after the state Γ_p is reached. Rather, as a dynamical generalization of Eq. (A.7), these parameters fluctuate about $\mathbf{A}^*$, executing random motions with microscopic amplitudes.

The fluctuation-regression hypothesis, rephrased in modern language, may now be stated as follows. To describe the dynamical fluctuations just mentioned, it is sufficient to use Onsager's purely macroscopic eqs. (A.38) and (A.40) modified to account for microscopic effects solely by the inclusion of random forces of the standard Brownian motion type; namely, zero mean white noise Gaussian forces that obey fluctuation-dissipation relations that ensure recovery of Eq. (A.45) as $t \to \infty$ [2].

Starting from the forms for Onsager's equations given in eqs. (A.43) and (A.44), the fluctuation-regression hypothesis yields the following equation of motion for the fluctuations $\mathbf{y}(t) = \tilde{\mathbf{M}}^{1/2}\alpha'(t) = \tilde{\mathbf{M}}^{1/2}[\mathbf{A}'(t) - \mathbf{A}^*]$ in state Γ_p:

$$\ddot{\mathbf{y}}(t) = -\Omega^2\mathbf{y}(t) - \beta\dot{\mathbf{y}}(t) + \tilde{\mathbf{f}}(t) \qquad (A.52)$$

and

$$\dot{\mathbf{y}}(t) = -\beta^{-1}\Omega^2\mathbf{y}(t) + \mathbf{f}(t) \qquad (A.53)$$

where $\tilde{\mathbf{f}} = \beta\mathbf{f}(t)$ and where the random force $\mathbf{f}(t) = [f_{p+1}(t)\ldots f_n(t)]$.

Eqs. (A.52) and (A.53) were used by Onsager to prove his reciprocal relations and to generate dynamical extensions of Boltzmann's Eq. (A.7) [3,4]. However, our interest in these equations arises because (a) they yield a bit more insight into the physical content of Onsager's theory: and (b) they provide the link between Onsager's model and Mori's identity [5].

We next briefly comment on point (a) above; merely noting the following.

1. Eq. (A.52) is identical to the *Langevin equation* for a set of $n - p$ coupled harmonic oscillators each of unit mass; with coordinates $\mathbf{y}(t)$, dynamical matrix Ω^2, and friction matrix β.

2. Eq. (A.53) is identical to the Langevin equation for the same system when Eq. (A.29) applies.

3. Eq. (A.52) is thus equivalent to the *Fokker-Planck equation* for the coupled oscillator system [2].

4. Similarly Eq. (A.53) is equivalent to the *diffusion equation* for the coupled oscillator system [2].

We next summarize our overview of Onsager's theory.

13. Summary

In summary, Onsager did succeed in finding a nonequilibrium extension of equilibrium thermodynamics. However, to resolve the dual problem of formulating a thermodynamic equation of motion and of choosing the thermodynamic forces, he was obliged to make limiting slow variable assumptions. Thus his central Eq. (A.53) models actual macroscopic parameters motions in a highly simplified way, namely, as *coupled diffusive motions* in the *equilibrium potential* $\propto S(\Gamma_p; \mathbf{A})$ of Eq. (A.36).

Relatedly, because of the slow variable assumption Onsager's theory applies only to irreversible processes of the "nearly reversible" type. However, many irreversible processes proceed via relaxation dynamics that differ sharply from nearly reversible dynamics [43]. Thus, Onsager's theory, while a significant step forward, is far narrower in scope than classical thermodynamics.

From the standpoint of liquid phase chemical reaction dynamics, the key points discussed so far are as follows:

1. The slow variable models are ultimately based on purely macroscopic measurements. For example Onsager's diffusional model of Eq. (A.53) derives from Eq. (A.15), which in turn is based on macroscopic phenomenology. Similarly, Onsager's Langevin model of Eq. (A.52) derives from Eq. (A.28), which in turn is suggested by the empirical friction law $-\zeta \dot{x}(t)$ for a macroscopic particle.

2. For slow variable models to be valid, throughout the process being considered the system must remain in near thermodynamic equilibrium.

3. Diffusional descriptions of irreversible motion, in addition to requiring the slow variable assumption, only validly describe the dynamics of strongly overdamped (weak force, high friction) systems, and then only over coarse-grained time intervals much longer than the system's acceleration decay times τ_A.

We next touch on questions of the validity of Onsager's theory.

14. The Validity of Onsager's Theory

As discussed in Section II the macroscopic roots of slow variable equations like Onsager's clearly render such equations questionable as frameworks for studying microscopic processes.

Here we note that Onsager's theory can fail even for fully macroscopic processes; if these are far removed from its empirical base.

For example, Bridgman's [43] work shows that Onsager's model can fail for thermodynamic processes $\Gamma_n \to \Gamma_p$ occurring in systems that display hysteresis or that are near a phase transition. We next paraphrase Bridgman's argument for the phase transition case.

Consider a macroscopic system so close to a phase transition $\alpha \to \beta$, that for state Γ_n the thermodynamically stable phase is α whereas for state Γ_p it is β. Then the qualitative kinetics of the process $\Gamma_n \to \Gamma_p$ depends on the relative occurrence times $\tau_{n \to p}$ and $\tau_{\alpha \to \beta}$ for the changes $\Gamma_n \to \Gamma_p$ and $\alpha \to \beta$. For example, if $\tau_{\alpha \to \beta} \gg \tau_{n \to p}$ the system will remain in phase α throughout $\Gamma_n \to \Gamma_p$. In contrast, if $\tau_{\alpha \to \beta} \ll \tau_{n \to p}$ the $\alpha \to \beta$ phase transition will occur during $\Gamma_n \to \Gamma_p$.

Moreover, in both limits Onsager's model can fail qualitatively. Suppose first that $\tau_{\alpha \to \beta} \gg \tau_{n \to p}$. Then $\Gamma_n \to \Gamma_p$ will be driven by the nonequilibrium entropy function $S_\alpha(\Gamma_p; \mathbf{A})$ for phase α, rather than by the equilibrium entropy function $S(\Gamma_p; \mathbf{A})$; thus invalidating Onsager's basic eqs. (A.15) and (A.28). Next, assume $\tau_{\alpha \to \beta} \ll \tau_{n \to p}$. Then $\Gamma_n \to \Gamma_p$ will be driven by $S(\Gamma_p; \mathbf{A})$ as is assumed by Onsager. However, this function will change its qualitative form during $\Gamma_n \to \Gamma_p$ due to the $\alpha \to \beta$ transition, thus invalidating Onsager's expansion Eq. (A.36) of $S(\Gamma_p; \mathbf{A})$ about the β state Γ_p and hence his linear eqs. (A.52) and (A.53).

Additionally for intermediate cases, for example $\tau_{\alpha \to \beta} \approx \tau_{n \to p}$, even more complex nonclassical behavior is expected.

In summary, the macroscopic nature of the unconstrained A's does not guarantee that they relax like classical slow variables. Thus, even for fully macroscopic processes $\Gamma_n \to \Gamma_p$, open questions about kinetics remain.

This concludes our outline of Onsager's theory. We next turn to Mori's [5] generalized Langevin equation.

15. Mori's Generalized Langevin Equation

The finite memory extension of Onsager's Eq. (A.53) is

$$\dot{\mathbf{y}}(t) = -\int_0^t \beta^{-1}(t - \tau)\Omega^2 \mathbf{y}(\tau)d\tau + \mathbf{f}(t) \tag{A.54}$$

where $\beta(t)$ and $\mathbf{f}(t)$ are standard generalized Langevin memory kernels and random forces [5]. Eq. (A.54) is equivalent to Mori's generalized Langevin equation [44].

Mori's eq. (A.54) has been well-discussed in many reviews, texts, and so on [45].

Thus, here we note only a few points that are pertinent to the themes of this review:

1. The form of Eq. (A.54) is set by that of Eq. (A.53).

2. Thus, use of Eq. (A.54) is problematic for processes whose physics is radically different from the overdamped slow variable physics responsible for the form of Eq. (A.53).

3. For example, from eqs. (A.36)–(A.39), it follows that Ω^2 determines the harmonic restoring forces produced by the equilibrium thermodynamic potential $\propto S(\Gamma_p; \mathbf{A})$; and thus governs kinetics only for near equilibrium (nearly reversible) processes and, moreover, only for those which conform to Eq. (A.36). Thus, to apply Eq. (A.54) to other types of processes, for example Bridgman's [43]; is to describe these processes in terms of driving force parameters Ω^2 that are unrelated to the actual thermodynamic forces experienced by the system.

REFERENCES

1. L. Onsager, *Phys. Rev.* **37**, 405 (1931); **38**, 2265 (1931).

2. S. Chandrasekhar, *Rev. Mod. Phys.* **15**, 1 (1943).

3. L. Onsager and S. Machlup, *Phys. Rev.* **91**, 1505 (1953).

4. S. Machlup and L. Onsager, *Phys. Rev.* **91**, 1512 (1953).

5. H. Mori, *Prog. Theor. Phys.* **33**, 423 (1965); **34**, 399 (1965).

6. For recent reviews of liquid phase chemical dynamics see, for example, P. J. Rossky and J. D. Simon, *Nature* **370**, 263 (1994); and R. M. Stratt and M. Maroncelli, *J. Phys. Chem.* **100**, 12981 (1996).

7. H. A. Kramers, *Physica* **7**, 284 (1940).

8. J. T. Hynes, *Ann. Rev. Phys. Chem.* **36**, 573 (1985).

9. B. J. Berne, M. Borkovec, and J. E. Straub, *J. Phys. Chem.* **92**, 3711 (1988).

10. P. J. Hänggi, P. Talkner, and M. Borkovec, *Rev. Mod. Phys.* **62**, 251 (1990).

11. S. D. Rosenthal, X. Xie, M. Du, and G. R. Fleming. *J. Chem. Phys.* **95**, 4715 (1991); R. Jimenez, G. R. Fleming, P. V. Kumar, and M. Maroncelli, *Nature* **369**, 471 (1994). These papers make clear the need for short timescale theories of dynamics and molecular models for friction in liquid phase chemistry.

12. See for example, S. Ruhman, L. R. Williams, A. G. Joly, B. Kohler, and K. A. Nelson, *J. Phys. Chem.* **91**, 2237 (1987); S-G Su and J. D. Simon, *J. Phys. Chem.* **91**, 2693 (1987); M. Maroncelli, J. MacInnis, and G. R. Fleming, Science **243**, 1674 (1989); W. Jarzeba, G. C. Walker, A. E. Johnson, and P. F. Barbara, *Chem. Phys.* **152**, 57 (1991); J. D. Simon and

R. Doolan, *J. Am. Chem. Soc.* **114**, 4861 (1992); M. J. Weaver, *Chem. Rev.* **92**, 463 (1992); D. Raftery, M. Iannone, C. M. Phillips, and R. M. Hochstrasser, *Chem. Phys. Letts.* **201**, 513 (1993); M. Maroncelli, *J. Molec.Liquids* **57**, 1 (1993); P. Cong, Y. J. Van, H. P. Deuel, and J. D. Simon, *J. Chem. Phys.* **100**, 7855 (1994); D. Raftery, E. Gooding, A. Romanovsky, and R. M. Hochstrasser, *J. Chem. Phys.* **101**, 8572 (1994); T. Joo and G. R. Fleming, *J. Chem. Phys.* **102**, 4063 (1995).

13. H. Sumi and T. Asano, *Chem. Phys. Letts.* **240**, 125 (1995).

14. For recent MD studies of the dynamics of model aqueous nucleophillic substitution reactions see (a) J. P. Bergsma, B. J. Gertner, K. R. Wilson, and J. T. Hynes, *J. Chem. Phys.* **86**, 1356 (1987) and B. J. Gertner, K. R. Wilson, and J. T. Hynes, *J. Chem. Phys.* **90**, 3537 (1989); and (b) W. P. Keirstead, K. R. Wilson, and J. T. Hynes, *J. Chem. Phys.* **95**, 5256 (1991); and (c) B. J. Gertner, R. M. Whitnell, K. R. Wilson, and J. T. Hynes, *J. Am. Chem. Soc.* **113**, 74 (1991).

15. For recent MD studies of vibrational energy relaxation in liquids see, for example, (a) R. M. Whitnell, K. R. Wilson, and J. T. Hynes, *J. Phys. Chem.* **94**, 8625 (1990) and R. M. Whitnell, K. R. Wilson, and J. T. Hynes, *J. Chem. Phys.* **96**, 5354 (1992); (b) I. Benjamin and R. M. Whitnell, *Chem. Phys. Letters* **204**, 45 (1993); (c) M. Bruehl and J. T. Hynes, *Chem. Phys.* **175**, 205 (1993); and (d) M. Ferrario, M. L. Klein, and I. R. McDonald, *Chem. Phys. Letts.* **213**, 537 (1993).

16. For recent MD studies of solvation dynamics see, for example, M. Maroncelli and G. R. Fleming, *J. Chem. Phys.* **89**, 5044 (1988); M. Maroncelli, *J. Chem. Phys.* **94**, 2084 (1991); and E. A. Carter and J. T. Hynes, *J. Chem. Phys.* **94**, 5961 (1991).

17. For miscellaneous MD studies of liquid phase chemical processes see, for example, I. Benjamin and K. R. Wilson, *J. Chem. Phys.* **90**, 4176 (1989); I. Benjamin, B. J. Gertner, N. J. Tang, and K. R. Wilson, *J. Am. Chem. Soc.* **112**, 524 (1990); G. Ciccotti, M. Ferrario, J. T. Hynes, and R. Kapral, *J. Chem. Phys.* **93**, 7137 (1990); B. B. Smith, A. Staib, and J. T. Hynes, *Chem. Phys.* **176**, 521 (1993); M. Ben-Nun and R. D. Levine, *J. Phys. Chem.* **97**, 2334 (1993); and P. A. Rejto, E. Bindewald, and D. Chandler, *Nature* **375**, 129 (1995).

18. See, for example, R. Zwanzig and M. Bixon, *Phys. Rev. A2*, 2005 (1970) and R. Zwanzig, *J. Chem. Phys.* **52**, 3625 (1970).

19. S. A. Adelman, R. Ravi, R. Muralidhar, and R. Stote, *Adv. Chem. Phys.* **84**, 73 (1993).

20. (a) The basic principles and equations of our theory of liquid phase chemical dynamics are developed for monatomic solvents in S. A. Adelman, *Adv. Chem. Phys.* **53**, 61 (1983) and references therein. (b) The theory is extended to molecular solvents in S. A. Adelman and M. W. Balk, *J. Chem. Phys.* **82**, 4641 (1985) and **84**, 1752 (1986).

21. The partial clamping model is developed for monatomic solvents in S. A. Adelman, *J. Chem. Phys.* **81**, 2766 (1984). The model is refined and extended to molecular solvents in S. A. Adelman, *Int. J. Quantum Chem. Symp.* **21**, 189 (1987).

22. Short time pictures of liquid phase vibrational energy relaxation, cage escape, and activated barrier crossing are described in S. A. Adelman, *J. Stat. Phys.* **42**, 37 (1986).

23. The short time treatment of liquid phase activated barrier crossing outlined in Section VB is described in detail in S. A. Adelman and R. Muralidhar, *J. Chem. Phys.* **95**, 2752 (1991).

24. The short time treatment of liquid phase vibrational energy relaxation overviewed in Section VC is developed and applied for monatomic solvents in S. A. Adelman and R. H. Stote, *J. Chem. Phys.* **88**, 4397 (1988); R. H. Stote and S. A. Adelman, *ibid.* **88**, 4415 (1988); and S. A. Adelman, R. Muralidhar, and R. H. Stote, *J. Chem. Phys.* **95**, 2738 (1991). The treatment is extended to molecular solvents in ref. 19 and in S. A. Adelman,

R. H. Stote, and R. Muralidhar, *J. Chem. Phys.* **99**, 1320 (1993) and S. A. Adelman, R. Muralidhar, and R. H. Stote *ibid.* **99**, 1333 (1993).

25. For brief reviews of the theory developed in refs. 20–24 see S. A. Adelman, *Rev. Chem. Intermed.* **8**, 321 (1987) and F. Patron and S. A. Adelman, *Chem. Phys.* **152**, 121 (1991).

26. For a preliminary account of some of the work presented here see R. Ravi and S. A. Adelman, *Trends Chem. Phys.* **3**, 1 (1994) and S. A. Adelman and R. Ravi, *Trends Stat. Phys.* **1**, 235 (1994).

27. (a) B. J. Berne, M. E. Tuckerman, J. E. Straub, and A. L. R. Bug, *J. Chem. Phys.* **93**, 5084 (1990); and (b) M. Tuckerman and B. J. Berne, *ibid.* **98**, 7301 (1993).

28. R. M. Stratt, *Acc. Chem. Res.* **28**, 201 (1995) and references therein.

29. J. Chandrasekhar, S. F. Smith, and W. L. Jorgensen, *J. Am. Chem. Soc.* **107**, 154 (1985).

30. R. F. Grote and J. T. Hynes, *J. Chem. Phys.* **73**, 2715 (1980).

31. See, for example, the discussion on p. 581 of ref. 8.

32. G. van der Zwan and J. T. Hynes, *J. Chem. Phys.* **76**, 2993 (1982).

33. For reviews of liquid phase vibrational energy relaxation see, for example, D. W. Oxtoby, *Adv. Chem. Phys.* **47**, 487 (1981) and *Ann. Rev. Phys. Chem.* **32**, 77 (1981); J. Chesnoy and G. M. Gale, *Ann. Phys.* (Paris) **9**, 893 (1984) and *Adv. Chem. Phys.* **70**, 297 (1988); C. B. Harris, D. E. Smith, and D. J. Russell, *Chem. Rev.* **90**, 481 (1990); J. C. Owrutsky, D. Raftery, and R. M. Hochstrasser, *Ann. Rev. Phys. Chem.* **45**, 519 (1994).

34. L. Landau and E. Teller, *Phys. Z. Sowjetunion* **10**, 34 (1936).

35. For a recent review of the Landau-Teller and related theories see D. W. Miller and S. A. Adelman, *Int. Rev. Phys. Chem.* **13**, 359 (1994).

36. For a comparison of the assumptions underlying our theory of liquid phase VER and those of the Landau-Teller theory see sec. 23 of ref. 19.

37. See, for example, F. Reif, *Fundamentals of Statistical and Thermal Physics* (McGraw-Hill, New York, 1965), and references therein.

38. See, for example, S. R. DeGroot and P. Mazur, *Nonequilibrium Thermodynamics* (Dover, New York, 1984); or J. Keizer, *Statistical Thermodynamics of Nonequilibrium Processes* (Springer-Verlag, New York, 1987).

39. See, for example, K. S. Pitzer, *Thermodynamics*, 3rd ed. (McGraw-Hill, New York, 1995).

40. J. W. Gibbs, *The Collected Works of J. Willard Gibbs* (Yale University, New Haven, 1948). See especially, pp. 55–57 and 353–356.

41. See, for example, the comments after eq. (15.18.7) on p. 595 of ref. 37.

42. See the first of eqs. (214) of ref. 2.

43. P. W. Bridgman, *Rev. Modern Physics* **22**, 56 (1950).

44. Mori's eq. (3.10) of ref. 5 and Eq. (A.54) in the present chapter are equivalent for the microscopically reversible A's assumed by Onsager.

45. See, for example, D. A. McQuarrie, *Statistical Mechanics* (Harper & Row, New York, 1976), chaps. 21 and 22 and references therein.

CHEMICAL REACTIONS AND REACTION EFFICIENCY IN COMPARTMENTALIZED SYSTEMS

JOHN J. KOZAK

Department of Chemistry
Iowa State University, Ames, IA 50011-3111

I. INTRODUCTION

In the introductory chapter of his treatise, "On Growth and Form," D'Arcy Thompson [1] begins with the statement, "Of the chemistry of his day and generation, Kant declared that it was a science, but not Science — eine Wissenschaft, aber nicht Wissenschaft — for that the criterion of true science lay in its relationship to mathematics." D'Arcy Thompson's primary purpose stated in the epilogue was, "to show that a certain mathematical aspect of morphology, to which as yet the morphologist gives little heed, is interwoven with his problems, complementary to his descriptive task, and helpful, nay essential, to his proper study and comprehension of Growth and Form."

The application of mathematics to the study of form is a problem of considerable interest in the physical and biological sciences and engineering. The title of the important 1977 monograph by Nicolis and Prigogine [2], "Self-Organization in Nonequilibrium Systems. From Dissipative Structures to Order through Fluctuations," already suggests the scope of this endeavor. As these and many other authors have shown [2,3], on studying chemical reactions the rate is generally a nonlinear function of the variables involved, with the consequence that a chemically reacting system is described by nonlinear differential equations having more than one solution. From the study of the stability of solutions to such equations for model systems, one finds that general principles emerge that govern the self-organized formation of structures (or functions) in both the animate and inanimate world. Indeed, as demonstrated by Haken [3] in his writings on "Synergetics," quite

Advances in Chemical Physics, Volume 115, edited by I. Prigogine and Stuart A. Rice.
ISBN 0-471-39331-2 © 2000 John Wiley & Sons, Inc.

diverse systems may evolve through a hierarchy of instabilities among solutions to such equations to yield more and more structured patterns.

In this review, we address the reciprocal question and explore the influence of system morphology on the efficiency of diffusion-controlled chemical reactions taking place within or on a compartmentalized system, that is, a system of finite spatial extent. One anticipates that geometrical factors such as the dimensionality of the system, its size and shape, and the number of distinct reaction pathways all contribute to influencing the reaction efficiency, but what this review shall be particularly concerned with, and document, is the interplay among these factors. In studying diffusion-reaction processes in compartmentalized systems, sometimes geometrical constraints are induced by the structure of the host medium (as in the case of clays or zeolites), and sometimes by inhomogenities in the microheterogeneous environment itself (the surface of a catalyst particle or a molecular assembly). For all but the simplest geometries, solution of a continuum (Fickian) equation of the form

$$\frac{\partial C(\vec{r}, t)}{\partial t} = D\nabla^2 C(\vec{r}, t) + f[C(\vec{r}, t)] \tag{4.1}$$

where $C(\vec{r}, t)$ is the concentration, D is the diffusion coefficient, ∇^2 is the Laplacian operator, and $f[C(\vec{r}, t)]$ is a (generally nonlinear) function of the concentration(s), can present formidable mathematical difficulties.

Accordingly, emphasis is placed here on the mobilization of an alternative approach, one based on the theory of finite Markov processes. In particular, the geometry of the reaction space is constructed explicitly using a lattice representation, and the survival probability of a reactant (random walker) diffusing on or within the defined, d-dimensional space before undergoing reaction at one or more reaction centers (traps) is determined by solving numerically the stochastic master equation for the problem. Although there is extensive literature dealing with random-walk models of diffusion-reaction processes in systems of infinite extent or for large lattices [4], considered here are processes taking place in domains of finite spatial extent which, in certain cases, may have an internal structure and/or boundaries of arbitrary shape.

In implementing the above approach, there are no limitations on the geometry of the compartmentalized system or on the nature of the boundary conditions imposed. The price paid for this generality is that the problem must be resolved for every choice of system geometry, and for each temporal boundary condition. One is then left with the task of extracting from the numerical data, in the spirit of an experimentalist, trends and correlations, rather than having at one's disposal an analytic solution from which the behavior of the system for various parameter choices can be extracted at a glance.

A further rationale for the approach developed here is that some of the most interesting diffusion-reaction problems in structured media involve processes taking place in or on spaces (sets) of fractal dimension. Strictly speaking, the Laplacian in Eq. (4.1) is defined rigorously only for spaces of integer dimension, thereby limiting the usefulness of an approach based on a Fickian diffusion equation for such systems. Even for media that are built up from planar lattice structures of integer dimension, for example, layered diffusion spaces, the dimensionality of the system does not change discontinuously with the addition of the first few overlayers; hence, the formulation of a diffusion-reaction problem using a Laplacian defined for spaces of integer dimension is necessarily approximate. Both of these situations can be dealt with quite straightforwardly using a lattice-based model, again subject to the caveat noted above.

A further spatial consideration arises if, in addition to short-range quantum chemical forces, reactant pairs interact via long-range multipolar potentials. It is expected that the strength and range of the governing potential function will play a critical role in influencing the efficiency of reaction, but this review will demonstrate that interesting new effects can arise when the size of the compartmentalized system and the range of the potential are of comparable magnitude. For systems of quite different topologies, there emerges a universality in the response of reactant pairs to changes in the system geometry, temperature and dielectric constant of the host medium.

The program followed in this review is now summarized. In Section II the theory that enables one to calculate the mean number $\langle n \rangle$ of displacements (or "hops") of the diffusing reactant before it terminates its walk in the presence of one or more partially or completely absorbing traps is outlined. The mean walklength $\langle n \rangle$ plays a central role in applications of the theory because it is related to the reciprocal of the smallest eigenvalue of the time-dependent solution to the stochastic master equation for the problem; its reciprocal is therefore a measure of the effective first-order rate constant and a signature of the long-time behavior of the system. Because, with two exceptions, the theory underlying Markovian models and the random walk problem has been discussed in detail elsewhere (e.g., refs. 4, 5, 6), proofs of theorems are omitted here. The two exceptions relate to formal techniques that can be used to determine numerically exact values of $\langle n \rangle$ (as well as higher moments of the probability distribution function) on time scales orders-of-magnitude faster than a full-scale Monte Carlo simulation. The reader who wants to get to the heart of the matter discussed in this review may, however, wish to defer reading about these methods until later.

Section III focuses on problems of system topology and documents, using results obtained from a series of model calculations, how the separate influences of system size, dimensionality, and reaction pathway(s) can be disentangled, and the principal effects on reaction efficiency quantified. With these factors clarified, Section IV demonstrates how these trends and correlations change when a multipolar potential is operative between reaction partners, both confined to a compartmentalized system. More general diffusion-reaction systems are described in Section V, where effects arising from nonrandom distributions of reaction centers and, secondly, the influence of multipolar potentials in influencing catalytic processes in crystalline and semiamorphous zeolites are explored. The conclusions drawn from these studies are then summarized in Section VI.

II. THEORETICAL BACKGROUND

Consider the encounter-controlled reactive process described by the equation

$$A + B \rightarrow C \tag{4.2}$$

where A is assumed to be a diffusing coreactant and B a stationary target molecule (reaction center or "trap"). In general, both short-range quantum-chemical and longer-range multipolar correlations will conspire to influence the efficiency of such a diffusion-reaction process, as will the topology of the compartmentalized system within or on which the above reaction takes place.

Suppose the geometry of a compartmentalized system is described by a lattice of integral or fractal dimension of given size and shape, and characterized by N discrete lattice points (sites) embedded in a Euclidean space of dimension $d = d_e$ and local connectivity or valency v. At time $t = 0$, assume that the diffusing coreactant A is positioned at a certain site j with unit probability. For $t > 0$ the probability distribution function $\rho(t)$ governing the fate of the diffusing particle is determined by the stochastic master equation

$$\frac{d\rho_i(t)}{dt} = -\sum_{j=1}^{N} G_{ij} \rho_j(t) \tag{4.3}$$

Specifically, $\rho_i(t)$ is the probability of the particle being at the site i at time t, with $\rho_i(t = 0) = \delta_{im}$. G_{ij} is the transition rate of the probability to the site i from a neighboring site j. The overall $\mathbf{G}$ matrix is linked to the $N \times N$ Markov transition probability matrix $\mathbf{P}$ with elements p_{ij} via the relation [6,7]

$$G_{ij} = [\delta_{ij} - p_{ij}]v_j \tag{4.4}$$

where p_{ij} is the probability that the diffusing particle, conditional on being at site i at any time t, will be at site j in its next displacement until terminating its walk (eventually) at one or more partially or completely absorbing traps. In general, the p_{ij} reflect all of the constraints (e.g., potential or concentration gradients, multipolar potentials) influencing the diffusional motion of the coreactant A. In the simplest case, a particle undergoing an unbiased, nearest-neighbor random walk,

$$p_{ii} = 0 \qquad (4.5a)$$

and

$$p_{ij} = 1/v_j \qquad i \neq j \qquad (4.5b)$$

where v_j is the valency of site j.

The master equation (4.3) can be solved numerically in various ways [6,7]. A symmetric $\mathbf{G}$ matrix guarantees that the solutions are of the form

$$\rho_i(t) = \sum_k a_{ik} \exp(-\lambda_k t) \qquad (4.6)$$

where the λ_k are the eigenvalues of the matrix $\mathbf{G}$ and the a_{ik} can be obtained from the matrix of the eigenvectors and the initial boundary condition. In the limit of large N, it can be shown that the first moment of the probability distribution function, which is just the mean walklength $\langle n \rangle$ of the Markovian theory, is related to the smallest eigenvalue λ_1 of the stochastic master equation via the relation

$$\langle n \rangle = v\lambda_1^{-1} \qquad (4.7)$$

In Section III the temporal behavior of diffusion-reaction processes occurring in or on compartmentalized systems of various geometries, as determined via solution of the stochastic master equation (4.3), is studied. Also, in Sections III–V, results are presented for the mean walklength $\langle n \rangle$. From the relation (4.7), and the structure of the solutions (4.6) to Eq. (4.3), the reciprocal of $\langle n \rangle$ may be understood as an effective first-order rate constant k for the process (4.2) or $\langle n \rangle$ itself as a measure of the characteristic relaxation time of the system; it is, in effect, a signature of the long-time behavior of the system.

Two strategies that can be used to simplify the calculation of the mean walklength $\langle n \rangle$ are now reviewed. In the first of these, it is shown that if the random walk is modeled by a stationary Markov process on a finite state space, the number of points in state space can often be substantially reduced by taking account of lattice symmetries [4,8–10]. A point in state space will then correspond to an entire symmetry class of nodes, or sites on the lattice.

To proceed, note that a random walk on a lattice with no traps continues forever. This property is expressed mathematically by the equation

$$\mathbf{P}\iota = \iota \qquad (4.8)$$

where ι is a vector (here, $N \times 1$) each element of which is 1. This equation says that the probability is 1 that, starting from any lattice site, the next step leads somewhere on the lattice; in particular, the step may not lead into a trap in which the walker would disappear. Equation (4.8) implies that ι is an eigenvector of $\mathbf{P}$ with eigenvalue 1, from which it follows that ι is an eigenvector of $[\mathbf{I} - \mathbf{P}]$ with eigenvalue 0, so that $[\mathbf{I} - \mathbf{P}]$ is a singular matrix.

The sites corresponding to state i on a lattice are traps if the sum of the elements $p_{ij}, j = 1, 2, \ldots, N$ of row i of $\mathbf{P}$ is less than 1. There is then a nonzero probability that the walk will end if it reaches state i. The most usual case is that of a deep trap; all the elements p_{ij} are zero, so that the walk ends with certainty whenever it reaches state i. Provided that there is a positive probability of reaching such a trap from any starting point on the lattice, any random walk will end in finite time with probability 1. In that case, one may inquire as to the average number $\langle n \rangle_{kj}$ of visits to the nontrapping site j in the course of the random walk conditional on starting from the specific site k, the average number $\langle n \rangle_k$ of visits to the $j = 1, \ldots, N - 1$ nontrapping sites of the lattice conditional on starting at site k, and the average overall length $\langle n \rangle$ of the random walk before trapping (reaction).

The answers to these questions are furnished by the matrix $[\mathbf{I} - \mathbf{P}]^{-1}$, which is denoted here as $\mathbf{\Pi}$. $\mathbf{\Pi}$ exists if and only if the expected walklength from each possible starting point is finite. The (k, j)th element of this matrix is just the quantity $\langle n \rangle_{kj}$ referred to in the previous paragraph. Although there are some interesting questions not answered by knowledge of the matrices $\mathbf{P}$ and $\mathbf{\Pi}$ alone, such as the distributions of the numbers of visits to various sites, rather than just the expected values of these numbers, attention in this review is limited to the (numerous) properties of random walks that can be derived from knowledge of these two matrices.

To illustrate the advantages gained in considering lattice symmetries, consider a target molecule B (or trap) positioned at an arbitrary site on a finite, 5×5 square-planar lattice. Calculation of the mean walklength $\langle n \rangle$ before reaction (trapping) of a coreactant A diffusing on this lattice, and subject to specific boundary conditions, requires the specification of the matrix $\mathbf{P}$ and subsequent inversion of the matrix $[\mathbf{I} - \mathbf{P}]$. If the trap is anchored at the centrosymmetric site on the lattice and periodic boundary

5	4	3	4	5
4	2	1	2	4
3	1	T	1	3
4	2	1	2	4
5	4	3	4	5

Figure 4.1. 5×5 lattice with a centrosymmetric trap T. The integers classify the symmetry-distinct sites of the lattice.

conditions are imposed, the 25 sites (or nodes of the dual lattice) group themselves into six different symmetry classes, as shown in Figure 4.1. Given a conditional probability

$$\Pr[q/r]$$

where r and q are statements which are independent and neither a self-contradiction, the n-step transition probabilities for a Markov process, denoted by $p_{ij}(n)$, are

$$p_{ij}(n) = \Pr[f_n = s_j / f_{n-1} = s_i]. \tag{4.9}$$

Here, f_n is the "outcome function," the value of which is s_j if the outcome of the nth step is s_j. A finite Markov process is a finite Markov chain if the transition probabilities do not depend on n. Then, the transition matrix for a Markov chain is the matrix $\mathbf{P}$ with entries p_{ij}.

Consider now all possible transitions between sites s_i of the lattice diagrammed in Figure 4.1, and specified by the attendant probabilities p_{ij} of the process $s_i \rightarrow s_j$:

$$\mathbf{P} = \begin{bmatrix} s_T \rightarrow s_T & s_T \rightarrow s_1 & s_T \rightarrow s_2 & s_T \rightarrow s_3 & s_T \rightarrow s_4 & s_T \rightarrow s_5 \\ s_1 \rightarrow s_T & s_1 \rightarrow s_1 & s_1 \rightarrow s_2 & s_1 \rightarrow s_3 & s_1 \rightarrow s_4 & s_1 \rightarrow s_5 \\ s_2 \rightarrow s_T & s_2 \rightarrow s_1 & s_2 \rightarrow s_2 & s_2 \rightarrow s_3 & s_2 \rightarrow s_4 & s_2 \rightarrow s_5 \\ s_3 \rightarrow s_T & s_3 \rightarrow s_1 & s_3 \rightarrow s_2 & s_3 \rightarrow s_3 & s_3 \rightarrow s_4 & s_3 \rightarrow s_5 \\ s_4 \rightarrow s_T & s_4 \rightarrow s_1 & s_4 \rightarrow s_2 & s_4 \rightarrow s_3 & s_4 \rightarrow s_4 & s_4 \rightarrow s_5 \\ s_5 \rightarrow s_T & s_5 \rightarrow s_1 & s_5 \rightarrow s_2 & s_5 \rightarrow s_3 & s_5 \rightarrow s_4 & s_5 \rightarrow s_5 \end{bmatrix}$$

$$\tag{4.10}$$

For the case where no sites on the lattice are deep traps, the transition matrix **P** reads

$$
\mathbf{P} = \begin{bmatrix}
0 & 1 & 0 & 0 & 0 & 0 \\
1/4 & 0 & 1/2 & 1/4 & 0 & 0 \\
0 & 1/2 & 0 & 0 & 1/2 & 0 \\
0 & 1/4 & 0 & 1/4 & 1/2 & 0 \\
0 & 1/4 & 1/4 & 1/4 & 1/4 & 0 \\
0 & 0 & 0 & 0 & 1/2 & 1/2
\end{bmatrix} \tag{4.11}
$$

whereas if we arbitrarily assign the centrosymmetric site T to be a deep trap the matrix **P** is

$$
\mathbf{P} = \begin{bmatrix}
1 & 0 & 0 & 0 & 0 & 0 \\
1/4 & 0 & 1/2 & 1/4 & 0 & 0 \\
0 & 1/2 & 0 & 0 & 1/2 & 0 \\
0 & 1/4 & 0 & 1/4 & 1/2 & 0 \\
0 & 1/4 & 1/4 & 1/4 & 1/4 & 0 \\
0 & 0 & 0 & 0 & 1/2 & 1/2
\end{bmatrix} \tag{4.12}
$$

The accessible states are classified by distinguishing between two types of sets, transient sets and ergodic sets. Ergodic sets are sets that, once entered, are never left again. Thus, if the centrosymmetric site T of the lattice diagrammed in Figure 4.1 is a deep trap, the site T comprises a (one-state) ergodic set, and the sites $\{1,\ldots,5\}$ comprise the transient set; taken together, the set $\{T, 1,\ldots,5\}$ defines an absorbing Markov chain.

The information in **P** can be reorganized in terms of ergodic sets and transient sets. The ergodic sets are united as one block division of **P** with the transient sets comprising the remaining blocks. Supposing there are s transient states and $r - s$ ergodic states, **P** may then be written as

$$
\mathbf{P} = \begin{bmatrix} \mathbf{L} & \mathbf{0} \\ \mathbf{R} & \mathbf{Q} \end{bmatrix} \begin{matrix} r-s \\ s \end{matrix} \tag{4.13}
$$

The $(r - s) \times s$ submatrix **0** consists entirely of zeros. The $s \times s$ submatrix **Q** codes the fate of the random walker as long as it remains in the transient states. The $s \times (r - s)$ submatrix **R** describes transitions that carry the random walker from transient states to ergodic states. Finally, the $(r - s) \times (r - s)$

submatrix $\mathbf{L}$ codes the fate of the random walker after it enters an ergodic set. In the deep-trap case illustrated by the transition matrix $\mathbf{P}$ in Eq. (4.12),

$$\mathbf{L} = (1) \qquad \mathbf{R} = \begin{bmatrix} 1/4 \\ 0 \\ 0 \\ 0 \\ 0 \end{bmatrix} \qquad \mathbf{0} = [0 \quad 0 \quad 0 \quad 0 \quad 0]$$

and

$$\mathbf{Q} = \begin{bmatrix} 0 & 1/2 & 1/4 & 0 & 0 \\ 1/2 & 0 & 0 & 1/2 & 0 \\ 1/4 & 0 & 1/4 & 1/2 & 0 \\ 1/4 & 1/4 & 1/4 & 1/4 & 0 \\ 0 & 0 & 0 & 1/2 & 1/2 \end{bmatrix} \qquad (4.14)$$

For finite Markov processes, it can be proved that for any finite Markov chain, no matter where the walker starts, the probability that the walker is in an ergodic state after n steps tends to unity as $n \rightarrow \infty$. Thus, powers of $\mathbf{Q}$ in the above aggregated version of $\mathbf{P}$ tend to $\mathbf{0}$ and consequently for any absorbing Markov chain, the matrix $\mathbf{I} - \mathbf{Q}$ has an inverse $\mathbf{N}$, called the fundamental matrix. In the problem defined by Eq. (4.12) the matrix $\mathbf{N}$ is

$$(\mathbf{I} - \mathbf{Q})^{-1} = \mathbf{N} = \begin{bmatrix} 4 & 4 & 4 & 8 & 4 \\ 4 & 28/5 & 24/5 & 52/5 & 26/5 \\ 4 & 24/5 & 32/5 & 56/5 & 28/5 \\ 4 & 26/5 & 28/5 & 64/5 & 32/5 \\ 4 & 26/5 & 28/5 & 64/5 & 42/5 \end{bmatrix} \qquad (4.15)$$

The interpretation of the elements and row sums of the matrix $\mathbf{N}$ follows directly from the definitions given earlier. For example, if a walker starts in state s_3, it will visit state s_3 an average of 6.4 times before being trapped eventually at site T. Further, the total number of times a diffusing particle, starting in state s_3 will be in state s_1, state $s_2, \ldots$, state s_5 before being trapped irreversibly in the ergodic state s_T will be given by the row sum:

$$4 + 4.8 + 6.4 + 11.2 + 5.6 = 32.$$

In terms of the notation introduced above, $\langle n \rangle_{3,3} = 6.4$, $\langle n \rangle_3 = 32$, and the overall mean walklength of the random walker before being trapped is given by

$$\langle n \rangle = \tfrac{1}{24}\{4\langle n \rangle_1 + 4\langle n \rangle_2 + 4\langle n \rangle_3 + 8\langle n \rangle_4 + 4\langle n \rangle_5\} = 31\,2/3.$$

The irreversible process described by Eq. (4.2) corresponds to the situation where the diffusing particle is trapped with unit probability upon reaching site T for the first time. One can also consider processes where the diffusing coreactant A, upon reaching the target molecule B (situated, say, at site T), forms an excited state complex $[AB]^*$, which may either proceed to the product C, or fall apart with the coreactant A resuming its motion in diffusion space. Provided there is a nonzero probability that the reaction proceeds eventually to the product C,

$$A + B \rightleftarrows [AB]^* \to C \qquad\qquad (4.16)$$

the reaction efficiency of this process can also be studied using the theory of finite Markov processes. Further, the case of competing reaction centers can be considered by positioning coreactants at any (or all) of the $N - 1$ satellite sites of the host lattice, and specifying that the diffusing coreactant reacts with probability $p_i \leq 1$ (with some or all of the p_i different) at site(s) i. In each of these situations, the classification of sites in diffusion space according to their spatial relationship to the reaction center (as was illustrated in Fig. 4.1) leads to a significant simplification in the calculation of the random walklength $\langle n \rangle$ and higher moments of the probability distribution function $\rho(t)$. As will be shown in later examples, the advantage of this contraction in state space for large, complex geometries is that, rather than studying the processes described by Eqs. (4.2) and (4.16) using full-scale Monte Carlo simulations, which are very computer intensive, or inverting large $N \times N$ matrices to obtain the matrix Π (or $\mathbf{N}$), numerical results can be obtained very efficiently (and economically) for a wide class of diffusion-reaction problems.

A second strategy to simplify the calculation of the mean walklength (and higher moments of the distribution function ρ) can be implemented if one is interested only in how Π changes in response to certain changes in $\mathbf{P}$ [11,12]. As noted above, the Π that corresponds to any given $\mathbf{P}$ can always be calculated by the inversion of a $N \times N$ matrix. But, again, if N is large, this may be a costly operation if it has to be performed many times. It is now shown that, if changes to $\mathbf{P}$ can be expressed as a matrix of rank $v < N$, then the corresponding changes to Π can be computed by inverting a $v \times v$ matrix instead of an $N \times N$ one. Consider a lattice with a random walk characterized by $N \times N$ matrices $\mathbf{P}$ and Π. Let $\mathbf{P}$ be changed to

$$\mathbf{P}' \equiv \mathbf{P} + \Delta\mathbf{P}$$

Let

$$\mathbf{\Pi}' = (\mathbf{I} - \mathbf{P}')^{-1}$$

and

$$\Delta \mathbf{\Pi} = \mathbf{\Pi}' - \mathbf{\Pi}$$

Then,

$$
\begin{aligned}
\Delta \mathbf{\Pi} &= (\mathbf{I} - \mathbf{P} - \Delta \mathbf{P})^{-1} - (\mathbf{I} - \mathbf{P})^{-1} \\
&= (\mathbf{I} - \mathbf{P})^{-1}[\mathbf{I} - \mathbf{P} - (\mathbf{I} - \mathbf{P} - \Delta \mathbf{P})](\mathbf{I} - \mathbf{P} - \Delta \mathbf{P})^{-1} \\
&= (\mathbf{I} - \mathbf{P})^{-1}\Delta \mathbf{P}(\mathbf{I} - \mathbf{P} - \Delta \mathbf{P})^{-1}
\end{aligned}
$$

Now,

$$\mathbf{I} - \mathbf{P} - \Delta \mathbf{P} = [\mathbf{I} - \Delta \mathbf{P}(\mathbf{I} - \mathbf{P})^{-1}](\mathbf{I} - \mathbf{P})$$

from which it follows that

$$
\begin{aligned}
\Delta \mathbf{\Pi} &= (\mathbf{I} - \mathbf{P})^{-1}\Delta \mathbf{P}(\mathbf{I} - \mathbf{P})^{-1}[\mathbf{I} - \Delta \mathbf{P}(\mathbf{I} - \mathbf{P})^{-1}]^{-1} \\
&= \mathbf{\Pi}\Delta \mathbf{P}\mathbf{\Pi}(\mathbf{I} - \Delta \mathbf{P}\mathbf{\Pi})^{-1}
\end{aligned}
\tag{4.17}
$$

If $\Delta \mathbf{P}$ is of rank v, say, then it is possible to find two matrices $\mathbf{A}$ and $\mathbf{B}$, of dimension $N \times v$, such that

$$\Delta \mathbf{P}\mathbf{\Pi} = \mathbf{A}\mathbf{B}^{T} \tag{4.18}$$

In the applications considerd in this review, there are simple natural choices for $\mathbf{A}$ and $\mathbf{B}$. From Eq. (4.17),

$$\Delta \mathbf{\Pi} = \mathbf{\Pi} \, \mathbf{A}\mathbf{B}^{T} \, (\mathbf{I} - \mathbf{A}\mathbf{B}^{T})^{-1} \tag{4.19}$$

Now,

$$(\mathbf{I}_{v} - \mathbf{B}^{T}\mathbf{A})\mathbf{B}^{T} = \mathbf{B}^{T} - \mathbf{B}^{T}\mathbf{A}\mathbf{B}^{T} = \mathbf{B}^{T}(\mathbf{I}_{N} - \mathbf{A}\mathbf{B}^{T})$$

whence

$$\mathbf{B}^{T}(\mathbf{I}_{N} - \mathbf{A}\mathbf{B}^{T})^{-1} = (\mathbf{I}_{v} - \mathbf{B}^{T}\mathbf{A})^{-1}\mathbf{B}^{T}$$

Thus, from Eq. (4.19) we obtain,

$$\Delta\Pi = \Pi A (I_v - B^T A)^{-1} B^T \qquad (4.20)$$

It is clear that computation of the right-hand side of Eq. (4.20) requires the inversion of a $v \times v$ matrix rather than of an $N \times N$ one, resulting in a concomitant decrease in the time required to obtain numerical results in specific problems.

This completes the survey of the methods that will be used in exploring the problems addressed in this review. As noted previously, no attempt was made to provide formal proofs of the theorems that are the basis of the theory of finite Markov processes, or the methods that have been developed to obtain numerical solutions to equations of the form, Eq. (4.3); the reader is directed to refs. 4–7 for a more thorough discussion of these matters. Taken together, these theorems and the "shorthand" theoretical methods described above (which result in a contracted description of the Markovian problem) allow the calculation of numerically exact values of the moments of the distribution function describing the underlying diffusion-reaction system. For example, in the simple problem described earlier, the matrix Q, whose elements are ratios of integers, results in the matrix N, whose elements are ratios of integers, from which the mean walklength $\langle n \rangle$ could be computed exactly, viz. $\langle n \rangle = 31\ 2/3$. Thus, the results reported in this review for specific problems, and the more extensive body of results in the literature generated using these methods, are results to which full-scale Monte Carlo results must converge and against which the quality of results obtained from analytic theories can be compared.

III. GEOMETRICAL EFFECTS

A. Regular Lattices

In this section are displayed graphically the numerically exact results that have been obtained for unbiased, nearest-neighbor random walks on finite $d = 2, 3$ dimensional regular, Euclidean lattices, each of uniform valency v, subject to periodic boundary conditions, and with a single deep trap. These data allow a quantitative assessment of the relative importance of changes in system size N, lattice dimensionality d, and/or valency v on the efficiency of diffusion-reaction processes on lattices of integral dimension, and provide a basis for understanding processes on lattices of fractal dimension or fractional valency.

In Figure 4.2 are the results calculated for $\langle n \rangle$ for finite hexagonal ($v = 3$) [10], square planar ($v = 4$) [9], and triangular ($v = 6$) [13] lattices in $d = 2$,

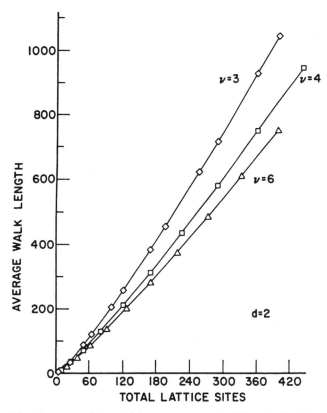

Figure 4.2. Average walklength $\langle n \rangle$ versus system size N for $d = 2$ and $v = 3, 4, 6$.

all subject to periodic boundary conditions. The implementation of periodic boundary conditions requires the specification of a unit cell. The N for a unit cell in a given dimension d may be different for different valencies; indicated in the figure (by diamonds, squares, and triangles) are the specific unit cells for which calculations were performed. Smooth curves through these points allow a comparison of the change in $\langle n \rangle$ with respect to changes in v for a given (N,d), and it is on the basis of these (sometimes) interpolated results that the conclusions presented later will be quantified. Similarly, in Figure 4.3 are recorded the results for finite, $d = 3$ tetrahedral $(v = 4)$ [14], and simple cubic $(v = 6)$ [9] lattices, these once again subject to periodic boundary conditions. It is also convenient to organize the data in Figures 4.2 and 4.3 so that one can study the consequences of changes in the profile of $\langle n \rangle$ versus N when one fixes the valency and changes the dimensionality and these results are presented in Figures 4.4 and 4.5.

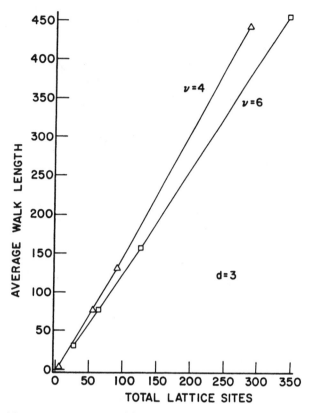

Figure 4.3. Average walklength $\langle n \rangle$ versus system size N for $d = 3$ and $v = 4, 6$.

Note that for finite d-dimensional lattices of valency v with a centrosymmetric trap, if a random walker upon reaching a boundary site is "reset" at that site (as one of its v possible "next steps"), the value of $\langle n \rangle$ calculated under this convention, the case of "confining" or "passive" boundary conditions, is exactly the same as for the case of periodic boundary conditions. In particular problems, other spatial boundary conditions may be more appropriate (see later text); a comparison of results obtained using periodic, confining and reflecting boundary conditions is presented in refs. 9,10,13, and 14.

To get a preliminary handle on the kind of differences that arise when changes in dimensionality and/or valency are made, consider a fivefold expansion in the total number N of sites for a lattice of given (d,v), that is, consider values of the average walklength $\langle n \rangle$ at $N = 50$ and $N = 250$. Suppose one fixes the dimensionality of the lattice and considers the change

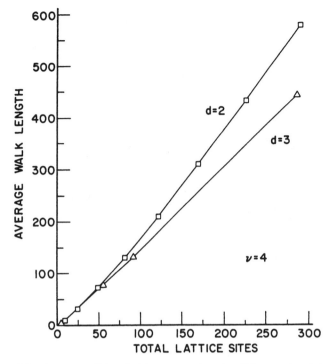

Figure 4.4. Average walklength $\langle n \rangle$ versus system size N for $v = 4$ and $d = 2, 3$.

in $\langle n \rangle$ at $N = 50$ and the change in $\langle n \rangle$ at $N = 250$ arising from a unit change in the valency of the lattice ($\Delta v = 1$); let these quantities be denoted as $\Delta\langle n \rangle(N = 50)$ and $\Delta\langle n \rangle(N = 250)$, and construct the factor

$$F = \Delta\langle n \rangle(N = 250)/\Delta\langle n \rangle(N = 50)$$

to calibrate the change $\Delta v = +1$ for given d. Similar factors can be constructed for other changes ($\Delta v, \Delta d$), and the composite results are presented in Table III.1. From these data, the following conclusions can be drawn:

1. The effect of changing the valency (Δv) in a lattice of given dimensionality becomes less significant the higher the dimensionality of the system ($F = 7.42$ vs $F = 4.92$).
2. The effect of changing the dimensionality (Δd) in a lattice of given valency v becomes less significant the higher the valency of the system ($F = 22.9$ vs $F = 9.02$).

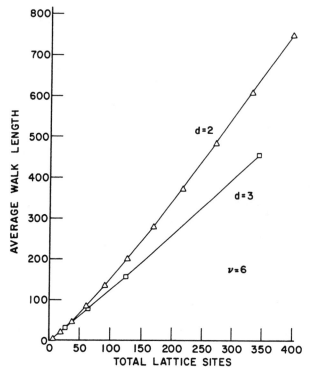

Figure 4.5. Average walklength $\langle n \rangle$ versus system size N for $v = 6$ and $d = 2, 3$.

3. For a lattice of given dimensionality, changes in the valency $\Delta v = +1$ or $\Delta v = +2$ produce (essentially) the same change in the factor $F (F = 7.50$ vs $F = 7.42)$.

4. Changes in $\langle n \rangle$ arising from changes in the dimensionality of the system are of greater consequence than changes in $\langle n \rangle$ arising from

TABLE III.1
Changes in the factor F corresponding to a five-fold increase
in the total number of lattice sites N (N: $50 \rightarrow 250$) for selected
changes in the lattice valency (Δv) or dimensionality (Δd)

Variation	$F = \Delta n(N = 250)/\Delta n(N = 50)$
$d = 2; \Delta v = +1$	7.50
$\Delta v = +2$	7.42
$d = 3; \Delta v = +2$	4.92
$v = 4; \Delta d = +1$	22.9
$v = 6; \Delta d = +1$	9.02

changes in the valency of the lattice (the factors $F = 22.9$, 9.02 vs the factors $F = 7.50$, 7.42, 4.92).

One can go beyond these preliminary results and examine two special cases. Consider first the valency effect and the data presented in Table III.2 for the case $d = 2$, a common system size $N = 169$, and a common metric

TABLE III.2

Site specific walklengths $\langle n \rangle_i$ for $N = 169$ lattices of a common dimensionality $(d = 2)$ but different valencies (see text)

$\langle n \rangle_i$		$v = 3$			$v = 4$			$v = 6$	
		ℓ/σ_i	m/p		ℓ/σ_i	m/p		ℓ/σ_i	m/p
$\langle n \rangle$	381.8			310.6			280.2		
$\langle n \rangle_1$	168.0	1/3	3/1.73	168.0	1/4	4/2.38	168.0	1/6	6/3.57
$\langle n \rangle_2$	250.5	2/6	9/5.36	213.2	2/4		218.1	2/6	
$\langle n \rangle_3$	276.0	3/3		241.6	2/4	12/7.14	229.9	2/6	18/10.71
$\langle n \rangle_4$	324.1	4/3		256.4	3/8		253.2	3/6	
$\langle n \rangle_5$	321.0	4/3		279.1	4/4		253.2	3/6	
$\langle n \rangle_6$	304.5	3/6	18/10.71	281.6	3/4	24/14.23	262.9	3/6	36/21.43
$\langle n \rangle_7$	338.9	4/6	30/11.86	287.6	4/8		273.3	4/6	
$\langle n \rangle_8$	346.7	5/6		299.8	5/8		276.0	4/6	
$\langle n \rangle_9$	374.1	6/6		313.0	6/4		276.1	4/6	
$\langle n \rangle_{10}$	385.1	7/3		305.7	4/4	40/23.81	282.8	4/6	60/35.71
$\langle n \rangle_{11}$	404.1	8/3		308.6	5/8		287.6	5/6	
$\langle n \rangle_{12}$	351.0	5/3		315.7	6/8		287.7	5/6	
$\langle n \rangle_{13}$	364.5	6/6		324.1	7/8		290.4	5/6	
$\langle n \rangle_{14}$	362.5	5/6	45/26.79	331.8	8/4		290.5	5/6	
$\langle n \rangle_{15}$	381.1	6/6	63/37.50	319.8	5/4	60/35.71	294.8	5/6	90/53.57
$\langle n \rangle_{16}$	387.4	7/6		321.6	6/8		295.7	6/6	
$\langle n \rangle_{17}$	404.1	8/6		326.1	7/8		296.5	6/6	
$\langle n \rangle_{18}$	412.1	9/6		331.9	8/8		296.7	6/6	
$\langle n \rangle_{19}$	425.1	10/6		337.5	9/8		298.8	6/6	
$\langle n \rangle_{20}$	431.2	11/3		341.9	10/4		299.0	6/6	
$\langle n \rangle_{21}$	440.5	12/3		326.5	6/4	84/50.00	301.3	6/6	126/75.00
$\langle n \rangle_{22}$	378.6	8/3		327.8	7/8	108/64.29	299.9	7/6	
$\langle n \rangle_{23}$	377.1	7/6		331.3	8/8	128/76.13	300.1	7/6	
$\langle n \rangle_{24}$	385.2	8/6		335.9	9/8	144/85.71	300.9	7/6	
$\langle n \rangle_{25}$	390.3	7/6	84/50.00	340.6	10/8	156/92.86	301.4	7/6	
$\langle n \rangle_{26}$	401.6	8/6	108/64.29	344.2	11/8	164/97.62	302.6	7/6	
$\langle n \rangle_{27}$	409.8	9/6	120/71.43	346.2	12/4	168/100.0	303.0	7/6	
$\langle n \rangle_{28}$	420.9	10/6	132/78.57				304.0	7/6	168/100.0
$\langle n \rangle_{29}$	428.9	11/6	141/83.33						
$\langle n \rangle_{30}$	437.8	12/6	150/89.29						
$\langle n \rangle_{31}$	443.6	13/6	156/92.86						
$\langle n \rangle_{32}$	449.6	14/6	162/96.43						
$\langle n \rangle_{33}$	452.6	15/6	165/98.21						
$\langle n \rangle_{34}$	455.6	16/3	168/100.0						

(or distance) between adjacent lattice points. For this fixed system size and metric, results for the hexagonal, square planar, and triangular lattices can be compared directly. Listed in Table III.2 are the site specific walklengths $\langle n \rangle_i$ and the overall walklength $\langle n \rangle$, where the site specifications are given explicitly in fig. 5 of ref. 10 for $v = 3$, in fig. 3 of ref. 9 for $v = 4$ and in fig. 1 of ref. 13 for $v = 6$. The second column under each valency heading is a couple (ℓ/σ_i), where the first integer ℓ gives the minimum number of lattice displacements required to pass from site i to the centrosymmetric trap, and the second integer σ_i denotes the number of such sites in the overall symmetry classification. In the third column under each valency classification is listed a couple (m, p), where m gives the total number of sites distributed at or within the ℓth nearest neighbor about the centrosymmetric target, and p is the percentage of the total number of nontrapping sites at or within that nearest-neighbor distance.

To see at once the principal result, note that in order to encompass (at least) 50% of the $N = 168$ surrounding, nontrapping sites, one must be at an effective nearest-neighbor distance $\ell = 7$ for $v = 3$, at $\ell = 6$ for $v = 4$, but only at $\ell = 5$ for $v = 6$. As this observation and the attendant $\langle n \rangle_i$ show, for lattices of fixed N and a common metric, increasing the valency of the lattice allows the clustering of more sites closer to the central trap and leads to a decrease in the average number of steps required for trapping. Trends associated with an increase in the dimensionality of the system can also be understood using this insight. Reported in Table III.3 are results for the site specific walklengths $\langle n \rangle_i$ and overall mean walklength $\langle n \rangle$ for lattices characterized by a common valency $(v = 6)$, a common metric, but two different dimensionalities $(d = 2, 3)$. The symmetry classifications for the sites i of the $d = 3$ cubic lattice are given in fig. 4 of ref. 9. What makes the results in Table III.3 striking is that the $d = 3$ lattice has a larger number of sites $(N = 343)$, yet has over 50% of these clustered at or within the shell $\ell = 5$, whereas a similar percentage is realized for the (here, smaller) $d = 2$ lattice only when $\ell = 7$. The increased "concentration" of nontrapping sites about the centrosymmetric trap with increase in dimensionality leads to a systematic reduction in the values of the $\langle n \rangle_i$ and, consequently, the overall walklength $\langle n \rangle$. A combined analytic and numerical study has also been carried out for random walks on finite, high-dimensional Cartesian lattices $(3 < d \leq 8)$, and the results found there as one increases the dimensionality of the system are even more pronounced [15,16].

It is of interest to compare the numerically exact results presented in refs. 9,10, and 13 for dimension $d = 2$ with the asymptotic analytic results obtained by Montroll and Weiss [17–19]. In this classic study, these authors showed that the overall average walklength $\langle n \rangle$ can be expressed explicitly

TABLE III.3

Site specific walklengths $\langle n \rangle_i$ for lattices of a common valency ($v = 6$) but different dimensionalities (see text)

$\langle n \rangle_i$	$d = 3, N = 343$			$d = 2, N = 331$				$d = 2, N = 331$		
		ℓ/σ_i	m/p		ℓ/σ_i	m/p			ℓ/σ_i	m/p
$\langle n \rangle$	455.3			608.9						
$\langle n \rangle_1$	341.9	1/6	6/1.8	330.0	1/6	6/1.8	$\langle n \rangle_{29}$	636.8	8/6	
$\langle n \rangle_2$	404.6	2/12		429.9	2/6		$\langle n \rangle_{30}$	637.7	8/6	
$\langle n \rangle_3$	427.7	3/8		454.1	2/6	18/5.5	$\langle n \rangle_{31}$	637.9	8/6	
$\langle n \rangle_4$	427.3	2/6	24/7.0	502.7	3/6		$\langle n \rangle_{32}$	640.6	8/6	
$\langle n \rangle_5$	441.1	3/24		502.7	3/6		$\langle n \rangle_{33}$	640.8	8/6	
$\langle n \rangle_6$	455.2	4/12		523.5	3/6	36/10.9	$\langle n \rangle_{34}$	644.6	8/6	
$\langle n \rangle_7$	448.8	4/24		546.8	4/6		$\langle n \rangle_{35}$	644.8	8/6	
$\langle n \rangle_8$	458.8	5/24		553.1	4/6		$\langle n \rangle_{36}$	648.8	8/6	216/65.5
$\langle n \rangle_9$	464.6	6/8		553.1	4/6		$\langle n \rangle_{37}$	645.1	9/6	
$\langle n \rangle_{10}$	451.4	3/6	62/18.1	569.1	4/6	60/18.2	$\langle n \rangle_{38}$	645.2	9/6	
$\langle n \rangle_{11}$	455.9	4/24	122/35.7	581.7	5/6		$\langle n \rangle_{39}$	646.4	9/6	
$\langle n \rangle_{12}$	462.7	5/24		581.7	5/6		$\langle n \rangle_{40}$	646.8	9/6	
$\langle n \rangle_{13}$	467.1	6/12		588.9	5/6		$\langle n \rangle_{41}$	648.9	9/6	
$\langle n \rangle_{14}$	459.2	5/24	194/56.7	588.9	5/6		$\langle n \rangle_{42}$	649.4	9/6	
$\langle n \rangle_{15}$	646.7	6/48	262/76.6	600.8	5/6	90/27.3	$\langle n \rangle_{43}$	651.9	9/6	
$\langle n \rangle_{16}$	468.4	7/24		605.7	6/6		$\langle n \rangle_{44}$	652.3	9/6	
$\langle n \rangle_{17}$	468.5	7/24	310/90.6	608.1	6/6		$\langle n \rangle_{45}$	654.6	9/6	270/81.8
$\langle n \rangle_{18}$	471.1	8/24	334/97.7	608.1	6/6		$\langle n \rangle_{46}$	649.0	10/6	
$\langle n \rangle_{19}$	473.1	9/8	342/100.0	614.5	6/6		$\langle n \rangle_{47}$	649.4	10/6	
$\langle n \rangle_{20}$				614.5	6/6		$\langle n \rangle_{48}$	649.8	10/6	
$\langle n \rangle_{21}$				623.3	6/6	126/38.2	$\langle n \rangle_{49}$	650.9	10/6	
$\langle n \rangle_{22}$				624.3	7/6		$\langle n \rangle_{50}$	651.7	10/6	
$\langle n \rangle_{23}$				624.4	7/6		$\langle n \rangle_{51}$	653.1	10/6	
$\langle n \rangle_{24}$				627.3	7/6		$\langle n \rangle_{52}$	654.0	10/6	
$\langle n \rangle_{25}$				627.4	7/6		$\langle n \rangle_{53}$	655.5	10/6	
$\langle n \rangle_{26}$				632.5	7/6		$\langle n \rangle_{54}$	656.1	10/6	
$\langle n \rangle_{27}$				632.6	7/6		$\langle n \rangle_{55}$	657.1	10/6	330/100.0
$\langle n \rangle_{28}$				638.8	7/6	168/50.9				

in terms of the variable N for each of the $d = 2$ lattices $v = 3$, 4, and 6. The expression derived takes the general form

$$\langle n \rangle = \frac{N}{N - 1}[A_1 N \ell n N + A_2 N + A_3 + A_4/N] \qquad (4.21)$$

Presented in Table III.4 are the coefficients A_i for the three lattice valencies $v = 3$, 4, and 6. den Hollander and Kasteleyn [20] reexamined the

TABLE III.4

Coefficients in the expression $\langle n \rangle = (N/(N-1))\{A_1 N \ln N + A_2 N + A_3 + A_4 \,(1/N)\}$ for $d=2$ finite lattices of valency $v=3$, 4, and 6, subject to periodic boundary conditions

Lattice structure	Ref	A_1	A_2	A_3	A_4
		$1/\pi$			
Square planar	a	0.318309886	0.195056166	−0.11696481	−0.05145660
	b	0.318309886	0.195062532	−0.116964779	0.484065704
	c	0.318309886	0.195062532	−0.116964779	0.481952
		$3\sqrt{3}/4\pi$			
Hexagonal	a	0.413496672	0.06620698	−0.25422279	
	c	0.413496672	0.06620698	−0.25422279	6.063299
		$\sqrt{3}/2\pi$			
Triangular	a	0.275664448	0.235214021	−0.251407596	
	c	0.275664448	0.235214021	−0.251407596	−0.0444857

[a] Ref. 19.
[b] Ref. 20.
[c] Ref. 13.

calculation of Montroll for square planar lattices [19] and found that although their results for A_1 through A_3 were essentially the same as Montroll's, a discrepancy was found between the values reported for the coefficient A_4. Calculations have been performed [13] using the numerically exact data on $\langle n \rangle$ cited above that lend support to the analysis of den Hollander and Kasteleyn. The data on square planar lattices were fitted with an expression of the above form. The coefficients A_1 through A_3 were set at the values given by den Hollander and Kasteleyn (which are essentially the Montroll values), but the coefficient A_4 was allowed to "float" so as to give the best overall representation of the data. The result of this calculation is given in Table III.4 and it is seen that the coefficient A_4 determined analytically by den Hollander and Kasteleyn is excellent agreement with the "fitted" coefficient A_4.

Inasmuch as Montroll did not report values of the coefficient A_4 for triangular or hexagonal lattices, the above procedure was repeated for $v=3$ and $v=6$ (viz., the coefficients A_1 through A_3 were set at the Montroll values and A_4 was determined to give the best fit of the data). The results are also given in Table III.4. These tabulations may be taken as the "state of the art" representation of $\langle n \rangle$ as a function of N for finite lattices in $d=2$ subject to periodic boundary conditions.

In dimension $d=3$, Montroll showed that for simple cubic lattices the analytic asymptotic representation of $\langle n \rangle$ as a function of N is of the form

$$\langle n \rangle = 1.516386059\, N + O(N^{1/2}) \qquad (4.22)$$

The success of this expression in reproducing the numerically exact data on $\langle n \rangle$ is less good than is the case for $d = 2$. In fact, if one simply wants an accurate representation of the data, the data on simple cubic lattices $(v = 6)$ can be fitted to an expression of the above functional form [14] viz.

$$\langle n \rangle = 1.474131 \, N - 2.812810 \, N^{1/2} \qquad (4.23)$$

Table III.5 gives a comparison of these two representations of the data for $\langle n \rangle$ for nearest-neighbor random walks on finite, cubic lattices with a centrosymmetric trap and subject to periodic boundary conditions. A similar analysis [14] shows that for $d = 3$, tetrahedral lattices $(v = 4)$

$$\langle n \rangle = 1.683449 \, N - 2.165826 \, N^{1/2}. \qquad (4.24)$$

These representations, the ones cited above for dimension $d = 2$, and Montroll's exact result in $d = 1$

$$\langle n \rangle = \frac{N(N+1)}{6} \qquad (4.25)$$

will be used later on in this work in developing comparisons.

B. Finite Domains

1. $d = 2$ Arrays

Owing to physical and/or chemical interactions, the defining constituents (atoms, molecules) of a surface are often organized, at least locally, into

TABLE III.5
Comparison of two analytic representations of the data for the mean walklength $\langle n \rangle$ on a finite, cubic lattice with a centrosymmetric trap and subject to periodic boundary conditions

N	$\langle n \rangle$	$\langle n \rangle_M{}^a$	(Percent)$_M$	$\langle n \rangle_P{}^b$	(Percent)$_P$
27	30.46	40.94	34.42	25.19	17.32
64	77.05	97.05	25.96	71.84	6.76
125	157.32	189.55	20.49	152.82	2.86
343	455.27	520.12	14.24	453.53	0.38
729	997.4	1105.4	10.83	988.70	−0.13
1331	1856.1	2018.3	8.74	1859.4	−0.18
2197	3104.2	3351.5	7.32	3106.8	−0.08
3375	4814.7	5117.8	8.98	4811.8	+0.06

[a] Calculated assuming Montroll's expression (Eq. 4.22) for cubic lattices $(v = 6)$ with (percent)$_M = [(\langle n \rangle - \langle n \rangle_M)/\langle n \rangle] \times 100$.

[b] Calculated using the best polynomial fit of the data $\langle n \rangle \simeq 1.474141 \, N - 2.812810 N^{1/2}$. Here, (percent)$_P = [(\langle n \rangle - \langle n \rangle_P)/\langle n \rangle] \times 100$.

well-defined constellations having hexagonal, square-planar, or triangular symmetry. Although such geometrical structures can persist globally, on extended length scales it is more usual to find that surfaces are broken up into domains, each of finite extent, with boundaries separating these domains. For isolated domains of definite symmetry, the efficiency of diffusion-reaction processes taking place locally on such arrays will be critically dependent on the geometrical organization of the system (its overall size, shape, and internal connectivity or valency). Moreover, inasmuch as the juxtaposition of m such domains can, as well, form patterns, lattice-like arrays that may be regular, fractal, or random, the expectation is that the experimentally observed reaction efficiency will be a composite function of the geometrical factors defining not only the local but the overall organization of the system.

Consider a molecule diffusing in free space or a solute molecule diffusing in solution. Upon colliding with a surface, assume that the molecule is sufficiently entrained by surface forces that there results a reduction in dimensionality of its diffusion space from $d = 3$ to $d = 2$, and that in its subsequent motion the molecule is sterically constrained to follow the pathways defined by the lattice structure of the surface (or, perhaps, the boundary lines separating adjacent domains). If at some point in its trajectory the molecule becomes permanently immobilized, either because of physical binding at a site or because an irreversible reaction has occurred at that site, then, qualitatively, this sequence of events is descriptive of many diffusion-reaction processes in biology, chemistry and physics.

The above surface-diffusion problems can be translated into a lattice-statistical one that can be studied systematically [21]. The objective is to calculate the mean walklength $\langle n \rangle$ of a molecule constrained to move along sterically allowed pathways (reaction channels) on a structured surface until immobilized at a receptor or target site. As noted earlier, although there is an extensive literature dealing with random walk models of diffusion-reaction processes in systems of infinite extent or for large lattices [4], considered here are processes taking place on domains of *finite* extent having boundaries of *arbitrary* shape. In particular, in formulating and then solving the stochastic problem of determining the first moment $\langle n \rangle$ of the distribution function, the simplifying feature of periodic boundary conditions is not imposed.

Recall that in the studies of Montroll and Weiss [17–19] on nearest-neighbor random walks on an infinite, periodic lattice of unit cells, the mean walklength $\langle n \rangle$ is completely determined once the dimensionality d, the system size (number of lattice sites) N, and the connectivity (or valency) v of the unit cell are specified. For the class of $d = 2$ problems considered here, there is, not unexpectedly, a more subtle dependence of $\langle n \rangle$ on the lattice parameters N and v, and a further, pronounced dependence on the number of sites N_b comprising the boundary of the domain. For a given setting of N_b, trends in the values of $\langle n \rangle$ calculated for a wide variety of finite lattices of given symmetry but of various shapes can be systematically organized in terms of the overall root-mean-square distance $\langle r^2 \rangle^{1/2}$ of the N lattice sites from the center of the array.

The number of microlattices characterized by a given setting of $[N, N_b, v]$ obviously escalates with increase in N; the several lattices considered below were designed to allow the separate influence of the variables $[N, N_b, v]$ on the dynamics to be disentangled. These lattices are characterized by a common (unit) distance separating lattice points, that is, although lattices of different $[N, N_b, v]$ and different overall geometries are considered, a common metric for all lattices is imposed.

Using the integer j to label the jth nearest-neighbor site to a trap positioned at site i on a *finite* lattice, the site-specific mean walklength $\langle n \rangle_j$ is found to depend on the location of the trap on the lattice. This result stands in contrast to the one found when nearest-neighbor random walks on an infinite, periodic lattice of unit cells are studied. In the latter case, owing to the imposition of periodic boundary conditions, the value calculated for $\langle n \rangle_j$ is invariant regardless of the positioning of the trap. This distinction provides the first indication that one's intuition might fail in interpreting results on finite lattices, if that intuition were based solely on results obtained in lattice-statistical studies in which a unit cell was identified and periodic boundary conditions imposed.

Consider now the lattices diagrammed in Figures 4.6 and 4.7. Whereas the N dependence of $\langle n \rangle$ will be found to be qualitatively correct, the fact that these finite lattices are not characterized by a uniform valency v limits the usefulness of Eq. (4.21). In fact, it is necessary to introduce composite (average) lattice parameters in order to be able to compare trends in the data. Specifically, one parameter is needed to account for the nonuniform valency and one parameter to reflect and/or characterize the variety of geometrical shapes. Consistent with the identification of an overall (average) walklength

$$\langle n \rangle = \frac{1}{N} \sum_i \langle n \rangle_i \qquad (4.26)$$

where $\langle n \rangle_i$ is the average walklength for a trap assumed to be situated at site i on a given lattice, and the sum is over all lattice sites i, one defines

$$\langle v \rangle = \frac{1}{N} \sum_i v_i \qquad (4.27)$$

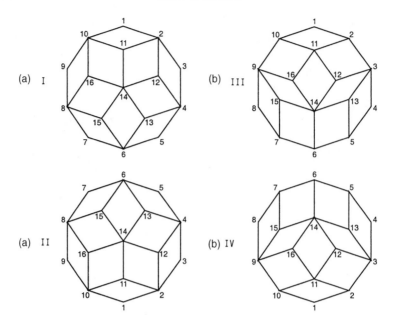

Figure 4.6. (a) I and II; lattice characteristics:
$$\{N, N_b, \langle r^2 \rangle^{1/2}, \langle v \rangle\} = \{16, 10, 1.395984, 3.125\}.$$
(b) III and IV; lattice characteristics:
$$\{N, N_b, \langle r^2 \rangle^{1/2}, \langle v \rangle\} = \{16, 10, 1.512821, 3.125\}.$$

Figure 4.7. (a) Lattice characteristics:
$$\{N, N_b, \langle r^2 \rangle^{1/2}, \langle v \rangle\} = \{16, 12, 1.500000, 4.125\}.$$
(b) Lattice characteristics:
$$\{N, N_b, \langle r^2 \rangle^{1/2}, \langle v \rangle\} = \{16, 12, 1.581139, 4.125\}.$$
(c) Lattice characteristics:
$$\{N, N_b, \langle r^2 \rangle^{1/2}, \langle v \rangle\} = \{16, 12, 1.581139, 3.000\}.$$
(d) Lattice characteristics:
$$\{N, N_b, \langle r^2 \rangle^{1/2}, \langle v \rangle\} = \{16, 12, 1.581139, 3.000\}.$$
(e) Lattice characteristics:
$$\{N, N_b, \langle r^2 \rangle^{1/2}, \langle v \rangle\} = \{16, 14, 1.802776, 2.375\}.$$

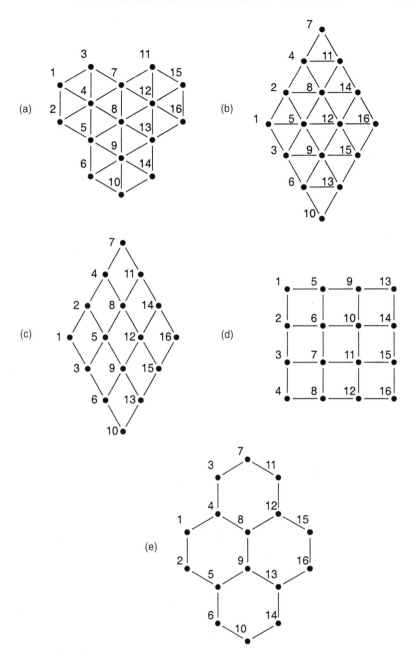

where v_i is the valency of site i of the lattice, the sum again being over all lattice sites. To distinguish among the various geometrical shapes of the finite lattices, one constructs

$$\langle r^2 \rangle^{1/2} = \left[\frac{1}{N} \sum_i r_i^2 \right]^{1/2} \tag{4.28}$$

the root-mean-square distribution of lattice sites with respect to the "center" of a given lattice. For those lattice structures characterized by two (or more) axes of bilateral symmetry, the intersection of these axes is taken as the "center" of the lattice. For lattices with only one axis of bilateral symmetry, one identifies a "stochastic center," defined as that lattice point on the bilateral axis for which the calculated $\langle n \rangle_i$ has the minimum value.

For the $\langle n \rangle_i$ data calculated for each of the lattices diagrammed in Figures 4.6 and 4.7, and from a consideration of their respective geometrical "shapes," values of $\langle n \rangle$, $\langle v \rangle$ and $\langle r^2 \rangle^{1/2}$ can be calculated for each lattice. These values are recorded in Table III.6, along with the values of N, N_b and A, where A is the area encompassed by the given figure.

From Table III.6, it is evident for the case $N = 16$, that the primary "order parameter" is N_b. Once, N_b has been specified, the distance $\langle r^2 \rangle^{1/2}$ provides a systematic organization of the data. Then, for a common setting of N, N_b, and $\langle r^2 \rangle^{1/2}$, the data may be further organized in terms of the average valency $\langle v \rangle$ of the lattice being considered. In particular, as N_b, increases from $N_b = 10$ to 12 to 14, the values of $\langle n \rangle$ (taken as a group) systematically increase. For $N_b = 10$, $\langle n \rangle$ increases as $\langle r^2 \rangle^{1/2}$ increases (while $\langle v \rangle$ remains constant); for $N_b = 12$, a similar increase of $\langle n \rangle$ with $\langle r^2 \rangle^{1/2}$ is observed. For fixed $[N_b = 12, \langle r^2 \rangle^{1/2} = 1.581139]$, $\langle n \rangle$ increases with respect to a decrease in the average valency $\langle v \rangle$. Thus, for fixed $[N_b, \langle r^2 \rangle^{1/2}]$, the qualitative dependence of the average walklength $\langle n \rangle$ on the valency (here $\langle v \rangle$) is the same as that predicted by Eq. (4.21).

TABLE III.6
Lattice characteristics and stochastic results: see figures 4.6 and 4.7

Lattice	N	N_b	$(r^2)^{1/2}$	$\langle n \rangle$	$\langle v \rangle$	A
Fig. 4.6(a)	16	10	1.395984	23.325121	3.125	7.694
Fig. 4.6(b)	16	10	1.512821	23.697968	3.125	7.694
Fig. 4.7(a)	16	12	1.500000	24.527863	4.125	7.794
Fig. 4.7(b)	16	12	1.581139	25.194569	4.125	7.794
Fig. 4.7(c)	16	12	1.581139	25.257143	3.000	9.000
Fig. 4.7(d)	16	12	1.581139	25.257143	3.000	9.000
Fig. 4.7(e)	16	14	1.802776	27.989569	2.375	10.392

Finally, examination of Figures 4.7c and 4.7d and the corresponding data in Table III.6 reveals that for fixed $[N, N_b, \langle r^2 \rangle^{1/2}, \langle v \rangle]$, the value of $\langle n \rangle$ remains constant, despite the fact that the shape of the lattice "looks" different. This last case reflects the fact that finite lattices subject to a diffeomorphic transformation leave invariant the value of $\langle n \rangle$. Since $\langle n \rangle$ summarizes the net result of considering all possible flows initiated from all possible sites on the lattice, $\langle n \rangle$ plays the role of a topological invariant.

It is in the dependence of $\langle n \rangle$ on the (average) valency $\langle v \rangle$ that the results here stand in contrast to the analytic and numerical results obtained for lattices subject to periodic boundary conditions. From studies on periodic lattices, $\langle n \rangle$ should decrease systematically with increase in the uniform valency v. This result pertains as well to random walks on $d = 3$ dimensional periodic lattices of unit cells and can also be demonstrated analytically and numerically for walks on higher-dimensional $(d \leq 8)$ cubic lattices [15,16]. In these problems, $v = 2d$ and hence the higher the dimensionality of the space, the greater the number of pathways to a centrally located deep trap in a periodic array of (cubic) cells; the decrease in $\langle n \rangle$ is found to be quite dramatic with increase in d, and hence v. However, an increase in v will also result in a greater number of pathways that allow the random walker to move away from the trap. For periodic lattices, this latter option positions the random walker closer to the trap in an adjacent cell. For finite lattices, moving away from the trap does not position the walker closer to a trap in an adjacent unit cell; it positions the walker closer to the finite boundary of the lattice from whence it must (eventually) work itself back. It is evident, therefore, why the v dependence for periodic lattices is modified when one studies the same class of nearest-neighbor random-walk problems on finite lattices.

Discussion of the results obtained on the finite lattices diagrammed in Figures 4.6 and 4.7 leads to the conclusion that the distance $\langle r^2 \rangle^{1/2}$, calculated with respect to the "center" of the lattice, is a parameter second in importance only to $[N, N_b]$ in organizing the data on $\langle n \rangle$. A further, concrete illustration of this conclusion is provided by considering in more detail the Penrose decagons diagrammed in Figure 4.6. Notice that I and II are "optical isomers," as are III and IV. No differences in the $\langle n \rangle$ values calculated for the structures I and II are found, nor are the $\langle n \rangle$ values for III and IV different. However, the $\langle n \rangle$ values for I and II are different from those for III and IV. The only lattice characteristic that distinguishes I(II) from III(IV) is the value of $\langle r^2 \rangle^{1/2}$, the values of $\langle v \rangle$ and A being exactly the same (see Table III.6). With respect to the calculation of $\langle r^2 \rangle^{1/2}$ in these two cases, whereas the "geometric center" and the "stochastic center" for the figure I(II) coincide, the two "centers" are different for III(IV), a consequence of the fact that III(IV) has only a single axis of bilateral symmetry. However, choosing the "stochastic center" in calculating $\langle r^2 \rangle^{1/2}$ for a finite lattice with a single axis of bilateral symmetry, the results calculated for the structures I(II) and III(IV), as well as for the more extensive range of structures studied in ref. 21, fall into place.

A further interesting feature of the four $N = 16$ Penrose decagons is that, for $N = 16$, the value of N_b is smaller than any other "regular" lattice that can be constructed, subject to the constraint that the "bond length" connecting all lattice points be fixed. Given that the $\langle n \rangle$ values calculated for the $N = 16$ Penrose decagons are smaller than the $\langle n \rangle$ values for the other $N = 16$ lattices diagrammed in Figure 4.7 again points to the importance of N_b as a principal organizing parameter for this class of random walk problems. When $N = 48$ lattices are considered, it is possible to construct a triangular lattice (Fig. 4.8a) with the same number $N_b = 22$ of boundary sites as the Penrose platelet, Figure 4.8b, the latter figure constructed by a conjunction of the simpler structures, I, II, III, and IV. In this case $\langle r^2 \rangle^{1/2}$ is smaller for the lattice shown in Figure 4.8a, with the consequence that the overall $\langle n \rangle$ is smaller for the more compact triangular lattice. Finally, a distinguishing feature of the Penrose platelet shown in Figure 4.8b is the presence of one site of valency $v = 7$; positioning a trap at that site leads to the smallest site-specific value of $\langle n \rangle_i \, (= 36.213)$ calculated for all other lattices with $N = 48$.

2. $d < 2$ Fractal Sets

The influence of geometry in modulating the efficiency of diffusion-controlled reactive processes can be broadened by considering spatial inhomogeneities and domains of fractal dimension. The difference in reaction efficiency when a reaction center is positioned at a regular versus defect site on a surface is easily studied. This kind of spatial inhomogeneity is local, that is, a defect site is a local imperfection in the surface that can be specified by the site valency. There can, however, also be present larger scale inhomogeneities. For example, the surface of a cell may have transmembrane (and other) proteins that would break the translational symmetry and interfere significantly with the lateral motion of a diffusing coreactant (and hence influence the reaction efficiency in a diffusion-controlled process). One strategy for dealing with such "excluded regions" is to consider fractal domains and formulate the problem so that the influence on the reaction efficiency of excluded regions of different spatial extent can be studied and contrasted with what would be expected if the surface were free of such inhomogeneities [22].

To proceed, consider two finite planar networks, a regular Euclidean triangular lattice (interior valence $v = 6$, but with boundary defect sites of valence $v = 4$ and $v = 2$) of dimension $d = 2$, and a fractal lattice

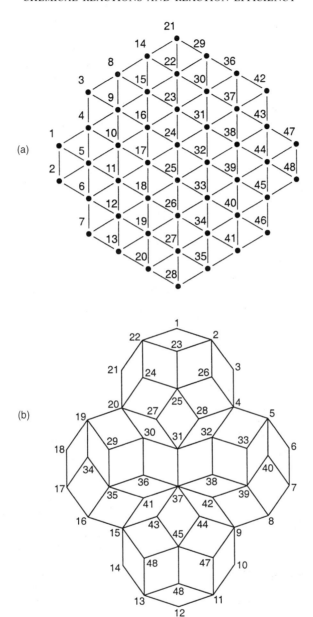

Figure 4.8. (a) Lattice characteristics:

$$\{N, N_b, \langle r^2 \rangle^{1/2}, \langle v \rangle\} = \{48, 22, 2.565801, 4.979\}.$$

(b) Lattice characteristics:

$$\{N, N_b, \langle r^2 \rangle^{1/2}, \langle v \rangle\} = \{48, 22, 2.576481, 3.458\}.$$

JOHN J. KOZAK

(Sierpinski gasket) [23–25] (principal valence $v = 4$ with $v = 2$ defect sites) of dimension $d = \ell n\ 3/\ell n\ 2 = 1.584962$. The punctuated structure of the second lattice (see Fig. 4.9) effectively restricts the diffusing coreactant to a subset of the overall (space filling) reaction space, thereby providing the opportunity to quantify the influence of longer-range spatial inhomogeneities on the dynamics. In the companion triangular lattice, there are no excluded regions and the surface interior to the boundary is fully connected by lattice points of valency $v = 6$.

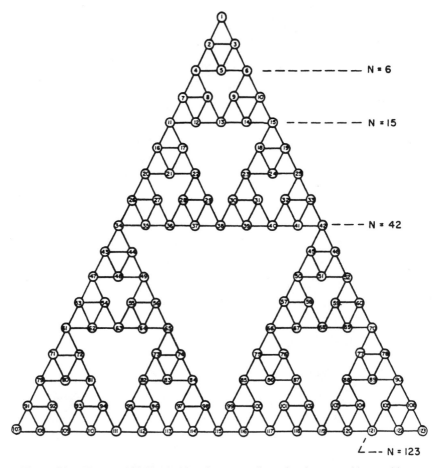

Figure 4.9. The $N = 123$ Sierpinski gasket, a two-dimensional uncountable set with zero measure and Hausdorff (fractal) dimension $\ell n\ 3/\ell n\ 2 = 1.584962\ldots$; the companion Euclidean lattice referred to in the text is a "space filling" triangular lattice of (interior valency $v = 6$.

Solution of the stochastic master equation, Eq. (4.3), allows one to monitor the efficiency of reaction by determining the time required for a diffusing molecule to reach a specific site on a lattice. Suppose the target molecule is anchored at the defect site 1 ($v = 2$) on the lattice diagrammed in Figure 4.9, and the motion of the diffusing coreactant is initialized at the lattice site farthest removed from the target site. On the discrete (regular and fractal) spaces considered here, any point on the base of Figure 4.9 will satisfy this condition and for definiteness the midpoint site 115 ($v = 4$) is chosen. One can then determine the probability $\rho(t)$ for an unconstrained random walk on both regular and fractal lattices and compare the differences. More specifically owing to the presence of a uniform concentration gradient or external field, the motion of the diffusing molecule may be constrained in such a way that it can move only laterally or toward the reaction center, and this can be dealt with by imposing a uniform bias on the motion of the molecule.

Some simple, formal results can be developed for the "bias" case. Inspection of Figure 4.9 shows that every vertex has either two upward connections and two sideways connections or one upward connection and one sideways connection. In either case, the ratio of upward to sideways connections is always equal to unity. As a result, at each time step the molecule either transits one level closer to the vertex site 1 or stays at the same level. The ratio of these two probabilities of these two motions is always the same, a conclusion valid both for the Sierpinski gasket and for the associated triangular lattice. The migration of a molecule through the lattice can then be regarded as a classical random walk relative to a fixed point. In this new frame of reference, the particle moves (on average) half a level spacing in a single step i; a step forward of half a level spacing and a step backward of half a level spacing are equally likely.

Since the above analysis is valid both for regular and fractal lattices, the probability of surviving to time t should be the same for both lattice geometries. Specifically, for a molecule positioned initially at the midpoint of the base, if the particle moves (on average) a half step upward in a single time step and there are m levels separating the particle initially from the trap (defined to be at level $m = 0$), the average number $\langle n \rangle$ of steps required for trapping (reaction) in either case is

$$\langle n \rangle = 2m$$

This result can be checked numerically for the sequence of fractal lattices defined by $N = 15, 42, 123$, and for the sequence of triangular lattices defined by $N = 15, 45, 153$. The results of these calculations are presented

TABLE III.7
Mean walklength data, $\langle n \rangle$ ($\langle n \rangle_i$)

	Case I [a]	Case II [b]	Case III [c]	Case IV [d]
$N = 6$	3.200	9.200	1.800	3.800
	(4.000) [e]	(10.000) [e]	(3.000) [f]	(5.000) [f]
		Sierpinski Gasket: $d = 1.584962 \cdots$		
$N = 15$	5.714	43.429	5.286	17.714
	(8.000) [e]	(50.000) [e]	(9.000) [f]	(25.000) [f]
$N = 42$	10.927	211.561	17.098	85.3414
	(16.000) [e]	(250.000) [e]	(28.000) [f]	(125.000) [f]
$N = 123$	21.508	1047.31	59.926	420.262
	(32.000) [e]	(1250.00) [e]	(92.000) [f]	(625.000) [f]
		Triangular Lattice: $d = 2$		
$N = 15$	5.714	44.674	5.091	17.723
	(8.000) [e]	(49.707) [e]	(8.300) [f]	(23.220) [f]
$N = 45$	10.9091	223.812	15.186	85.285
	(16.000) [e]	(247.581) [e]	(22.574) [f]	(108.936) [f]
$N = 153$	21.4737	1118.414	48.503	413.724
	(32.000) [e]	(1218.20) [e]	(61.875) [f]	(510.996) [f]

[a] Vertex site (1) target, biased flow (see text).
[b] Vertex site (1) target, unconstrained flow.
[c] Midpoint-base target, biased flow (see text).
[d] Midpoint-base site target, unconstrained flow.
[e] Initialized at midpoint-base site.
[f] Initialized at vertex site (1).

in Table III.7; examination of the data listed in parenthesis under Case I shows that the above relationship is satisfied (exactly).

The above can be generalized for both lattices and the mean number of steps $\langle n \rangle$ before trapping when the motion of the diffusing coreactant is initialized at *any* nontrapping site can be calculated, again assuming that the flow is guided (biased) by a uniform concentration gradient or external field. For a triangular lattice, the number of vertices on level m is $(m + 1)$. Thus, the total number N of vertices for this geometry is

$$N = \sum_{m=0}^{m} (m + 1) = (m + 1)(m + 2)/2$$

The total number N' of nontrapping (satellite) vertices is

$$N' = N - 1 = m(m + 3)/2$$

Averaging over all nontrapping sites, the number $\langle n \rangle$ of steps taken by a diffusing coreactant before trapping is

$$\langle n \rangle = \sum_{m=1}^{m} \frac{(m+1)}{N'}(2m) = \frac{4}{3}\frac{(m+1)(m+2)}{(m+3)} \tag{4.29}$$

The validity of this result can be checked against the numerical evidence presented in Table III.7 under "triangular" Case I. In particular, for $m = 2, 4, 8, 16$ this equation gives 3.2, 5.714, 10.9091, and 21.4737, respectively.

A similar sort of analysis shows that, for the same initial condition [diffusion of the coreactant proceeding from any nontrapping (satellite) site in the network with the subsequent flow biased in the direction of the reaction center, site 1], the average number $\langle n \rangle$ of steps before trapping on the Sierpinski gasket is

$$\langle n \rangle = \sum_{m=1}^{m} \frac{(\text{number of vertices on level } m)}{N'}(2m) \tag{4.30}$$

The results obtained for $N = 6, 15, 42$, and 143 (respectively, for $m = 2, 4, 8, 16$, one finds 3.2, 5.714, 10.92682, and 21.508191) are in accord with the numerical evidence presented under "Sierpinski" in Case I of Table III.7.

The above analytic/numerical results for $\langle n \rangle$ were calculated assuming that the diffusing coreactant initiated its motion at a single site (the midpoint of the base for each geometry considered) or at any site i (then averaging over all nontrapping (satellite) sites of the regular/fractal lattice). In both calculations, the subsequent motion of the coreactant was biased in the direction of the reaction center. The stochastic master equation, Eq. (4.3), can be solved numerically for this problem. For the largest networks considered here (the $N = 123$ Sierpinski gasket and the $N = 153$ triangular lattice), the results are displayed in Figure 4.10 for the two cases noted above. The profiles in the upper right of this figure correspond to a flow initiated from the single (midpoint base) site while the ones in the lower left are those for which initialization is possible at any satellite site. In both figures, the circles and triangles specify results generated for the Sierpinski gasket and the triangular lattice, respectively.

Two features of these evolution curves are immediately noticeable. First, for both initial conditions, trapping (reaction) on the fractal lattice is distinctly slower than reaction on the triangular one. At first sight, this result would appear to be anomalous inasmuch as the space-filling triangular

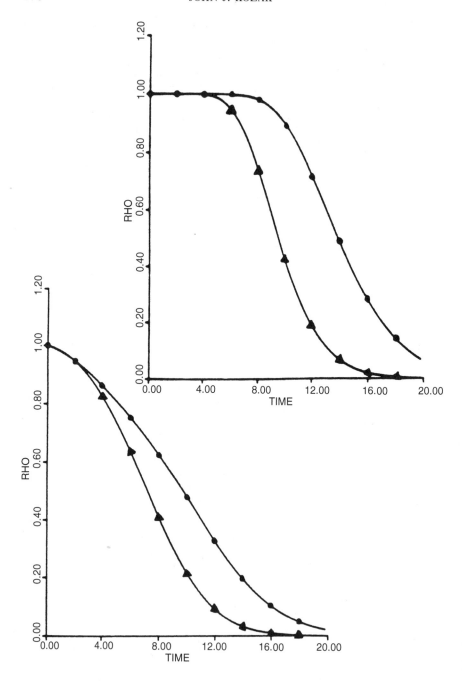

lattice has more lattice sites than does the Sierpinski gasket ($N = 153$ vs $N = 123$, respectively) and one would expect [17–19] the average walklength $\langle n \rangle$ (and hence the characteristic relaxation time) to increase with increase in the number N of lattice sites. However, all interior sites of the fractal lattice ($d = 1.58$) have valency $v = 4$ whereas all interior sites of the triangular one ($d = 2$) have valency $v = 6$. Increasing both d and v places more satellite sites closer to the reaction center and counteracts the increase in $\langle n \rangle$ arising from the greater number of vertices at each level $m \, (> 2)$ (and hence the overall N) on the triangular versus fractal lattice. These results are consistent with the theoretical insight of Gefen, Aharony, and Alexander [26] that diffusion on percolation networks is slower than on Euclidean ones; they showed that the mean-square displacement of a random walker is given by

$$\langle r^2(t) \rangle \sim t^{2/(2+\theta)} \tag{4.31}$$

where $\theta = 0.8$ in dimension $d = 2$ for percolating networks versus $\theta = 0$ for the Euclidean case.

The second notable feature of these evolution curves is the pronounced "shoulder" effect seen on short time scales, particularly for the case where the flow is initiated from a site farthest removed from the reaction center. The appearance of "shoulders" is related to the fact that, for a particle initiating its motion at a specific site somewhere in the lattice, there is a minimum time required for the coreactant to reach the reaction center; this time is proportional to the length of the shortest path, and hence the reactive event cannot occur until (at least) that interval of time has expired. This effect is analogous to the one observed in computer simulations of Boltzmann's H function calculated for two-dimensional hard disks [27]. Starting with disks on lattice sites with an isotropic velocity distribution, there is a time lag (a horizontal shoulder) in the evolution of the system owing to the time required for the first collision between two hard particles to occur.

Whereas in the above the consequences of imposing a directionality on the motion of the diffusing coreactant were explored, the data given in Case

---

Figure 4.10. Survival probability $\rho(t)$ versus (reduced) time t for a reactive event monitored at the vertex site 1 in Figure 4.9. The motion of the diffusing coreactant is initialized either at a single site (here, midpoint of base) (upper right figure), or averaged over all nontrapping (satellite) sites, subject to the constraint that the molecule moves either "right/left" or toward site 1. The upper (circles) and lower (triangles) curves in each of these figures refer to reaction spaces modeled as a Sierpinski gasket and a regular triangular lattice, respectively.

II of Table III.7 and in the companion profiles in Figure 4.11 describe the situation where, following initiation of the flow, no motional constraints are placed on the subsequent migration of the diffusing molecule. From an examination of these data vis-à-vis those presented in Case I and Figure 4.10 two conclusions follow. First, the mean walklength $\langle n \rangle$ and time scale characterizing the (eventual) reaction at the target site for the case of an unconstrained flow are significantly longer than for the case of a focused flow. The diffusion-reaction process is much less efficient. Second, on the extended time scale characterizing the case of unconstrained flow, the pronounced shoulder effect noted in Figure 4.10 is suppressed for both choices of initial conditions; the evolution profiles displayed in Figure 4.11 have a canonical exponential form. Thus, both the qualitative and quantitative characteristics of the decay profiles change vis a vis the case of a biased flow.

The calculations reviewed thus far refer to the case where the target molecule is anchored at a defect site $(v = 2)$ of the reaction space. It is of interest to assess the generality of the conclusions drawn from these data when the location of the target molecule is switched to a site of valency $v = 4$. Consider positioning the target molecule at the midpoint base site. Entries for Case III in Table III.7 can be compared directly with corresponding entries for Case I and the evolution profiles displayed in Figure 4.12 can be compared with those in Figure 4.10; in both cases it is assumed that the motion of the diffusing coreactant is biased, that is, the molecule suffers either a lateral displacement or migrates in the direction of the reaction center. The calculation shows that, if the diffusing coreactant is injected at the vertex (defect) site and migrates toward a reaction center on the opposite boundary, the $\langle n \rangle$ is systematically longer than if the coreactant is injected at the midpoint base site and migrates toward a reaction center placed at the vertex (compare data in parentheses under Case III versus Case I). On the other hand, for unconstrained (unbiased) flow, exactly the opposite conclusion pertains (compare data in parentheses under Case IV versus Case II and the evolution profiles displayed in Figure 4.13 versus Figure 4.11). These generalizations are valid for both the fractal and regular lattice.

Finally, if one relaxes the constraint that the coreactant is injected into the system at a specific site and instead assumes that the coreactant can initiate its motion at any site i of the reaction space, some interesting new effects arise. For the case of unconstrained motion, the results calculated for the overall $\langle n \rangle$ show that placing the reaction center at the midpoint base site (with $v = 4$) results in a much more efficient diffusion-reaction process than placing the trap at a defect (vertex) site (with $v = 2$). On the other hand, biasing the motion of the diffusing coreactant (e.g., by switching on a

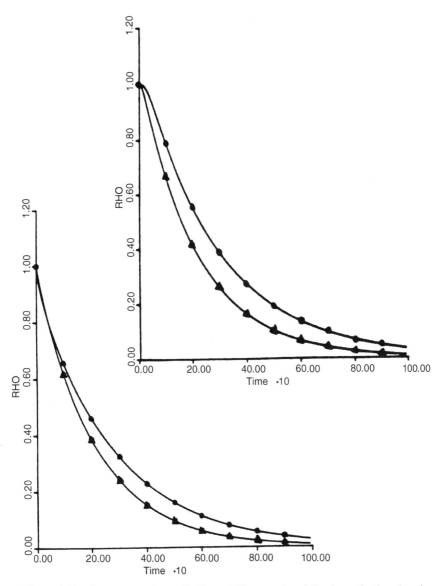

Figure 4.11. Same conventions as in Figure 4.10 except that, following activation, there is no restriction on the directionality of the coreactant's motion.

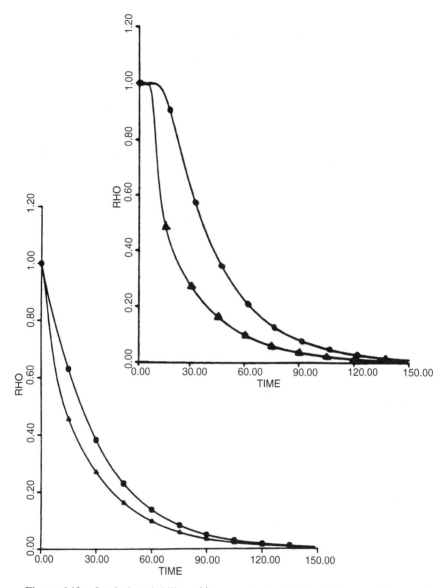

Figure 4.12. Survival probability $\rho(t)$ versus (reduced) time t for a reactive event monitored at the midpoint base site 115 in Figure 4.9. The motion of the diffusing coreactant is initialized either at a single site (top vertex) (upper right figure) or averaged over all nontrapping (satellite) sites (lower left figure), subject to the constraint that the molecule moves either "right/left" or toward site 115. The upper (circles) and lower (triangles) curves in each of these figures refer to reaction spaces modeled as a Sierpinski gasket and a regular triangular lattice, respectively.

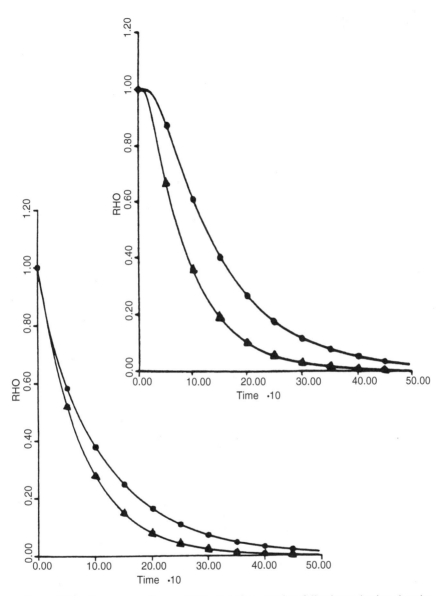

Figure 4.13.   Same conventions as in Figure 4.12 except that, following activation, there is no restriction on the directionality of the coreactant's motion.

uniform external field) leads to the same conclusion only if $N < 15$, that is, there is a "crossover" in reaction efficiency.

For the largest fractal and regular lattices considered here ($N = 123$ and $N = 153$, respectively), conclusions drawn from an examination of the time scales characterizing the evolution curves displayed in Figure 4.10 versus Figure 4.12 for biased flows and in Figure 4.11 versus Figure 4.13 for unconstrained ones, are consistent with those developed from an examination of the $\langle n \rangle$ (for $N > 15$), as they must. Specifically, for biased flow, the diffusing coreactant is trapped within times $t < 20$ when the target molecule is positioned at the vertex defect site (1) with $v = 2$, whereas when the target is at the midpoint base site ($v = 4$), the trapping is within $t < 150$. Conversely, for totally unconstrained flows the coreactant is trapped within $t < 100$ when the target molecule is anchored at the defect site 1, but is trapped within $t < 50$ when the target is positioned at a site of higher local coordination ($v = 4$).

The first of the conclusions that can be drawn from these calculations is that for finite planar lattices of integral or fractal dimension, in the absence of any external bias, the diffusion–reaction process is more efficient when the target molecule is localized at a site of higher valency. This conclusion is consistent with the trends noted in Section IIIA, and with the results reported in Section IIIB. Further, the conclusion is consistent with results obtained in a lattice-based study of reactivity at terraces, ledges, and kinks on a (structured) surface [28]. There the reaction

$$A + B \rightleftarrows [AB]^* \rightarrow C \qquad (4.32)$$

between a diffusing coreactant A and a stationary target molecule B was studied as a function of the degree of reversibility s of reaction. Changes in the reaction efficiency before reaction, as gauged by $\langle n \rangle$, were quantified for various settings of s, the probability that the above reaction proceeds at once to completion with formation of the product C [or, conversely, a probability $(1 - s)$ that the diffusing molecule A, upon confronting the target molecule B, forms an excited state complex, but one that eventually falls apart with regeneration of the species A (which subsequently continues its random walk on the surface)]. These calculations demonstrated that in the absence of specific energetic effects (e.g., those associated with the extra number of "free bonds" at a defect site of the lattice), the reaction efficiency was always higher the more highly coordinated the site to which the reaction center was anchored. Specifically, the reaction efficiency increased as the reaction center was positioned sequentially at sites of valency $v = 3$, $v = 4$, and $v = 5$ on a reaction space characterized overall by the valency $v = 4$. Also consistent with the above conclusion is the one that emerged in a lattice-based study of sequestering and the influence of domain structure on excimer formation in spread monolayers [13]. There, changes in lifetime of a diffusing, excited-state monomer (probe molecule) in a monolayer, studied experimentally by monitoring the excimer-monomer steady-state photo-excitation of the probe [29], were examined in a variety of situations, with the overall conclusion being that the lifetime of the probe molecule was always longer the lower the local coordination of the domain. And, finally, in Section IV, where the influence of symmetry-breaking potentials in influencing the efficiency of diffusion-controlled reactive processes in $d = 3$ and on surfaces will be examined, the efficiency of reaction will be found to decrease with decrease in the valency of the site at which the target molecule is positioned.

The second conclusion follows from the more global symmetry-breaking introduced above, that is, imposing a uniform bias on the motion of the diffusing species at each and every point of the reaction space. For system sizes $N < 15$, the results obtained were consistent with those reviewed above (see Table III.7). Both for fractal and Euclidean surfaces, upon averaging over all possible trajectories, placing the target molecule at a defect site $(v = 2)$ resulted in longer trapping times than if the target were positioned at a site of valency $v = 4$. However, a new feature appeared when a uniform bias was imposed on the motion of the diffusing species for processes taking place on reaction spaces, $N > 15$. Here, the focusing effect of this (global) constraint leads to an enhanced efficiency of reaction if the target is positioned at a vertex defect site, $v = 2$. This "crossover" effect in reaction efficiency with increase in system size emerged both for surfaces of fractal and integral dimension. Since this is the only case in lattice models studied to date where, in the absence of any specific energetic effects at a given defect site, one finds that localizing the target molecule at a site of lower coordination leads to an enhancement in the reaction efficiency, further discussion of this case is instructive.

In Case I two factors are at play in focusing the flow to the site (1) of lower valency. First, there is the bias imposed on the particle's motion and second, there is a geometrical focusing, that is, the fact that the reaction space available to the diffusing molecule contracts systematically as the molecule moves (irreversibly) from site to site closer to the trap. These two factors need to be disentangled in order to determine whether the enhanced efficiency at the site of lower valency is driven by the biasing effect, the geometrical effect or, perhaps, a synergetic interplay of both factors.

It is already known (see Cases II and IV in Table III.7) that, if one turns off the potential, the reaction efficiency is higher the greater number of channels intersecting at the reaction center. It remains to be determined whether, in the presence of a bias, positioning the target molecule at a site of lower valency $(v = 2)$ in a reaction space not characterized by the constrictive geometry of Case I, the reaction efficiency remains higher at a site of lower valency. To investigate this point, suppose that molecules were injected, one at a time, at the defect site 1 in the reaction space. If these molecules were regarded as "hard spheres" and if gravity were acting downward (see Fig. 4.9), one would simulate an apparatus known as a Galton board [30], after Francis Galton who constructed a physical apparatus based on hexagonal geometry. Such an apparatus is often displayed in museums to demonstrate visually how one produces a Gaussian distribution starting with $n(\sim 1000)$ balls initialized at the vertex site (here, site 1) and exiting at one of the sites defining the lower boundary of Figure 4.9. In the lattice model here, this corresponds to initializing $n \to \infty$ molecules at the vertex site 1 and determining $\langle n \rangle_i$ and the overall $\langle n \rangle$ for molecules being trapped at each basal site i. A plot can be constructed for the normalized quantity (probability)

$$P = \langle n \rangle \text{ (midpoint base site)}/\langle n \rangle \text{ (base site)} \qquad (4.33)$$

versus exit location, both for the fractal lattice and the "space filling" Euclidean one. As displayed in Figure 4.14, this plot yields the (anticipated) Gaussian profile of probability theory, with the maximum in the profile centered at the midpoint site on the base. Note that the relative width of the curve generated for the Sierpinski gasket is somewhat broader than for the triangular lattice, reflecting the longer residence time of the diffusing coreactant on the fractal surface. From an examination of Figure 4.14, one finds for either lattice that the $\langle n \rangle_i$ of a diffusing coreactant injected at the vertex site 1 and flowing downward is smallest at the midpoint base exit $(v = 4)$ and increases (P decreases) systematically as one moves away from that site. In particular, values of $\langle n \rangle_i$ calculated for the left/right vertex sites on the base yield values of $P = 0.31$ for the Sierpinski gasket and $P = 0.25$ for the triangular lattice. Hence, consistent with results obtained in all previous studies, the reaction efficiency is again found to be higher if the target molecule is positioned at a site of higher valency (here, $v = 4$). It is only when one "inverts" the process, that is, injects molecules at the base of Figure 4.9 and monitors the flows at the (top) vertex site 1, that an enhanced efficiency of reaction is determined for a target molecule anchored at a site of lower valency, and then only if $N > 15$. Thus, it is only when the motion of the diffusing particle is influenced simultaneously by a global focusing effect (the bias) and a systematic, geometrical contraction in the reaction space that the "crossover" effect noted above emerges.

As a final comment, consider the relationship between the theory of cellular automata and the geometrical unfolding of the two lattices studied

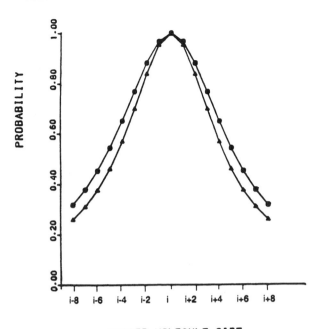

TARGET MOLECULE SITE

Figure 4.14.   A plot of the factor

$$P = \frac{\langle n \rangle \,(\text{midpoint base site})}{\langle n \rangle \,(\text{base site})}$$

versus exit location for the Sierpinski gasket (circles) and the "space filling" Euclidean triangular lattice (triangles). For the Sierpinski gasket as one moves pointwise from the midpoint base site 115 in Figure 4.9 (where $\langle n \rangle$ = 59.926) to the corner vertex site, the $\langle n \rangle$ values are 61.934, 67.959, 78.000, 92.057, 110.131, 132.221, 158.328, and 188.451. For the triangular lattice, the corresponding $\langle n \rangle$ values are 48.503 (midpoint), 50.842, 57.823, 69.342, 85.227, 105.250, 129.131, 156.660, and 187.154.

above. Given an initial configuration (state) defined by a sequences of sites with value 0 or 1 on a line, Wolfram [31] has demonstrated that both regular and fractal patterns can be generated, each site evolving according to definite rules involving the values of its nearest neighbors. For example, given the eight possible states of three adjacent sites (111, 110, 101, 100, 011, 010, 001, and 000) evolving via "rule 90" (i.e., the binary code for the number 90: 01011010), a Sierpinski gasket is generated; other rules (e.g., rule 50) generate regular (e.g., triangular) Euclidean patterns.

A diffusion-controlled reactive process can be designed wherein the development of the pattern, Figure 4.9, occurs in discrete stages. Let stage I be the triangular pattern defined by the sites 1, 2 and 3; stages II-IV are then the patterns generated by the set of sites $(1, \ldots, 15)$, $(1, \ldots, 42)$, and $(1, \ldots, 123)$, respectively. Assume that the onset of growth of each stage requires a chemically specific "trigger mechanism." That is, consider that a molecule or atom localized initially at site 1, upon activation, is released from site 1 and migrates randomly through the lattice, reacting (eventually) at a single, specific site (reaction center) on the far boundary of the system. Once the diffusing coreactant is transformed at the target site, the information carried by the code 90 is processed at each of the points defining the boundary, and the next stage in the evolution of the system unfolds. Similar mechanisms for pattern formation on Euclidean lattices can also be defined. For example, connecting sites 8 and 9 on Figure 4.9 produces a regular, triangular lattice with uniform valency, $v = 6$, and successive generations of this triangular structure can evolve through discrete stages following reaction at a boundary target site and the subsequent reading of a code (e.g., rule 50).

The overall conclusion which follows from solution of the stochastic master equation for each stage in the unfolding of the patterns noted above, viz., the Sierpinski gasket and the associated triangular lattice, is that pattern development triggered by a site-specific recognition event on a Sierpinski gasket is distinctly slower than the corresponding process on a regular triangular lattice [32,33]. This result is again consistent with what would be expected based on the profiles displayed in Figures 4.10 through 4.13. It has been suggested [33] that the evolution profiles for each of the stages described above may have relevance to the generation of neutral networks of the Purkinje type.

3. $d \geq 2$ Fractal Sets

A principal insight that follows from the work of Montroll and Weiss is that the mean walklength $\langle n \rangle$ decreases with increase in the integral dimension of the lattice. Studies of the random walk problem on the triangular lattice $(d = 2)$ versus the Sierpinski gasket $(d = \ell n \, 3 / \ell n \, 2 = 1.584962)$, reviewed in the previous subsection, also showed that trapping on this fractal lattice was distinctly slower (i.e., $\langle n \rangle$ was larger) than on the triangular one. That is, the result remains valid when $\langle n \rangle$ is calculated for flows on sets of fractal dimension, $1 < d < 2$, as well as those calculated for lattices of integral dimension $d = 1, 2$ and is consistent with many studies of phenomena that take place in fractal sets $(d < 2)$ or in sets of low dimensionality [34–40]. A natural question is whether this behavior holds when one considers random walks on fractal sets, $d \geq 2$. This can be addressed by considering the geometry defined by the Menger sponge [23–25,41], a fractal set of dimension $d = \ell n \, 20 / \ell n \, 3 = 2.7268$, and by comparing values of $\langle n \rangle$ for a given N with values of $\langle n \rangle$ calculated for lattices of integral dimension $d = 2$ and $d = 3$. There is also a physical motivation for considering random walks on the Menger sponge, viz. one can use this geometry to study the efficiency of diffusion-controlled reactive processes taking place within or on the surface of a porous catalyst.

Considering first the theoretical issue, diagrammed in Figure 4.15 is the $N = 72$ first-generation Menger sponge (top of figure), and several

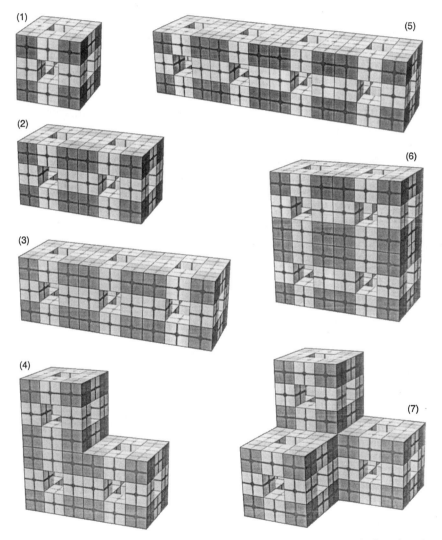

Figure 4.15. The $N = 72$ "first generation" Menger sponge, a symmetric fractal set in three dimensions of Hausdorff (fractal) dimension $\ln 20 / \ln 3 = 2.7268\ldots$, is configuration (1). Calculations based on (1) and the additional configurations (2–7) are discussed in the text.

configurations based on this "building block." The mean walklength $\langle n \rangle_i$ from each of the N sites of each configuration can be calculated [42] and the overall (average) walklength $\langle n \rangle$ determined using the expression, Eq. (4.26). As in the earlier discussion of random walks on simple, finite domains, it is useful to characterize the different configurations in Figure 4.15 by the (overall) root-mean-square distance of the N sites relative to the geometric center of the array [Eq. (4.28)]. Summarized in Table III.8 are the lattice characteristics for the configurations diagrammed in Figure 4.15, and the corresponding values of the overall walklength $\langle n \rangle$ for these geometries.

For all cases, the $\langle n \rangle$ calculated for random walks on the Menger sponge of given N are larger in magnitude than the $\langle n \rangle$ calculated using Montroll's asymptotic expression, Eq. (4.22), for random walks on a $d = 3$ simple cubic lattice ($v = 6$) of N sites with a centrosymmetric trap and subject to periodic boundary conditions. Second, with one exception, the $\langle n \rangle$ calculated for random walks on a Menger configuration of given N are (slightly) smaller than the $\langle n \rangle$ calculated using the Montroll-den Hollander-Kasteleyn asymptotic expression [see Eq. (4.21) and the coefficients in Table III.4] for

TABLE III.8
Lattice characteristics and stochastic results

		Menger Sponge		Torus			Square Planar and Simple Cubic Lattice (Hypothetical)[a]	
N	Symmetry distinct sites	$(r^2)^{1/2}$	$\langle n \rangle$	$N = 1 \times m$	$(r^2)^{1/2}$	$\langle n \rangle$	$d = 2$	$d = 3$
72	3 (Case 1)	1.708	115.937	8×9	3.452	113.740	113.525	109.180
128	12 (Case 2)	2.372	226.551	8×16	5.148	238.667	224.297	194.097
184	18 (Case 3)	3.133	354.612	8×23	7.018	396.369	343.075	279.015
	57 (Case 4)	2.844	348.072					
240	22 (Case 5)	3.954	500.051	8×30	8.954	586.740	467.337	363.933
224	18 (Case 6)	2.909	413.920	14×16	6.124	432.201	431.363	339.670
	42 (Case 7)	3.273	473.533	15×16	6.318	467.546		

[a] Calculated using Eqs. [21, 22].

random walks on a $d = 2$ square-planar lattice $(v = 4)$ of N sites with a centrosymmetric trap.

Comparisons using the asymptotic results in dimensions $d = 2, 3$ are somewhat artificial because for the configurations considered, $N^{1/2}$ and $N^{1/3}$ are not integers. A better comparison in the $d = 2$ case can be obtained by placing the results calculated for a Menger configuration of given N in correspondence with results calculated for walks on a $d = 2$ torus, $N = l \times m$ (where l, m are integers). From the data displayed in Table III.8, again with one exception, the values of $\langle n \rangle$ calculated for random walks on the Menger configurations shown in Figure 4.15 are smaller than $\langle n \rangle$ calculated for walks on a $d = 2$ torus with $N = 8 \times m$.

Thus, for every case $N > 72$ studied, the value of $\{\langle n \rangle, \langle r^2 \rangle^{1/2}\}$ calculated for a Menger configuration of given N is systematically smaller than the $\{\langle n \rangle, \langle r^2 \rangle^{1/2}\}$ calculated for random walks on the corresponding $d = 2$ torus, $N = 8 \times m$. The one exception is the $N = 72$ first-generation Menger sponge. Accordingly, a rigorous test [43] of the basic idea is to calculate $\langle n \rangle$ for the $N = 1056$ second-generation Menger sponge, the structure displayed at the top of Figure 4.16; from the data in Table III.9 the value of $\langle n \rangle$ for walks on the fractal set $d = \ell n\, 20/\ell n\, 3$ is intermediate between results calculated for walks on finite lattices of dimension $d = 2$ and $d = 3$, thereby confirming the lower bound.

A rather surprising result emerges when one plots $\langle n \rangle$ versus $\langle r^2 \rangle^{1/2}$ for the sequence of "linear" Menger configurations, cases 1, 2, 3, and 5 in Figure 4.15, and for the corresponding tori $N = 8 \times m$. Since $\langle n \rangle$ gauges the efficiency of an underlying diffusion-reaction process, the regularity seen in Figure 4.17 suggests seeking correlations between values of $\langle n \rangle$ and turnover numbers for porous catalysts of, here, increasing linear dimension.

A more detailed analysis of diffusion-reaction processes taking place within/on a porous catalyst can be carried out using the geometry specified by the first- and second-generation Menger sponge [44]. Three different cases can be considered. First, one can assume the surface is free of defects and model the catalyst as a Cartesian shell [45] (Euler characteristic, $\chi = 2$) of dimension $d = 2$ and uniform site valency $v = 4$. Second, one can consider processes in which the diffusing reactant confronts areal defects (excluded regions on the surface); in this case, both d and χ remain unchanged, but there is a constriction of reaction space and the site valencies are no longer uniform. Finally, the case of a catalyst with an internal pore structure can be considered explicitly by modeling the system as a fractal solid, here the Menger sponge with fractal dimension $d = 2.73$. Since it is of interest to contrast and distinguish results obtained for surface sites of different symmetry, the symmetry-distinct sites for the configurations diagrammed in Figure 4.16 are labeled.

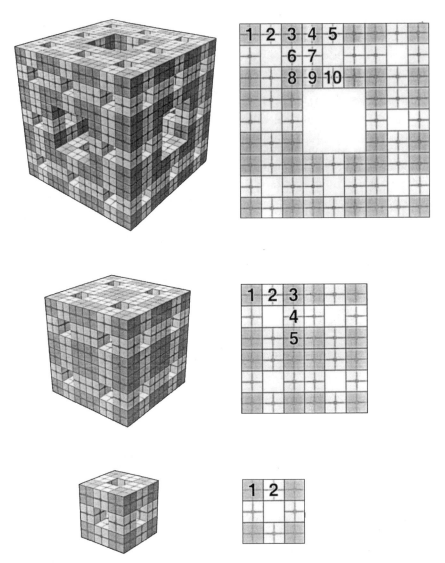

Figure 4.16. The $N = 72$ "first generation" Menger sponge (bottom of figure) and the $N = 1056$ "second generation" Menger sponge (top of figure), are referred to as I and III, respectively, in the text. The configuration in the middle of the figure is II. The integers denote the symmetry-distinct sites on the surface of each configuration (figures on right).

TABLE III.9
Comparison of $\langle n \rangle$ values for integral and fractal lattices

---

$d = 2$

(1) $N = 32 \times 32 = 1024$ square-planar lattice subject to periodic boundary conditions:
   $\langle n \rangle = 2461.339718$ calculated
   $\langle n \rangle = 2461.339738$ Eq. (4.21)

(2) $N = 33 \times 33 = 1089$ square-planar lattice subject to periodic boundary conditions:
   $\langle n \rangle = 2638.784709$ calculated
   $\langle n \rangle = 2638.784732$ Eq. (4.21)

(3) $N = 1056$ square-planar lattice subject to periodic boundary conditions:
   $\langle n \rangle = 2548.538292$ Eq. (4.21)

$d \sim 2.7268$

(1) $N = 1056$ "second generation" Menger sponge:
   $\langle n \rangle = 2161.923358$

$d = 3$

(1) $N = 10 \times 10 \times 10 = 1000$ simple cubic lattice subject to periodic boundary conditions:
   $\langle n \rangle = 1382.581089$ calculated
   $\langle n \rangle = 1516.386$ Eq. (4.22)
   $\langle n \rangle = 1385.192138$ Eq. (4.23)

(2) $N = 11 \times 11 \times 11 = 1331$ simple cubic lattice subject to periodic boundary conditions:
   $\langle n \rangle = 1856.063$ calculated
   $\langle n \rangle = 2018.310$ Eq. (4.22)
   $\langle n \rangle = 1859.452282$ Eq. (4.23)

(3) $N = 1056$ simple cubic lattice subject to periodic boundary conditions:
   $\langle n \rangle = 1601.304$ Eq. (4.22)
   $\langle n \rangle = 1465.287390$ Eq. (4.23)

(4) $N = 1056$ tetrahedral lattice subject to periodic boundary conditions:
   $\langle n \rangle = 1707.341130$ Eq. (4.24)

Three diffusion-reaction scenarios can be specified by referring to the geometry of the $N = 72$ first-generation Menger sponge (bottom of figure). First, imagine that all the surface holes are covered, and at the center of each "cover" is placed a lattice point; this site will be four-coordinated to its first nearest neighbors, and the surface lattice as a whole will have $N = 54$ sites, each of valency $v = 4$. In the second case, imagine that the surface "holes" are covered, but now assume that the covered regions are inaccessible to the diffusing coreactant. In this case, not all sites accessible to the coreactant will have the same valency; the site labeled "1" will have a valency $v = 4$, while the site labeled "2" will have valency $v = 3$. Consideration of the excluded surficial regions leads to a constriction in reaction space, that is, fewer lattice sites ($N = 48$). Finally, changes in the reaction efficiency when

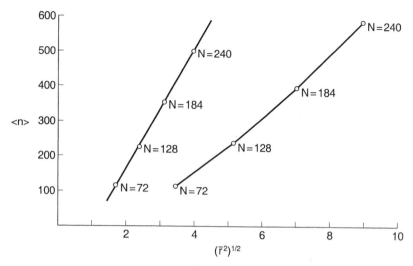

Figure 4.17. A plot of $\langle n \rangle$ versus $\langle r^2 \rangle^{1/2}$ for the configurations 1, 2, 3, and 5 in Figure 4.15 and for the $8 \times m$ tori with $m = 9$, 16, 23, and (30) (line on right).

the diffusing species has access to the inner pore structure of the catalyst can be considered. In this case, a particle situated at site 2 can find its way into the interior of the system, with each lattice site (both in the interior and on the surface) characterized by a uniform valency, $v = 4$. Here, the total number of sites accessible to the diffusing particle is $N = 72$.

Three geometries are displayed in Figure 4.16. The simplest structure (at the bottom of the figure and referred to in the previous paragraph) is denoted as I. The structure at the center of the figure is denoted as II and is characterized by the fact that the number of internal lattice sites is exactly equal to the number of surface sites. Finally, the $N = 1056$ second-generation Menger sponge (at the top) is denoted as III. Listed in Table III.10 are the values calculated for the mean walklength $\langle n \rangle$ for each of these cases. The location of the reaction center is specified in column 2, with $\{1, 2, \ldots\}$ in each case referring to the sites coded in the diagrams accompanying the structures I, II, III. The entries $\langle n \rangle_S$, $\langle n \rangle_D$, and $\langle n \rangle_M$ then denote, respectively, the $\langle n \rangle$ values for the case of coreactant trajectories confined to the surface (only), for the case where the diffusing particle confronts areal defects on the surface, and for the case where the reactant has access to the interior pore structure of the catalyst particle.

Listed in Table III.11 are values of $\langle n \rangle$ calculated for lattices of integer dimensional and specified (uniform) valency, corresponding to the N values in Table III.10; these values were determined from the Montroll-den

TABLE III.10

Values of the mean walklength $\langle n \rangle$ for the three structures, I, II, III

Geometry	Reaction Center Site	$\langle n \rangle_S$	$\langle n \rangle_D$	$\langle n \rangle_M$
I		$N = 54$	$N = 48$	$N = 72$
	1	90.343	83.021	121.705
	2	86.206	99.585	116.479
II		$N = 216$	$N = 192$	$N = 384$
	1	482.136	460.684	823.861
	2	453.821	511.888	773.717
	3	441.177	413.838	738.276
	4	436.689	483.536	706.693
	5	432.876	402.161	704.485
III		$N = 486$	$N = 384$	$N = 1056$
	1	1248.687	1118.344	2542.604
	2	1180.086	1216.327	2397.092
	3	1140.343	1035.285	2270.147
	4	1119.995	1025.430	2228.183
	5	1113.686	1156.746	2236.462
	6	1129.120	1194.568	2170.784
	7	1114.203	1228.874	2152.696
	8	1115.831	1064.126	2104.022
	9	1106.322	1254.162	2090.323
	10	1102.955	1538.772	2127.889

TABLE III.11

Values of the mean walklength $\langle n \rangle$ for lattices of integer dimension

N	$d = 2; v = 4$	$d = 3; v = 4$
48	69.859	65.800
54	80.481	74.991
72	113.525	102.831
192	360.530	293.212
216	413.510	331.794
384	804.236	604.003
384	804.236	604.003
486	1053.851	770.410
1056	2548.538	1707.341

Hollander-Kasteleyn expressions or from polynomial representations of numerical data on $\langle n \rangle$ for lattices of various $[N, d, v]$. It was noted previously that when calculations are performed in which all symmetry-distinct sites, both internal and external in the geometries represented are considered, values of $\langle n \rangle$ for the Menger sponge lie between the corresponding values for $\langle n \rangle$ in $d = 2$ and $d = 3$ for all $N > 72$. The data in Table III.11 show that this correlation holds even if one restricts consideration to surface-only locations of the reaction center, except for a few special site locations (viz., the corner sites in all cases, and the additional site 2 in I).

From studies on regular lattices, the expectation is that the larger the lattice, the longer the mean walklength $\langle n \rangle$ before trapping. Considering the case of a catalyst particle with areal defects (with $d = 2$, $\chi = 2$), the behavior of $\langle n \rangle_D$ with respect to $\langle n \rangle_S$ is found to be more subtle, and depends on the location of the reaction center. If the reaction center is anchored at a site of valency $v = 4$, the N-dependence is as expected, that is, $\langle n \rangle_D < \langle n \rangle_S$. However, positioning the reaction center at a site of valency $v = 3$ or $v = 2$ results in a $\langle n \rangle$ value for the defect case which is greater than that calculated for the defect-free Cartesian shell, even though on the basis of the theoretical N dependence, one would expect exactly the opposite. This behavior, which pertains for all three structures, demonstrates again the sensitivity of the reaction efficiency to the number of nearest-neighbor pathways leading from the surrounding lattice to the reaction center.

Finally, there is an interesting "rule of the thumb" that can be deduced from the data. Define the ratios

$$R_D = \frac{\langle n \rangle_D}{\langle n \rangle_S} \qquad (4.34)$$

and

$$R_M = \frac{\langle n \rangle_M}{\langle n \rangle_S}. \qquad (4.35)$$

As seen from Table III.12, the ratios R_M are in approximate accord with the ratio of sites, that is,

$$R_M \sim \frac{N_M}{N_S} \qquad (4.36)$$

where, N_M is the total number of sites (exterior and interior) accessible to the diffusing coreactant, and N_S is the number of surface-only sites. For the case of unbiased, nearest-neighbor random walks, the site ratio provides a useful,

TABLE III.12
The ratios R_D and R_M

Geometry	Reaction Centre Site	$\dfrac{N_D}{N_S}$	v_D	R_D	$\dfrac{N_M}{N_S}$	v_M	R_M
I		0.888			1.333		
	1		4	0.919		4	1.347
	2		3	1.155		4	1.351
II		0.888			1.753		
	1		4	0.956		4	1.709
	2		3	1.128		4	1.705
	3		4	0.938		4	1.673
	4		3	1.107		4	1.618
	5		4	0.929		4	1.627
III		0.790			2.173		
	1		4	0.896		4	2.036
	2		3	1.031		4	2.031
	3		4	0.908		4	1.991
	4		4	0.916		4	1.989
	5		3	1.039		4	2.008
	6		3	1.058		4	1.923
	7		3	1.103		4	1.932
	8		4	0.954		4	1.886
	9		3	1.134		4	1.889
	10		2	1.395		4	1.929

first-order estimate of the dilation in time scale when a diffusion-controlled reactive process takes place on the surface of a solid with an internal pore structure. The relation (4.36) may be thought of as a counterpart to the kind of power-law scaling relation

$$\text{interaction property} \sim r^d \tag{4.37}$$

(where r is the size of the object and d is an empirical exponent related to a "fractal dimension") often used in the interpretation of phenomena [45–50] in problems as diverse as physisorption, chemisorption, thermal decomposition, noncatalytic reactions, catalytic reactions with supported metal catalysts, and photochemical reactions with semiconductors.

C. Consequences of Dimensional Changes

1. Reduction of Dimensionality

The timing and efficiency of a diffusion-controlled process in a compartmentalized system can be enhanced by reducing the dimensionality of the reaction space of the system. This idea, introduced by Adam and Delbruck [51] and referred to as "reduction of dimensionality," can be explored using stochastic models of chemical reactions in finite, discrete (lattice) systems.

Stating their idea within its original context, Adam and Delbruck proposed that "organisms handle some of the problems of timing and efficiency, in which small numbers of molecules and their diffusion are involved by reducing the dimensionality in which diffusion takes place from three-dimensional space to two-dimensional surface diffusion." In the continuum, diffusion-reaction approach taken by these authors, a target (trap) is fixed at the origin of a coordinate system and one considers the diffusion of a molecule to the target. To assess the importance of reduction of dimensionality, they solved the corresponding (Fickian) diffusion equation for the three-, two-, and one-dimensional problems, keeping the number of traps fixed. For the case of a coreactant diffusing to a centrosymmetric trap via purely random jumps, Monte Carlo simulations showed [52] that the enhancement in reaction efficiency with reduction of dimensionality was indeed borne out, provided the number of traps remained constant as d decreased. However, it was also shown that just the *opposite* behavior pertained if the concentration of traps was kept constant as the dimensionality of the system was reduced. These distinctions could also be inferred from the asymptotic, analytic expressions, Eqs. (4.21) and (4.22), for the number of steps required for trapping on finite, d-dimensional lattices subject to periodic boundary conditions. The approach described in Section II can be mobilized to extend significantly the above conclusions.

Thus far in this review, the focus has been on diffusion-reaction processes that are strictly encounter controlled. Chemical intuition suggests that if the diffusing molecule enters the reaction sphere of the target molecule, short-range quantum-chemical effects are likely to take over and, on such a length scale, the reaction in this situation will become chemically controlled. Alternatively, in condensed phases one often appeals to the so-called "cage effect" to describe aspects of reaction kinetics, and argues that two reactants may remain for some time together in a cage formed by the surrounding solvent, thereby enhancing encounters between the species involved. Clearly, this is also an effect operative in the immediate vicinity of the reaction center.

In the problem discussed in this subsection, as well as those dealt with in subsequent ones, it will be useful to quantify the influence of a (nonspecific) short-range quantum chemical perturbation or a cage effect when such constraints are imposed on the motion of the diffusing particle in the immediate vicinity of the reaction center [53]. Within the context of a lattice model, a first estimate of the importance of these two effects can be obtained by formulating the problem mathematically in such a way as to place special emphasis on the role of the sites that are nearest neighbors to the reaction center; specifically, one imposes the constraint that, upon arriving at one of these nearest-neighbor sites, the diffusing molecule reacts with the target molecule in the next step, that is, migration of the coreactant away from the neighborhood of the target molecule is prohibited. Let $\langle n \rangle$(C) denote the average number of steps before trapping for a molecule undergoing an unbiased, nearest-neighbor random walk in the presence of such a nonspecific "chemical" or "cage" effect.

The quantity $\langle n \rangle$(C) can be related, exactly, to the average number $\langle n \rangle = \langle n \rangle$(RW) of steps before trapping for a molecule undergoing an unbiased, nearest-neighbor random walk on a lattice of N sites, of dimension d, with uniform valence v, and with a centrosymmetric trap. A random walker can be trapped only by passing through a site which is a nearest-neighbor to the central trap. The walklength before trapping can, therefore, be expressed as the walklength until the first passage to a nearest-neighbor site, plus the subsequent walklength until trapping. The first-passage time is, by definition, $\langle n \rangle$(C) $- 1$ (with -1 because one does not count the last step into the trap). The expectation of the subsequent walklength is the expected walk length conditional on starting from a site nearest-neighbor to the trap, which Montroll [17] has shown to be $(N - 1)$. Thus,

$$\langle n \rangle (\text{RW}) = \langle n \rangle (\text{C}) - 1 + (N - 1)$$

or

$$\langle n \rangle (\text{C}) = \langle n \rangle (\text{RW}) - (N - 2). \tag{4.38}$$

To examine this local, short-range effect within the context of the Adam-Delbruck formulation (fixed number of reaction centers in dimension d), consider the results obtained for a 1×13 lattice, a 13×13 lattice, and a $13 \times 13 \times 13$ lattice with a single, centrosymmetric trap [54]. In $d = 1$, the relative enhancement in the efficiency of the process resulting from a short-range chemical or cage effect is $\sim 62\%$, that is, the number of steps required before the diffusing particle reacts irreversibly at the reaction center is only $\sim 62\%$ of what would have been expected were the process a purely random one. In $d = 2$, this percentage changes to $\sim 45\%$, while in $d = 3$ the percentage is $\sim 30\%$. These results show that the higher the dimensionality of the system, the more pronounced are short-range focusing effects in enhancing the efficiency of the diffusion-reaction process. The same general trends are seen when the problem is posed within the context of the Montroll formulation (fixed concentration of reaction centers in dimension d). The results for a 1×729 lattice (i.e., 728 locations specifying the $d = 1$ reaction space surrounding the target molecule), a 27×27 $(= 729)$ lattice and a

$9 \times 9 \times 9$ $(= 729)$ lattice are $> 95\%$ in $d = 1$, $\sim 57\%$ in $d = 2$, and $\sim 25\%$ in $d = 3$. Thus, regardless of whether one considers the number or the concentration of traps fixed, the consistent picture which emerges as the dimensionality of the system is changed, is that short-ranged, chemical or cage effects are relatively more important the higher the dimensionality of the system.

Consider now the pair of curves displayed in each of the plots in Figure 4.18. The lower line corresponds to the situation where the random walk is initialized at (any) one of the $N - 1$ sites surrounding the reaction center. The upper curve corresponds to the case where the migration of the diffusing coreactant is initialized at (any) one of the surface sites of the finite, d-dimensional lattice (only). Although allowing access initially to locations in the near neighborhood of the reaction center should lead to a more efficient diffusion-reaction process than one in which the reactant is confined initially to surface sites only, the results show that for compartmentalized systems the differences in efficiency are not all that great. Overall, in $d = 1$ there is a $\sim 10\%$ enhancement in efficiency when all sites (as opposed to surface sites only) are accessible to the coreactant initially; in $d = 2$ the advantage drops to $\sim 5\%$, whereas in $d = 3$ the estimate is $\sim 1\%$.

A variety of constraints can be imposed on the migration of the diffusing coreactant A as it encounters the boundary of the system. For example, one can impose the condition that the diffusing species A simply returns to the site from whence it came; for the special case that the reaction center is situated at the centrosymmetric site of the defining lattice, adopting this class of boundary conditions yields the same results as for periodic boundary conditions. To explore the concept of "reduction of dimensionality," the so-called "tracking" boundary condition can be invoked; here it is assumed that the walker moves randomly in a space of d dimensions until it encounters the boundary for the first time, after which its trajectory is restricted entirely to the lower dimensional boundary of the compartmentalized system. Considering first the case of symmetrical geometries, the

---

Figure 4.18. Influence of a short-range chemical/cage effect on the efficiency of an irreversible, encounter-controlled reaction at a centrally located reaction center, relative to a purely random process. Plotted is the ratio $\langle n \rangle (C) / \langle n \rangle (RW)$ versus system size for a d-dimensional Euclidean lattice of valency $v = 2d$. The abscissa in the figure is the edge length ℓ; overall, the reaction space has $N - 1$ sites where $N = \ell^d$. The lower curve (Δ) corresponds to the case where migration of the coreactant is initialized at (any) one of the surrounding $N - 1$ sites; the upper curve (0) corresponds to the situation where the random walk is initialized at (any) one of the surface sites of the compartmentalized system. (a) $d = 1$, $v = 2$. (b) $d = 2$, $v = 4$. (c) $d = 3$, $v = 6$.

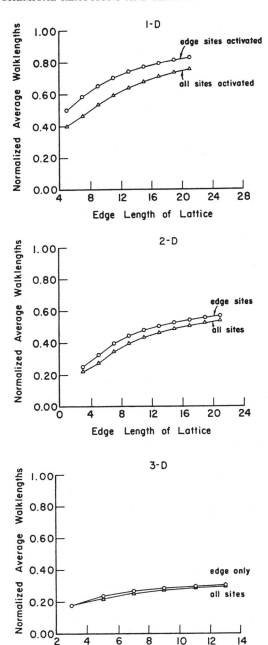

results obtained for the case of an unrestricted ($d = 3$) random walk to a trap situated at the centrosymmetric site of the reaction space are designated in Figure 4.19 by the filled circles; those obtained assuming the tracking boundary condition, with the target molecule situated at the centrosymmetric site on one face of the defining $N \times N \times N$ cubic lattice, are coded by the open squares. The curves drawn through these two set of points cross in the neighborhood of $N = 6$. These results show that upon increasing (symmetrically) the geometrical size of the compartmentalized system, there is a system size beyond which it is clearly more efficient to move the target molecule from an internal, centrosymmetric site in the reaction space to a position on the boundary, provided this shift is accompanied by a concomitant reduction in the dimensionality of the flow of the diffusing particle, here from $d = 3$ to $d = 2$.

Consider next the three curves displayed in Figure 4.20. The curve through the filled circles has the same interpretation as in the previous figure; it represents the results of calculating $\langle n \rangle$ versus edge length for a diffusing particle undergoing a $d = 3$ random walk in a compartmentalized system, with the target located at the centrosymmetric site of the reaction space. The curve through the open squares gives the results obtained using the tracking boundary condition, with the target molecule situated on the boundary of the compartmentalized system, but now one site removed from the centrosymmetric site on the surface. Slight differences are found between the latter curve and the curve through the open squares. If, however, the target molecule is displaced to a relatively inaccessible site on the boundary, the advantages gained in reducing the dimensionality of the flow in the manner suggested by the work of Adam and Delbruck (implemented here using the tracking boundary condition) are compromised. This can be illustrated by placing the target molecule at a surface site characterized not by a valency $v = 4$, but at a site of lower valency, $v = 3$. The curve through the open circles in Figure 4.20 shows clearly that the advantages of reducing the dimensionality of the diffusion space of the coreactant are lost if the number of nearest-neighbor paths to the target is restricted even slightly (here, from $v = 4$ to $v = 3$). Since the number of nearest-neighbor pathways to the reaction center is not a natural parameter in a continuum diffusion theory of the sort implemented in ref. 51, it is not surprising that this local effect had not been assessed. Rather, what is surprising is that a decrease in the local valency for a target molecule positioned on the boundary (from $v = 4$ to $v = 3$) should change the qualitative conclusions drawn from the continuum diffusion-reaction theory. As seen in Figure 4.20, when the number of nearest-neighbor paths to the reaction center decreases by one, it becomes more efficient once again to position the target molecule at the center of the compartmentalized system and assume that the diffusing

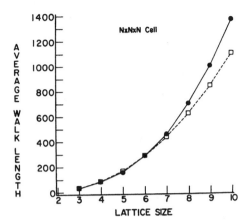

Figure 4.19. A plot of the average walklength $\langle n \rangle$ versus system size (normalized edge length ℓ) for $N \times N \times N$ simple cubic lattices. The curve through the filled circles gives the results for a $d = 3$ walk to a central trap. The curve through the open squares displays the results obtained using the tracking boundary condition (see text), with the trap anchored at a centrosymmetric site on the boundary of the compartmentalized (lattice) system.

coreactant undergoes a random walk in $d = 3$ until it hits the trap (eventually).

It is next of interest to assess the consequences of considering geometries other than the symmetrical ones studied above. Elongation or compression of a symmetrical lattice structure to a $M \times M \times N$ tubule $(N > M)$ or platelet $(N < M)$, respectively, amounts to introducing the simplest sort of symmetry-breaking in the underlying diffusion-reaction problem. In Figures 4.21 and 4.22, changes in the efficiency of the trapping process are studied as a function of changes in N relative to a fixed setting of $M \times M$, again for two choices of boundary conditions. The curve through the filled circles gives the results for an unrestricted $(d = 3)$ random walk to an internal, centralized trap, subject only to the constraint that upon encountering the boundaries of the compartmentalized system, the particle remains confined within the system. The curve through the open squares gives the results for a $d = 3$ random walk to a target molecule positioned at a central site on the boundary of the largest face of the asymmetrical compartmentalized system. Finally, the curve through the open triangles gives the results for walks to the same centrally disposed site on the largest boundary face of the system, but subject to the constraint that when the diffusing particle encounters any boundary for the first time, its subsequent motion is restricted to motion on the boundary (only).

From the results displayed in Figures 4.21 and 4.22, one finds an interesting interplay between the geometry of the compartmentalized system

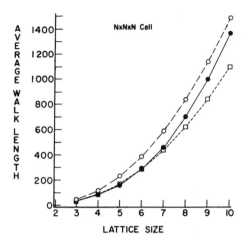

Figure 4.20. A plot of the average walklength $\langle n \rangle$ versus system size (normalized edge length ℓ) for $N \times N \times N$ simple cubic lattices. The curve through the filled circles has the same interpretation as in Figure 4.19. The curve through the open squares gives the results obtained using the tracking boundary condition (see text), for a trap located at a position one site removed from a centrosymmetric site on the boundary. The curve through the open circles shows the consequences of using the tracking boundary condition but shifting the trap to a corner site of the compartmentalized (lattice) system.

and the efficiency of the irreversible reaction, $A + B \rightarrow C$. For relatively small compartmentalized systems, placing the target molecule at a centrally disposed interior site and assuming unrestricted $(d = 3)$ motion of the diffusing species A is somewhat more efficient than positioning the target at a site of maximal accessibility on the boundary of the compartmentalized system and invoking the tracking boundary condition. However, with a distortion in the symmetry and size of the system, in particular along one dimension (the one coded by the integer N), the advantages gained of placing the target molecule on the boundary and invoking the tracking boundary condition become apparent and pronounced. As N increases for a given setting of $M \times M$, a crossover in reaction efficiency occurs, with the critical length ℓ_c shifting to smaller and smaller values of N with increase in size of the $M \times M$ face of the system. The crossover for $3 \times 3 \times N$, $4 \times 4 \times N$, and $5 \times 5 \times N$ occurs between $N = 7$ and $N = 8$, for $6 \times 6 \times N$ lattices between $N = 5$ and $N = 6$, and for the remaining $M \times M \times N$ lattices, between $N = 3$ and $N = 4$. Note that it is only when the tracking boundary condition is imposed that such a crossover in reaction efficiency occurs, relative to the case of unrestricted $d = 3$ diffusion to a centrally disposed, internal reaction center. Note further that if the target molecule is displaced to a centrally disposed position on the largest face of the

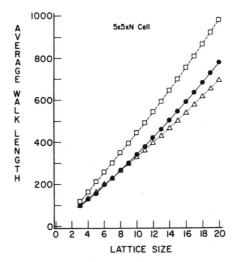

Figure 4.21. A plot of the average walklength $\langle n \rangle$ versus system size (normalized edge length ℓ) for a series of $5 \times 5 \times N$ lattices. The curve through the filled circles gives the results for a $d = 3$ walk to a central trap. The curve through the open squares gives the results for a $d = 3$ walk to a trap positioned at the centrosymmetric site on the boundary of the largest face of the compartmentalized (lattice) system. The curve through the open triangles gives the results for walks to the same centrally disposed site on the boundary of the largest face of the system, but calculated using the tracking boundary condition.

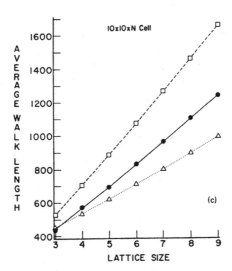

Figure 4.22. A plot of the average walklength $\langle n \rangle$ versus system size (normalized edge length ℓ) for a series of $10 \times 10 \times N$ lattices. The conventions here are the same as in Figure 4.21.

compartmentalized system, but unrestricted $d = 3$ motion of the diffusing coreactant is assumed (subject only to confining boundary conditions), in none of the cases studied is this process competitive with either of the processes described above.

These results highlight once again the point made in previous sections concerning the accessibility of the reaction center to the diffusing coreactant. Statistically, the most accessible site in a d-dimensional system is the site (or set of sites) nearest the geometric center of the system. If one moves the reaction center away from this centralized, internal site, the calculated value of $\langle n \rangle$ increases, with the consequent deterioration in reaction efficiency becoming more pronounced the farther removed the reaction center is from the geometric center of the system. Only by focusing in some way the motion of the diffusing coreactant can the effect of this displacement be counterbalanced and minimized. The proposal of Adam and Delbruck, the concept of "reduction of dimensionality," is one way of inducing such a bias, and their argument that organisms may handle problems of timing and efficiency by collapsing the dimensionality of the flow is simply understood in this context.

Having confirmed that the concept of reduction of dimensionality can play an important role in determining the efficiency of diffusion-controlled reactions in both symmetrical and asymmetrical compartmentalized systems, one may ask: How does a substrate "know," upon first encounter with the boundary, to move along the interior surface of a cellular unit, to react (eventually) with (say) a membrane-bound enzyme? While substrate-specific, surface binding or association forces can conspire to keep the substrate in immediate vicinity of the boundary, once the latter has been encountered for the first time, certain (chemically) nonspecific, statistical factors are also likely to play a role in reduction of dimensionality.

By the very nature of compartmentalized systems, once a diffusing particle hits the boundary, it is positioned on (or near) a site for which the valency v is less than the valency of sites interior to the boundary. For motion interior to the boundary, a diffusing particle can move (randomly) to any of the $v = 2d = 6$ nearest-neighbor sites in the reaction space. If the particle finds itself on the boundary, it can move (randomly) to any of the $v = 5$ nearest-neighbor sites, but of these accessible sites, four are on the boundary. Therefore, in a completely random process, there is an 80% statistical probability that the diffusing particle in its next step will remain on the boundary, and only one chance in five that a displacement into the interior of the system will take place. Using this statistical argument, it is reasonable to suppose that a substrate upon encountering the (inner) surface of a compartmentalized system would tend to remain in the vicinity of this surface in its subsequent motion. Moreover, one would have to invoke only very weak surface association forces to ensure that the 20% statistical probability of moving away from the boundary is suppressed further.

The geometrical fact that the surface sites of a compartmentalized system comprise an ever-increasing fraction of the total number of sites (or, in effect, the volume) of the system as a system increases in size suggests that the concept of reduction of dimensionality is not an exotic alternative implemented only by biological organisms [51]. Rather it emerges as a necessary consequence of the interplay between statistics and geometry, and should be considered in unraveling the kinetics of processes taking place in all organized molecular assemblies (e.g., micelles, vesicles, and micro-emulsions).

A closer inspection of the results reported in Figures 4.19 through 4.22 reveals that the type of symmetry breaking associated with a shift of the reaction center from a (more or less) centrosymmetric, internal position to a boundary site, coupled with an attendant reduction in the dimensionality of the flow, casts light on diffusion-reaction processes in tubules versus platelets. In these systems, one can study the interplay between two geometrical characteristics of the reaction space, the surface-to-volume ratio and the eccentricity and, by constructing a certain phase portrait, display regimes of parameter space where reduction of dimensionality takes over as the governing statistical process. With respect to the first of these characteristics, the surface to volume ratio, S/V, for $M \times M \times N$ lattices is

$$\frac{S}{V} = \frac{2M^2 + 4(M-1)(N-2)}{M^2 N}.$$

(4.39)

The following limits for an infinitely long tubule or for an infinitely extended platelet pertain:

$$\lim_{\substack{M \text{ fixed} \\ N \to \infty}} \frac{S}{V} = \frac{4(M-1)}{M^2} \quad \text{for tubules}$$

$$\lim_{\substack{N \text{ fixed} \\ M \to \infty}} \frac{S}{V} = \frac{2}{N} \quad \text{for platelets.}$$

Table I in ref. 54 shows that the tubule limit is nearly achieved for $3 \times 3 \times N$ lattices with $N = 15$ and the platelet limit is nearly realized for $M \times M \times 3$ lattices with $M = 10$.

A second geometrical parameter that characterizes the reaction space of a compartmentalized system is the eccentricity ε, defined here as

$$\varepsilon = \frac{N}{M} \qquad (4.40)$$

for both tubules and platelets.

The interplay between the choice of boundary conditions (confining or tracking) and the two geometrical factors S/V and ε, in influencing the efficiency of diffusion-controlled reactive processes in compartmentalized systems is illustrated in Figure 4.23. There the region denoted "internal target" shows the regimes of parameter space $(S/V, \varepsilon)$ for which unrestricted, random diffusion in $d = 3$ of the migrating coreactant A to a target molecule B situated at a centrosymmetric, internal site is the more efficient process. Then, the region designated "surface target" indicates regimes of $(S/V, \varepsilon)$ for which unrestricted, random motion in $d = 3$ of the diffusing particle, followed by reduction of dimensionality in the flow to $d = 2$ upon first encounter with the boundary, is the preferred strategy. The

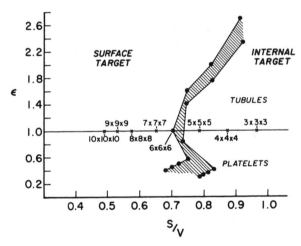

Figure 4.23. Cross-over in reaction efficiency as a function of system geometry for $M \times M \times N$ lattices. The vertical axis calibrates the eccentricity $\varepsilon = N/M$ and the horizontal axis calibrates the surface-to-volume ratio S/V (see text). To the right of the hatched area, random $d = 3$ diffusion to an internal, centrosymmetric reaction center in the compartmentalized system is the more efficient process. To the left of the hatched area, reduction of dimensionality in the $d = 3$ flow of the diffusing coreactant to a restricted $d = 2$ flow upon first encounter with the boundary of the compartmentalized system is the more efficient process. The lines delimiting the hatched region give upper and lower bounds on the critical crossover geometries.

hatched area in the figure corresponds to upper and lower bounds on the extent of the transition zone between the two kinds of behavior. For example, the crossover for the $3 \times 3 \times N$ lattice occurs somewhere between the integer N values of 7 and 8; the values of S/V and ε for these two settings of N are different and this fact is reflected in two different points on the graph, viz., the two (nearly aligned) points in the upper right-hand corner. Since, from Eq. (4.40), values above the abscissa $\varepsilon = 1$ correspond to tubular geometries, whereas those below $\varepsilon = 1$ correspond to platelet geometries, one has a clear picture of those cellular geometries for which reduction of dimensionality is expected to play a significant role in determining the efficiency of diffusion-controlled kinetic processes. The concept of "reduction of dimensionality" will be revisited in a later section in discussing the case where there exist multipolar potential interactions between the coreactants A and B.

2. Continuum Versus Lattice Theories

In classical, continuum theories of diffusion-reaction processes based on a Fickian parabolic partial differential equation of the form, Eq. (4.1), specification of the Laplacian operator is required. Although this specification is immediate for spaces of integral dimension, it is less straightforward for spaces of intermediate or fractal dimension [47,55,56]. As examples of problems in chemical kinetics where the relevance of an approach based on Eq. (4.1) is open to question, one can cite the avalanche of work reported over the past two decades on diffusion-reaction processes in microheterogeneous media, as exemplified by the compartmentalized systems such as zeolites, clays and organized molecular assemblies such as micelles and vesicles (see below). In these systems, the (local) dimension of the diffusion space is often not clearly defined.

The question arises, therefore, whether in the study of diffusion-reaction processes in heterogeneous media one can sensibly use a continuum approach based on the use of a Laplacian operator specified for spaces of integer dimension. Notice that in an approach based on the formulation and solution of a stochastic master equation this question does not arise; one simply defines the geometry of the diffusion space, however complicated, using a lattice representation, and then studies the time evolution of the distribution function for the problem using Eq. (4.3). In fact, the great generality of this approach provides an avenue to explore the above question. By establishing the effective Euclidean dimension for a layered array of finite regular lattices using an approach based on the stochastic master equation, the regime of applicability of a continuum approach to the study of diffusion-reaction processes in compartmentalized systems can be quantified.

To begin, recall that, in general, spaces can be characterized by three quantities: d_e, the dimension of the embedding Euclidean space, d_f the Hausdorff or fractal dimension, and d_s the spectral or fracton dimension. A key to what follows is that for Euclidean spaces, these three dimensions are equal [57,58].

Since the spectral dimension plays an important role in determining dynamic properties on a lattice, considerable efforts have been directed toward its numerical evaluation [59]. These methods are based on a random-walk model: If R_n is the number of distinct sites visited during an n-step walk, then

$$\tfrac{1}{2} d_s = \ell n[R_n]/\ell n[n]. \qquad (4.41)$$

It has also been established that the spectral dimension can be evaluated using the definition

$$\tfrac{1}{2} d_s = -\sum_k P_k \ell n P_k / \ell n[n] \qquad (4.42)$$

where P_k is the probability of visiting the site k in an n-step walk: if site k is visited i_k times, then

$$P_k = i_k/n. \qquad (4.43)$$

This definition of the spectral dimension agrees with that of Eq. (4.41) in the limit as n goes to infinity. The numerical simulation of the spectral dimension based on random-walk models requires enormously large lattices and long walklengths to obtain reliable results [59].

Instead of mobilizing large scale Monte Carlo simulations of the visitation probability P_k as a function of walklength n, P_k can be evaluated (as a function of time) using the stochastic master equation [60]. Suppose at time $t = 0$ a random walker is positioned with unit probability at a site m in the interior of the lattice (away from the boundary of the system). For $t > 0$ this probability evolves among the lattice sites as determined by Eq. (4.3). An entropy-like quantity

$$S(t) = -\sum_i P_i(t)\ell n[P_i(t)] \qquad (4.44)$$

constructed from the solutions obtained to Eq. (4.3) increases as a function of time. Initially, $S(0) = 0$ and, after a long time when uniform probability characterizes each point in the lattice, S approaches the value $\ell n\, N$, where N is the total number of lattice sites. In the limit $N \to \infty$, $S(t)$ increases continuously with t. The growth of $S(t)$ with t here is analogous to the behavior of the numerator of Eq. (4.42) with respect to n. With this identification, it was proposed and verified (see ref. 60 and later text) that the spectral dimension of an arbitrary lattice can be determined from the temporal behavior of the entropy $S(t)$, determined via solution of the stochastic master equation. In particular,

$$\tfrac{1}{2}d_s = S(t)/\ell n(t). \tag{4.45}$$

The evolution of $S(t)$ was studied for $d = 1$ lattices, $d = 2$ square-planar and triangular lattices, and the Sierpinski gasket. In each case, it was observed that after an initial interval of time (which is of the order of one), $S(t)$ grows linearly with $\ell n(t)$. For a finite lattice, on a time scale which depends on the size of the lattice, the evolution curve deviates from strictly linear behavior. The linear regime persists longer the larger the size of the lattice.

Displayed in Figure 4.24 are the results of calculations on a linear lattice of 21 sites, a 21×21 square-planar lattice, a triangular lattice of 361 sites,

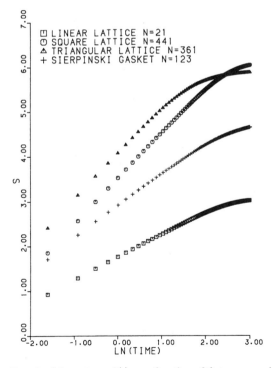

Figure 4.24. Growth of the entropy $S(t)$ as a function of $\ell n \, t$ on several finite lattices.

JOHN J. KOZAK

and a Sierpinski gasket of 123 sites. For regular lattices, the initial boundary condition is imposed at the site symmetrically positioned at the center of the lattice; for the Sierpinski gasket, a site at the middle of one of the sides of the gasket was chosen. Even for these small lattices, there is a pronounced linear regime in the growth of $S(t)$ versus $\ell n(t)$. That this linear regime becomes more pronounced with increase in lattice size can be seen from the profiles displayed in Figure 4.25, where plots are shown for a linear lattice of 101 sites, a 31×31 square-planar lattice, a triangular lattice of 1027 sites, and a Sierpinski gasket of 366 sites. Results obtained from the best (least squares) fit of the slope of $S(t)$ versus $\ell n(t)$ are reported in Table III.13, and it is seen that very accurate estimates of the spectral dimension of each lattice are obtained via the use of Eq. (4.45).

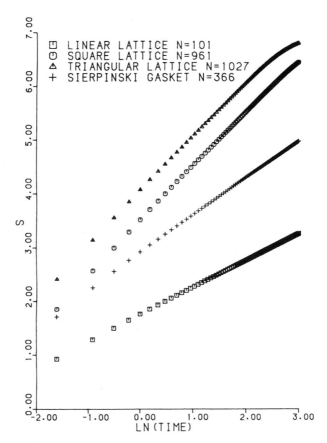

Figure 4.25.   Same as Figure 4.24, but for larger lattices.

TABLE III.13
Results from the least-squares fits to the linear growth regimes of $S(t)$

| Lattice | Present Estimate | | Theoretical Value of d_s |
	Slopes	d_s	
Linear	0.500428 ± 0.000003	1.00	1.00
Square	1.000259 ± 0.000121	2.00	2.00
Triangular	1.000588 ± 0.000113	2.00	2.00
Sierpinski	0.683745 ± 0.000103	1.367	1.365 [a]

[a] From ref. 58.

The above approach, based on the solution of Eq. (4.3), is numerically superior and more accurate than one based on conventional Monte Carlo simulations. For comparison, Pitsianis et al. [59b] performed Monte Carlo simulations using 100,000 random walks on a Sierpinski gasket of 29,526 sites and obtained a value of 1.354 for d_s (the exact value is 1.365). In the approach elaborated above, the value 1.367 was obtained by solving the stochastic master equation on a gasket of only 366 sites.

Now consider the behavior of $S(t)$ versus $\ell n(t)$ for a sequence of layered lattices, with the objective of determining the (minimal) size of a $k \times k \times k$ cubic lattice such that an effective Euclidean dimension $d_e = 3$ is realized [61]. The expectation is that one will then have a concrete means of gauging the validity of theoretical studies of diffusion-reaction processes occurring in compartmentalized systems of finite spatial extent when analyzed using an approach based on Eq. (4.1) with the Laplacian defined for integer dimension. In Table III.14 are displayed data on the slopes and correlation coefficients for a sequence of (one-layer) square-planar lattices determined from the linear regime of $S(t)$ versus $\ell n(t)$ (the range $1 < t < 20$). Note that although the correct asymptotic value of S is given by the limiting value $(\ell n \, N)$ for the 11×11, 21×21, and 31×31 lattices (see Table III.15), the values of the slope of S versus $\ell n(t)$ in the linear regime remain effectively the same for these lattices; in fact, from Table III.14, it is seen that the slopes of S versus $\ell n(t)$ for the latter two lattices are exactly the same (to eight significant figures). The data in Table III.16 then gives the results for the slopes and the correlation coefficients for a sequence of two-layer, square-planar lattices, with each site of valency $v = 5$. As is seen from these data, especially for the case of the 11×11 square-planar lattice with a companion 11×11 overlayer (a total of $N = 2(121) = 242$ sites), the Euclidean dimension of the layered diffusion space calculated using Eq. (4.45) remains effectively $d_e \sim 2$.

JOHN J. KOZAK

TABLE III.14

Slope of the linear regime [$S(t)$ versus ℓn [t] for one-layer systems[a]

Δt	m^{b}	7×7	9×9	11×11	21×21	31×31
1.0–2.6	9	0.447785	0.744491	0.906475	1.007620	1.007620
		(0.979862)	(0.995307)	(0.999056)	(0.999997)	(0.999997)
1.0–2.4	8	0.476082	0.769195	0.920830	1.008382	1.008382
		(0.984328)	(0.996545)	(0.999347)	(0.999997)	(0.999997)
1.0–2.2	7	0.507647	0.794571	0.935052	1.009188	1.009188
		(0.988556)	(0.997571)	(0.999588)	(0.999998)	(0.999998)
1.0–2.0	6	0.542050	0.821131	0.948087	1.009873	1.009873
		(0.992167)	(0.998473)	(0.999740)	(0.999998)	(0.999998)
1.0–1.8	5	0.579079	0.847115	0.960882	1.010909	1.010909
		(0.995017)	(0.999095)	(0.999857)	(0.999998)	(0.999998)
1.0–1.6	4	0.618940	0.873219	0.973131	1.012718	1.012718
		(0.997122)	(0.999542)	(0.999939)	(1.000000)	(1.000000)
1.0–1.4	3	0.664221	0.898363	0.983933	1.013416	1.013416
		(0.998880)	(0.999789)	(0.999983)	(1.000000)	(1.000000)
1.0–1.2	2	0.713110	0.927043	0.992869	1.014811	1.014811
		(1.000000)	(1.000000)	(1.000000)	(1.000000)	(1.000000)

[a] Listed under each lattice are the slope and associated correlation coefficient (in parentheses) for a linear least-squares fit of the data generated for S versus $\ell n\ t$.

[b] m is the number of points considered in the specified range of Δt; the grid used is uniformly 0.2.

Reported in tables I–V of ref. 61 are the results obtained for the effective Euclidean dimension d_e of the lattice spaces $3 \times 3 \times 3$, $5 \times 5 \times 5$, $7 \times 7 \times 7$, $9 \times 9 \times 9$, and $11 \times 11 \times 11$. Also recorded in these tables are the limiting values of $\ell n\ N$ for each case, calculated in the long-time regime; the correspondence of these latter data with the theoretically-predicted value of

TABLE III.15

Asymptotic values of the entropy $S(t)$

Lattice	Layer	N	$S(t \to \infty)$	$\ell n\ N$
3×3	1	9	2.197225	2.197225
5×5	1	25	3.218879	3.218879
7×7	1	49	3.891820	3.891820
9×9	1	81	4.394449	4.394449
11×11	1	121	4.795791	4.495791
21×21	1	441	6.089045	6.089045
31×31	1	961	6.867974	6.867974
7×7	2	98	4.584967	4.587967
9×9	2	162	5.087596	5.087596
11×11	2	242	5.488938	5.488938

TABLE III.16
Slope of the linear regime [$S(t)$ versus $\ln t$] for two-layer systems[a]

Δt	m[b]	7×7[c]	9×9[c]	11×11[c]
1.0–2.6	9	0.445778 (0.978614)	0.752670 (0.994794)	0.915005 (0.998873)
1.0–2.4	8	0.485163 (0.983219)	0.778677 (0.996118)	0.930686 (0.999209)
1.0–2.2	7	0.518088 (0.987598)	0.805783 (0.997259)	0.945738 (0.999452)
1.0–2.0	6	0.554189 (0.991355)	0.833497 (0.998159)	0.960578 (0.999635)
1.0–1.8	5	0.593240 (0.994289)	0.861920 (0.998.854)	0.975589 (0.999784)
1.0–1.6	4	0.637697 (0.996914)	0.890889 (0.999344)	0.990075 (0.999880)
1.0–1.4	3	0.685223 (0.998654)	0.922248 (0.999747)	1.004934 (0.999952)
1.0–1.2	2	0.740538 (1.000000)	0.954471 (1.000000)	1.020296 (1.000000)

[a] Listed under each lattice are the slope and associated correlation coefficient (in parentheses) for a linear least-squares fit of the data generated for S versus $\ln t$.

[b] m is the number of points considered in the specified range of Δt; the grid used is uniformly 0.2.

[c] Spatial extent of basal plane; total number of lattice sites is 2 (n $\times$ n).

$\ln N$ (where $N = 27, 125, 343, 729$, and 1331, respectively) is a critical check on the accuracy of the calculations. Summarizing these calculations, the results in Table III.17 show that the $3 \times 3 \times 3$ lattice is still "behaving" like a Euclidean lattice of dimension $d_e \sim 2$. However, already for the $5 \times 5 \times 5$ lattice, there is a distinct "jump" in the value of d_e; then, with the $7 \times 7 \times 7$, $9 \times 9 \times 9$, and $11 \times 11 \times 11$ lattice, one approaches asymptoti-

TABLE III.17
Values of $\langle n \rangle$ and d_e calculated for $k \times k \times k$ cubic lattices (see text)

N	$\langle n \rangle$	$\langle n \rangle_M$	%[a]	d_e (calc.)	d_e (integer)	%[b]
27	30.46	40.94	34.42	1.9465061	3	35.12
125	157.32	189.55	20.49	2.5772854	3	14.09
343	455.27	520.12	14.24	2.8031932	3	6.56
729	997.4	1105.4	10.83	2.8992408	3	3.36
1331	1856.1	2018.3	8.74	2.945152	3	1.83

[a] % $\equiv [(\langle n \rangle - \langle n \rangle_M)/\langle n \rangle]100$, with $\langle n \rangle_M$ calculated from Eq. (4.22).

[b] % $= [(d_e$ (integer)$-d_e$ (calc.)) $/d_e$ (integer)]100.

cally a spatial regime where the dimensionality is $d_e \sim 3$. This establishes a lower bound on the size of a compartmentalized system for which a continuum approach based on Eq. (4.1) with a $d_e = 3$ dimensional Laplacian is justified. These results will be used in interpreting trends found in the study of diffusion-reaction processes in pillared days (Section III.D.2).

D. Compartmentalized Systems

1. Molecular Organizates and Colloidal Catalysts

At the molecular level, the surface of a colloidal catalyst particle or molecular organizate (such as a cell, micelle, or vesicle) is not smooth and continuous, but rather differentiated by the geometry of the constituents and, if the surface composition is not homogeneous, often organized into clusters or domains. Thus, in many diffusion-controlled reactive processes in which the reactants are confined to the surface of a particle, the species are not free to diffuse freely over the surface. For example, in cellular systems, although lateral diffusion over the surface of the medium may be quite free, the diffusing species must nonetheless negotiate transmembrane (and other) proteins whose presence bifurcates the surface into domains connected by channels. In colloidally dispersed catalyst systems, randomly dispersed "islands" of catalytic activity on the surface of a particle are believed to govern the turnover and overall kinetic response of the system.

These and many other examples suggest the development of a model wherein the presence of individual sites connected by "pathways" or "reaction channels" to a reaction center (the consequence of having a surface broken up into domains or differentiated into clusters) is built into the formulation of the problem from the very outset. That is, rather than assuming that all sites are accessible to a diffusing coreactant so that a rotational diffusion equation can be written down

$$\frac{\partial C(\theta, t)}{\partial t} = \frac{D}{\sin \theta} \frac{\partial}{\partial \theta} \left[\sin \theta \frac{\partial C(\theta, t)}{\partial t} \right] \qquad (4.46)$$

(where C is the concentration, θ is an angle defined with respect to a defining spherical coordinate system, and D is an associated rotational diffusion coefficient) and subject to a variety of initial conditions [62,63], a lattice model can be constructed whose elaboration allows one to explore some consequences of relaxing the free diffusion approximation.

In the simplest approach, a lattice can be wrapped on the surface of a particle homeomorphic to a sphere [64]. Consider the connectivity and attendant domain structure defined by the five Platonic solids [45]. If d denotes the dimensionality of the surface of the particle, N the number of sites on the surface, and v the number of channels or reaction pathways accessible to a diffusing coreactant at each site on the surface, then the surface topology of the five Platonic solids can be specified by the triple $[d, N, v]$; thus the tetrahedron would be [2,4,3], the octahedron [2,6,4], the hexahedron (cube) [2,8,3], the icosahedron [2,12,5], and the dodecahedron [2,20,3]. One can then compute, using the theory of finite Markov processes, the average number of displacements required before a particle diffusing on the surface reacts with a target molecule, the latter anchored at a fixed site on the surface. Since all sites N on the surface of a given Platonic solid are characterized by the same valency v and local symmetry (the defining angles and bond lengths), any site can be specified as the reaction center without loss of generality. Assuming that a common jump time $\hat{\tau}$ is assigned for the site-to-site transitions on each lattice, calculations show [64] that the mean walklength $\langle n \rangle$ increases with increase in the number of sites surrounding the target molecule. Whereas this trend seems intuitively reasonable, its generality is open to question for several reasons.

In calculating $\langle n \rangle$ for the five Platonic solids as a function of increasing N, the valency v in the triple $[d, N, v]$ also changes (irregularly) with increase in N (from $v = 3$ to 4 to 3 to 5 to 3 for the five members of the series). Recall that the relation between the mean walklength $\langle n \rangle$ and the smallest eigenvalue of the defining stochastic master equation is mediated by the valency (or, here, the number of reaction channels) of the underlying reaction network [see Eq. (4.7)]. Thus, there exists the possibility that the dynamics of the diffusion-controlled reaction process, as calibrated by a characteristic relaxation time and determined by solving explicitly the stochastic master equation for the problem, may not of necessity be characterized by the same regularity as that which is found by studying the N dependence of $\langle n \rangle$.

Accordingly, it is necessary to determine the exact dynamics of the diffusion-reaction lattice model described above and study the change in reactivity as a function of $[d, N, v]$. It is also worthwhile to broaden the class of surface domain structures and this can be done by studying the efficiency of encounter-controlled reactive processes on the surfaces of the so-called Archimedean solids [65]. These generalizations allow a more comprehensive evaluation of the interplay of the variables $[N, v]$ in influencing the dynamics of $d = 2$ surface-mediated processes.

In solving the stochastic master equation, two classes of initial conditions are specified. Consider first a coreactant diffusing in an ambient environment (e.g., a solution) and colliding with a colloidal particle (or cellular assembly) at some site k on its surface. From that moment ($t = 0$) on, the coreactant is assumed to diffuse randomly on the surface (only) from site to site until it reacts irreversibly with a target molecule anchored at one site.

The number of discrete steps in this trajectory will be some number, say k_1. However, the trajectory followed by the diffusing coreactant starting from that initial site k is not unique; a variety of possible paths on the surface will be accessible to the coreactant, and each will be characterized by a walklength k_i. The most probable or average walklength will be the statistical average of all possible paths from site k to the reaction center, and this number is denoted $\langle n \rangle_k$. Of course, there is nothing special about the particular site k; the coreactant may collide with the surface at any of the $N - 1$ satellite sites defining the polyhedral assembly. The corresponding, overall average walklength characterizing all possible flows from all possible satellite sites is denoted $\langle n \rangle$.

Following from the above discussion, the first class of initial conditions is specified by assigning $\rho_k(t = 0) = 1$ with $\rho_i(t = 0) = 0$ for all i not equal to k. The second class of conditions is characterized by assigning $\rho_k(t = 0) = (N - 1)^{-1}$ for all k (excluding the trap or target site). In more descriptive language, the first class of initial conditions corresponds to evolution from a "pure state," specifically from trajectories of the coreactant initiated from the site on the polyhedral surface farthest removed from the reaction center, while the second class relates to evolution from a "manifold of states," that is, from trajectories initiated from any (and all) satellite sites of the system.

Displayed in Figures 4.26 and 4.27 are the survival probabilities $\rho(t)$ of a coreactant diffusing on a polyhedral surface with a single target molecule anchored at a site of valency $v = 3$ (the $N = 4$ tetrahedron, the $N = 8$ hexahedron, the $N = 20$ dodecahedron, the $N = 24$ truncated octahedron), $v = 4$ (the $N = 6$ octahedron), or $v = 5$ (the $N = 12$ icosahedron). The two curves for the case $N = 14$ refer to a rhombic dodecahedron; this polyhedron has eight sites of valency $v = 3$ and six sites of valency $v = 4$, and hence an average (or "fractional") valency $v = 3.4286$. The two curves for $N = 14$ correspond to having the reaction center positioned at a site of valency $v = 3$ or $v = 4$.

The eigenvalue spectrum for each of the polyhedral geometries considered in Figures 4.26 and 4.27 is given in Table III.18. Displayed in Table III.19 is the correspondence between the smallest eigenvalue λ_1 and the average walklength $\langle n \rangle$. In the limit of large N, the formal relationship is given by Eq. (4.7). The extent to which the single eigenvalue λ_1 dominates the decay can be gauged by calculating higher-order moments of the distribution function, $\rho(t)$, that is, its relative width, skewness, and kurtosis, and also by fitting the profiles displayed in Figure 4.26 to a two-term polynomial of the form, $\ln \rho(t) = \ln \bar{a} - \bar{\lambda} t$, and determining $\bar{\lambda}$ via a least-squares procedure. Values of the moments are given at the bottom of Table III.18. The $\bar{\lambda}$ encompasses information on the set $\{\lambda_m\}$ for each polyhedra

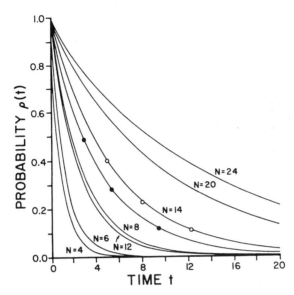

Figure 4.26. Survival probability $\rho(t)$ of a coreactant diffusing on a polyhedral surface with a single reaction center (target molecule) anchored at a site of valency $v = 3$ ($N = 4$ tetrahedron, $N = 8$ hexahedron, $N = 20$ dodecahedron, $N = 24$ truncated octahedron), $V = 4$ (the $N = 6$ octahedron), or $V = 5$ (the $N = 12$ icosahedron). The two curves for the case $N = 14$ refer to a rhombic dodecahedron ($\langle v \rangle = 3.4286$) with a reaction center positioned at a site of valency $v = 4$ (filled circles) or $v = 3$ (open circles). An equal a priori probability of the diffusing coreactant being at any of the $N - 1$ satellite sites at time $t = 0$ is taken as the initial condition, viz. $\rho(0) = (N - 1)^{-1}$.

(Table III.18) and the close correspondence between λ_1 and $\bar{\lambda}$ (Table III.19) shows clearly the dominance of the eigenvalue λ_1 in driving the decay of the system. From these calculations, two general features emerge: (1) the correspondence between the moment $\langle n \rangle$ and the parameter λ_1^{-1} is quite close, even for the smallest systems, and (2) the decays on all surfaces studied are effectively dominated by the single eigenvalue λ_1.

From the data on $\langle n \rangle$ for Platonic solids, the mean walklength increases with increase in the number N of sites on the polyhedral surface. Since the bonds (and angles) defining the surface of a given Platonic solid are all the same, if one assumes a common bond length (or metric) for all the Platonic solids and a common jump time $\hat{\tau}$ between adjacent sites on the surface, the trend in the data on $\langle n \rangle$ versus N is sensible. It is in analyzing the full dynamics of the decay that one realizes there is more to the story. While the evolution curves displayed in Figures 4.26 and 4.27 show a general slowing of the decay with increase in N, there is an inversion of order in the curves describing the evolution on the hexahedral ($N = 8$) versus the octahedral

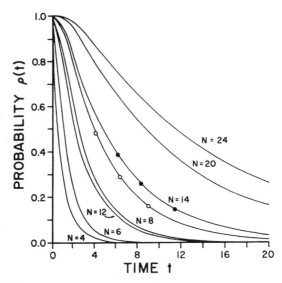

Figure 4.27. Same conventions as in Figure 4.26 except that at $t = 0$ the diffusing coreactant initiates its motion at the site farthest removed from the reaction center, viz., $\rho(0) = 1$ at that site.

($N = 12$) surface. The factor that has been overlooked in understanding the ordering of the decay curves is the valency v. It was this factor that was necessary to make precise the relationship between $\langle n \rangle$ and λ_1^{-1}, and hence it is of interest to quantify precisely how v influences the overall decay process.

First of all, it is clear that for fixed valency v the ordering of the decay profiles displayed in Figures 4.26 and 4.27, specified by λ_1^{-1} as a function of N, is in one-to-one correspondence with the order of increase of $\langle n \rangle$ with respect to N; for the tetrahedron $[N, v] = [4,3]$, the hexahedron [8,3], and the dodecahedron [20,3], polyhedra characterized by a common valency $v = 3$, both measures of the efficiency of the underlying diffusion-reaction process are consistent. The data for the truncated octahedron (with $N = 24$ and $v = 3$) are also consistent with the trends noted above for the three Platonic solids, although a qualification enters here. The angles of the truncated octahedron (second member of the second row in Figure 4.28) are not all the same, and hence results for this polyhedral surface can be compared with those for the Platonic surfaces only if one assumes that the nearest-neighbor jump times $\hat{\tau}$ are the same irrespective of this difference.

A second indicator that the channel structure of the surface is a critical determinant of the dynamics emerges upon examining the data on the rhombic dodecahedron. This figure has eight sites of valency $v = 3$ and six

TABLE III.18
Eigenvalue spectrum for specific polyhedral geometries

Figure	N	v	i	j^a	λ_i
Tetrahedron[*]	4	3	1	1	1.00000
			2	2	4.00000
Octahedron[†]	6	4	1	1	0.763932
			2	2	4.00000
			3	1	5.23607
			4	1	6.00000
Hexahedron[‡]	8	3	1	1	0.354249
			2	2	2.00000
			3	1	3.00000
			4	2	4.00000
			5	1	5.64575
Icosahedron[§]	12	5	1	1	0.388565
			2	2	2.76393
			3	1	3.80311
			4	4	6.00000
			5	1	6.81082
			6	2	7.23607
Dodecahedron[¶]	20	3	1	1	0.100647
			2	2	0.763932
			3	1	1.11238
			4	4	2.00000
			5	1	2.53758
			6	3	3.00000
			7	1	4.10625
			8	3	5.00000
			9	1	5.14314
			10	2	5.23607

Figure	N	v	i	j	λ_i^c	i	j	λ_i^d
Rhombic dodecahedron	14	3.4286[b]	1	1	0.180060	1	1	0.225533
			2	2	1.43845	2	2	1.43845
			3	1	1.92583	3	1	2.16663
			4	3	3.00000	4	4	3.00000
			5	2	4.00000	5	1	4.00000
			6	1	5.07417	6	1	4.83337
			7	2	5.56155	7	2	5.56155
			8	1	6.81994	8	1	6.77447
Truncated octahedron	24	3	1	1	0.0760907			
			2	2	0.585786			
			3	1	0.848668			
			4	1	1.26795			
			5	1	1.49030			
			6	2	2.00000			
			7	1	2.26288			
			8	2	2.58579			
			9	1	3.00000			

TABLE III.18   (*Continued*)

Figure	N	v	i	j	$\lambda_i{}^c$	i	j	$\lambda_i{}^d$
			10	2	3.41421			
			11	1	3.73712			
			12	2	4.00000			
			13	1	4.50970			
			14	1	4.73205			
			15	1	5.15133			
			16	2	5.41421			
			17	1	5.922391			

^{a} Degeneracy of the eigenvalue λ_i.
^{b} The rhombic dodecahedron has eight sites of valency $v = 3$ and six sites of valency $v = 4$.
^{c} Trap positioned at a site of valency $v = 3$.
^{d} Trap positioned at a site of valency $v = 4$.
* Relative width = 0.8165; skewness = 2.0412; kurtosis = 6.1667.
† Relative width = 0.9053; skewness = 2.0140; kurtosis = 6.0562.
‡ Relative width = 0.9584; skewness = 2.0124; kurtosis = 6.0516.
§ Relative width = 0.9762; skewness = 2.0089; kurtosis = 6.0376.
¶ Relative width = 1.0135; skewness = 2.0138; kurtosis = 6.0599.

sites of valency $v = 4$. As is seen from the data in Figures 4.26 and 4.27 and Tables III.18 and III.19, switching the target molecule from a site of valency $v = 3$ to a site of valency $v = 4$ decreases the average walklength $\langle n \rangle$ (from $\langle n \rangle = 18.8$ to 14.6) and decreases the time scale over which the diffusing reactant may be expected to react with the (fixed) target molecule (for both sets of initial conditions).

To summarize, the trends uncovered for the Platonic solids show that holding d and v constant in the triple $[d, N, v]$ while increasing N leads to stochastic and dynamic system responses that are both self-consistent and intuitively reasonable. Secondly, the data on the hexahedron and icosahedron indicate that, for fixed dimensionality d, the interplay between the variables N and v in the triplet $[d, N, v]$ can lead to results that are, at first sight, counterintuitive. And, thirdly, the data on the rhombic dodecahedron, where both d and N are held constant, argue that the variable v is responsible for the differences observed in calculations on the $N = 8$ and $N = 12$ Platonic surfaces; clearly, the influence of channel structure (as coded by the valency) on the reaction efficiency requires further study. Although the role of lattice valency in influencing $d = 2$ dimensional random walks might be inferred from Eq. (4.21) (and the coefficient values in Table III.4), it is important to stress that in deriving Eq. (4.21), periodic boundary conditions were imposed and the lattices studied were of uniform valency [17–19]. Here, the lattices are wrapped on a $d = 2$ surface of different topology [45] (Euler number $\chi = 2$ here versus $\chi = 0$ for planar lattices), and furthermore

TABLE III.19
Correspondence between $\langle n \rangle$, $v\lambda_1^{-1}$ and $v\bar{\lambda}_1^{-1}$

Figure	N	v	$\langle n \rangle$	λ_1	$v\lambda_1^{-1}$	$\%^d$	$\bar{\lambda}$	$v\bar{\lambda}_1^{-1}$	$\%^e$
Tetrahedron	4	3	3.0000	1.00000	3.00000	0.0	1.00391	2.9883	−0.39
Octahedron	6	4	5.200	0.76393	5.2361	0.69	0.76830	5.2063	0.12
Hexahedron	8	3	8.2857	0.35425	8.4686	2.21	0.35372	8.4813	2.36
Icosahedron	12	5	12.7273	0.38857	12.8679	1.10	0.39133	12.7770	0.39
Rhombic dodecahedron	14	3.4286^a	14.6154^b	0.22553	15.2022	4.01	0.22555	15.2009	4.01
			18.7692^c	0.18006	19.0414	1.45			
Dodecahedron	20	3	28.8417	0.10065	29.8071	3.35	0.10124	29.6340	2.74
Truncated octahedron	24	3	37.9752	0.076091	39.4266	3.82	0.076907	39.0079	2.72

^{a} The rhombic dodecahedron has eight sites of valency $v = 3$ and six sites of valency $v = 4$.

^{b} Trap positioned at a site of valency $v = 4$.

^{c} Trap positioned at a site of valency $v = 3$.

$^d \% = [(v\lambda^{-1}\langle n \rangle)/\langle n \rangle] \times 100$.

$^e \% = [(v\bar{\lambda}^{-1}\langle n \rangle)/\langle n \rangle] \times 100$.

323

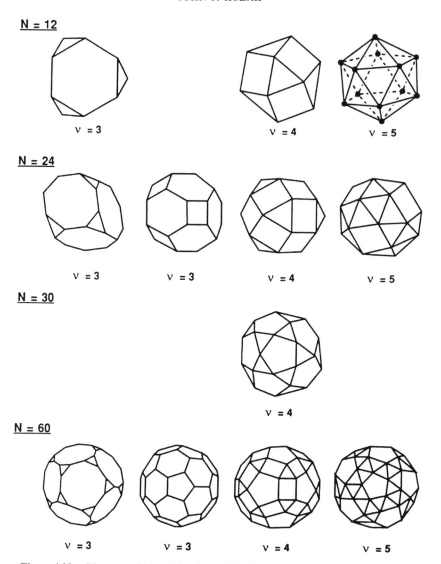

Figure 4.28. Diagrams of 11 Archimedean solids discussed in the text. The Platonic figure $(N, v) = (12, 5)$, the icosahedron, is also included.

the rhombic dodecahedron is a system of mixed valency. Consider, then, the 11 Archimedean solids displayed in Figure 4.28 (along with the icosahedron). Recorded in Table III.20 are the first four moments of the probability distribution function characterizing the overall decay, calculated assuming that class of initial conditions wherein the diffusing coreactant can

TABLE III.20
Characteristics of Archimedean solids

N	v		Site	$\langle n \rangle$	$(\sigma^2/\langle n \rangle)^{1/2}$	γ_1	γ_2	$\lambda^{-1} \equiv t_0$
12	3			16.418	1.0147	2.0241	6.1049	5.47
	4			13.455	0.9832	2.0105	6.0446	3.36
		5		12.727	0.9762	2.0089	6.0376	2.55
24	3 ^{a}			45.130	1.0385	2.0227	6.0995	15.0
	3			37.975	1.0221	2.0157	6.0681	12.7
	4			32.723	1.0105	2.0112	6.0482	8.18
		5		31.091	1.0072	2.0101	6.0435	6.22
30	4			44.690	1.0165	2.0114	6.0492	11.2
36	3		B	89.208	1.0608	2.0343	6.1521	29.7
			A	85.258	1.0569	2.0345	6.1542	28.4
48	3			95.211	1.0307	2.0147	6.0638	31.7
60	3 ^{a}			156.20	1.0397	2.0172	6.0749	52.1
	3			116.65	1.0256	2.0115	6.0498	38.9
	4			103.53	1.0220	2.0101	6.0436	25.9
		5		98.466	1.0205	2.0096	6.0410	19.7
108	3		D	474.10	1.0701	2.0354	6.1567	158
			C	467.89	1.0692	2.0354	6.1573	156
			E	459.37	1.0726	2.0381	6.1694	153
			B	456.27	1.0669	2.0351	6.1564	152
			A	433.01	1.0671	2.0374	6.1676	144
120	3			306.92	1.0304	2.0115	6.0497	102

^{a} Left-hand member in row in Figure 4.28.

initiate its random motion on the surface from any site with equal a priori probability. For purely exponential decay, the second moment [the relative width $(\sigma^2/\langle n \rangle)^{1/2}$], the third moment (the skewness, γ_1) and the fourth moment (the kurtosis γ_2) must have the values 1, 2, and 6, respectively. From the data recorded in Table III.20, it is seen that the evolution curves are essentially exponential, so we can use the mean $\langle n \rangle$ to determine the mean relaxation time τ (with, at most, a few percent error).

Upon correlating changes in the characteristics τ and $\langle n \rangle$ with changes in surface structure, the following trends emerge, again assuming that the jump time $\hat{\tau}$ between adjacent sites is the same regardless of the (small) differences in the length of the channel (or lattice bond length) separating the sites or the angular orientation of a local group of sites. Considering first the series $[N, v] = [12, 3], [12, 4],$ and $[12, 5]$, it is seen that both $\langle n \rangle$ and τ decrease with increase in v. This trend is also clearly displayed in the series $[N, v] = [24, 3], [24, 4],$ and $[24, 5]$, as well as in the series $[N, v] = [60, 3], [60, 4],$ and $[60, 5]$. Note that in the latter two series there are two structures consistent with the setting $[N, v] = [24, 3]$ and with the setting $[N, v] = [60, 3]$ but that, despite the differences in $\langle n \rangle$ observed between the members

v = 3

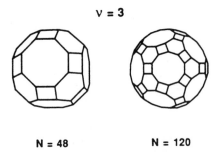

N = 48　　　　　　　　**N = 120**

Figure 4.29.　Two Archimedean solids characterized by valency $v = 3$.

of a pair with fixed N (see below), both structures induce dynamics that are consistent with the trend noted above.

The second trend is that, for a fixed setting of v, both the average walklength $\langle n \rangle$ and the relaxation time increase with increase in N. See the data on the three figures for which $v = 5$, the four figures for which $v = 4$, and the seven structures displayed in Figures 4.28 and 4.29 in the case $v = 3$.

Consider now the data for the pair of figures $[N, v] = [24, 3]$ and the pair $[N, v] = [60, 3]$. Differences in $\langle n \rangle$ and τ are seen in each case, and the feature common to the ordering observed is that the figure with triangular regions on its surface produces values of $\langle n \rangle$ and τ that are systematically larger than those for which no triangular regions are present. Since the presence of triangular domains (or clusters of sites) seems to drive the $\langle n \rangle$ to higher values (qualitatively because the diffusing particle can get trapped in a localized corner of reaction space, with fewer options to get out than for the cases $v > 3$), a situation might arise wherein the presence of multiple triangular domains might drive these $\langle n \rangle$ to values sufficiently high that the dominant role of the variable N in the couple $[N, v] = [N, 3]$, say, might be compromised. To explore whether, for fixed v, a compression or inversion in $[\langle n \rangle, \tau]$ values with increase in N can be realized, calculations were performed on three generically related structures where the number of triangular regions was systematically increased. The first is the figure $[N, v] = [12, 3]$ displayed in Figure 4.28, the singly truncated tetrahedron. Data for this figure, the doubly truncated tetrahedron $[N, v] = [60, 3]$ and the triply truncated tetrahedron $[N, v] = [108, 3]$ are given in Table III.20. Whereas the trend of increasing $[\langle n \rangle, \tau]$ with increasing N is preserved in these three figures, the values of $[\langle n \rangle, \tau]$ for $N = 108$ are actually larger than those calculated for $[N, v] = [120, 3]$, the figure displayed in Figure 4.29. This demonstrates that if the number N of the surface sites of two figures is roughly comparable (here about 10%), an inversion in the broad trend noted above can occur if the percent of triangular domains relative to all other surface domain structures increases. This percentage increases from 50% to 60% to 64% as one considers sequentially the singly, doubly, and triply truncated tetrahedron.

A second insight that follows from the data generated for the n-truncated tetrahedra is that, even for a given $[d, N, v]$, a reaction center situated at a site of valency $v = 3$ but at the intersection of regions of different symmetry (viz. triangular, hexagonal, or dodecahedral symmetry) can have $[\langle n \rangle, \tau]$ values that are slightly different. Data are given in Table III.20 for two distinct sites, A and B, for the figure $[N, v] = [36, 3]$ and for five distinct sites, A, B, C, D, and E for the figure $[N, v] = [108, 3]$. These differences are not great, viz., a difference of $\sim 4\%$ for the two sites A and B for the doubly truncated tetrahedron and a maximum difference of 8.7% for the five distinct sites of the triply truncated tetrahedron, but they suggest the important physical insight that the relative efficiency of reaction on the surface of a molecular organizate or colloidal catalyst can be fine tuned depending on the site valency to which the target molecule is attached.

2. Clays

The versatility of lattice models to describe encounter-controlled reactions in systems of more complicated geometries can be illustrated in two different applications. In this subsection layered diffusion spaces as a model for studying reaction efficiency in clay materials are considered and in the following subsection finite, three-dimensional lattices of different symmetries as a model for processes in zeolites are studied. Now that the separate influences of system size N, dimensionality d (integral and fractal), and valency v have been established, and the relative importance of $d = 3$ versus surface diffusion (and "reduction of dimensionality") has been quantified, the insights drawn from these studies will be used to unravel effects found in these more structured systems.

Many crystals can be considered as built up of parallel layers, especially if the forces operative between layers are of less consequence than those acting within layers. In some materials such as graphite the layers may be monotonic, but in others such as micas or clay materials they may be several atoms deep. Much attention has been directed toward the study of reactions enhanced by molecular catalysts intercalated in smectite clays [66,67]. Because of the ability of these minerals to imbibe water, a solution-like environment can exist between layers and, provided the layers are sufficiently swollen to permit rapid diffusion of reactants, many reactions involving cations or neutral molecules can be carried out. Photocatalytic and photochemical reactions in/on clays or clay-modified materials (electrodes) have been studied and discussions of these experiments are presented in the monographs of Thomas [68] and of Kalyanasundaram [69] and in the seminal papers of Bard et al. [70], DeSchrijver et al. [71], and Van Damme et al. [72].

Explored first is the extent to which the efficiency of reaction between a fixed target molecule and diffusing coreactant is affected when the interlamellar space separating the atomic layers defining the underlying host lattice is expanded [73] or "swollen." The second example illustrates how different spatial distributions of pillaring agents and different interlamellar spacings can influence the reaction efficiency in pillared clays, clays in which the intercalation of polynuclear metal cations or metal ion clusters are used as a means of keeping separate the silicate layers in a smectite clay in the absence of a swelling agent [74].

In the first application, consider a target molecule placed at the centrosymmetric site of a $d = 2$ dimensional lattice of valency $v = 3$. The consequent (hexagonal) lattice is assumed to be of finite extent containing, overall, N total sites. In general, the diffusing coreactant A can initiate its motion from any of the satellite sites surrounding the target molecule B and, after a certain time, can undergo, say, the irreversible reaction, Eq. (4.2). Averaged over all locations accessible to the diffusing coreactant, the mean walklength $\langle n \rangle$ characterizing the trajectories to the target molecule calibrates the efficiency of reaction on the host lattice.

Suppose now a new assembly is designed in which two hexagonal layers are stacked one above the other. The target molecule will be assumed to occupy the same centrosymmetric site in the basal layer of the array and the program of calculation described above repeated, only now with $2N - 1$ satellite sites accessible to the diffusing coreactant. Further "swelling" of the interlamellar space can be studied by aligning yet another finite, hexagonal layer above the previous array; again, the target molecule is positioned at the centrosymmetric site of the basal layer, and the coreactant can diffuse through a reaction space with $3N - 1$ satellite sites. In this construction, one is, in effect, exploring the consequences of changes in the (effective) dimensionality d of the reaction space; in the single-layer problem, diffusion is strictly two dimensional, but in the two- and three-layer problems the diffusing coreactant samples a $d > 2$ dimensional reaction space.

The efficiency of diffusion-controlled reactive processes on one-, two-, and three-layer hexagonal lattices is studied by calculating the mean walklength $\langle n \rangle_i$ of a coreactant diffusing from a given site i in the reaction space to a reaction center placed at the centrosymmetric site of the basal plane. The assembly is subject to nontransmitting (confining) boundary conditions such that if the diffusing particle attempts to exit the system from a boundary site, it is reset at that site (returning eventually to the interior of the assembly). For the one-layer hexagonal lattice, the common valency v of all lattices sites is $v = 3$; for the two-layer ("stacked") assembly, all sites are of valency $v = 4$; and for the three-layer assembly the lattice sites in the upper and lower layers are of valency $v = 4$, while the lattice sites in the middle layer are of valency $v = 5$. Both "in plane" and "out of plane" displacements of the diffusing coreactant are characterized by a common ("jump") length ℓ.

Presented in Table III.21 are the results for $\langle n \rangle$ as a function of N. Different initial conditions were examined. First, in the three-layer system, the notations "upper lattice layer," "lower lattice layer," and "entire lattice" refer respectively to walks initiated from sites on the upper layer only, the basal layer only, or all possible sites comprising the assembly. In Figure 4.30, the results obtained for the case where reactant trajectories are initiated from all possible nontrapping sites of the assembly for one-, two-, and three-layer lattices are compared for common values of N(total) in the range $N < 300$.

For a common setting of N (see Table III.21), and for the same choice of initial conditions, let $\langle n(\mathrm{I}) \rangle$, $\langle n(\mathrm{II}) \rangle$, and $\langle n(\mathrm{III}) \rangle$ designate the average walklength for the one-, two-, and three-layer lattices, respectively. A plot of the ratio $\langle n(\mathrm{II}) \rangle / \langle n(\mathrm{I}) \rangle$ and $\langle n(\mathrm{III}) \rangle / \langle n(\mathrm{I}) \rangle$ over the entire range of N values is presented in Figure 4.31. From this figure, it is seen that in the limit of large N, the ratio $\langle n(\mathrm{II}) \rangle / \langle n(\mathrm{I}) \rangle \rightarrow 1.7$ and the second ratio $\langle n(\mathrm{III}) \rangle / \langle n(\mathrm{I}) \rangle \rightarrow 2.4$. That is, in the limit of large N, the reaction efficiency decreases by a factor of 1.7 for a stacked two-layer hexagonal structure and by a factor of 2.4 for a three-layer structure. In expanding the interlamellar reaction space, the efficiency should decrease, and doubling or tripling the number of sites accessible to the diffusing coreactant might have led to decreases in the reaction efficiency, perhaps by simple factors of 2 or 3. The data here show that the actual decrease is somewhat smaller.

Turning next to the example of pillared clays, as noted by Pinnavaia [66], two principal advantages are realized when one considers the intercalation of polynuclear metal cations or metal ion clusters in smectites. First, in ordinary metal-ion-exchanged forms of smectite clays, there occurs dehydration at high temperatures ($>200°C$); the subsequent collapse of the interlayer region thereby precludes the study of a great variety of possible reactions. The introduction of pillaring cations, however, keeps the silicate layers separated in the absence of a swelling solvent and affords the study of reactions in the high-temperature regime. Second, the use of different pillaring cations allows the design of new catalytic materials with pore sizes that can be made much larger than those of available zeolite catalysts.

Consider now the consequences of assuming different distributions of pillaring cations and different interlamellar spacings on the efficiency of reaction. Since the pillaring cations are space filling, the interlamellar

TABLE III.21

Data for the average walklength as a function of system geometry for $v = 3$

| | One lattice layer | | | | Two lattice layers | | | | | | Three lattice layers | | | | |
| | | | | | | | | ⟨n⟩ | | | | | ⟨n⟩ | | |
N (layer)	N total	C_T	Entire lattice		N total	C_T	Entire lattice	Lower lattice layer	Upper lattice layer		N total	C_T	Entire lattice	Lower lattice layer	Upper lattice layer
4	4	0.25	3.000		8	0.125	8.000	9.000	6.6667		12	0.0833	16.813	21.079	10.771
16	16	0.0625	21.600		32	0.03125	43.148	43.800	42.453		48	0.02083	71.229	75.412	66.572
25	25	0.040	37.500		50	0.020	72.507	73.056	71.934		75	0.01333	116.07	120.19	111.65
49	49	0.02041	86.625		98	0.01020	159.95	160.31	159.57		147	0.00680	246.31	250.30	242.24
64	64	0.01562	119.62		128	0.00781	217.70	218.00	217.39		192	0.00521	331.16	335.10	327.21
100	100	0.0100	204.86		200	0.0050	364.42	364.60	364.24		300	0.00333	543.66	547.51	539.91
121	121	0.00826	256.96		242	0.00413	453.19	453.32	453.06		363	0.00276	671.08	674.90	667.41
169	169	0.00592	381.79		338	0100296	663.79	663.82	663.75		507	0.00197	970.68	974.43	967.15
196	196	0.00510	454.41		392	0.00255	785.47	785.47	785.48		588	0.00170	1142.70	1146.42	1139.23
256	256	0.00391	621.38		512	0.00195	1063.34	1063.26	1063.41		768	0.00130	1533.04	1536.71	1529.69
289	289	0.00346	715.64		578	0.00173	1219.40	1219.30	1219.51		867	0.00115	1751.25	1754.89	1747.94

TABLE III.22

Data for the average walklength as a function of system geometry for $v = 6$

One lattice layer				Two lattice layers		⟨n⟩			Three lattice layers		⟨n⟩		
N (layer)	N total	C_T	Entire lattice	N total	C_T	Entire lattice	Lower lattice layer	Upper lattice layer	N total	C_T	Entire lattice	Lower lattice layer	Upper lattice layer
7	7	0.14286	6.000	14	0.07143	15.256	12.444	17.667	21	0.04762	30.522	19.525	38.736
19	19	0.05263	20.750	38	0.02632	43.877	41.401	46.222	57	0.01754	75.006	64.765	83.356
37	37	0.02703	46.562	74	0.01351	92.145	89.827	94.399	111	0.00901	149.67	139.76	157.98
61	61	0.01639	84.635	122	0.00820	161.07	158.85	163.25	183	0.00546	255.11	245.41	263.39
91	91	0.01099	135.83	182	0.00549	251.51	249.37	253.63	273	0.00366	391.90	382.33	400.13
127	127	0.00787	200.82	254	0.00394	364.23	362.15	366.30	381	0.00262	560.61	551.15	568.81
169	169	0.00592	280.17	338	0.00296	499.87	497.84	501.89	507	0.00197	761.85	752.46	770.01
217	217	0.00461	374.37	434	0.00230	658.99	657.00	660.97	651	0.00154	996.14	986.83	1004.3
271	271	0.00369	483.82	542	0.00185	842.07	840.12	844.02	813	0.00123	1264.0	1254.7	1272.1

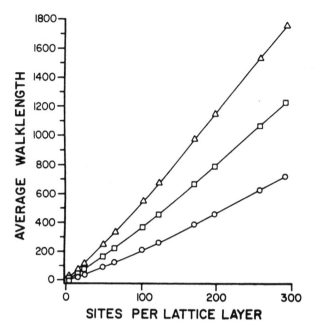

Figure 4.30. A comparison of the average walklength $\langle n \rangle$ versus the total number of lattice sites for a one-layer (circles), two-layer (squares), and three-layer (triangles) hexagonal lattice assembly subject to confining boundary conditions (see text), with the reaction center positioned at centrosymmetric site of the lowest (basal) layer of the compartmentalized system.

reaction space will be bifurcated into an interconnected set of lateral and vertical channels through which the diffusing coreactant can migrate. In the stacked hexagonal structure described above, rather than regarding the "holes" in the assembly to be "empty space," these regions can be regarded as volumes excluded to the diffusing coreactant owing to the presence of the space-filling, pillaring cations. The links (or bonds) of the lattice then represent channels through this array that are accessible to a migrating species in its random motion through the partially ordered lattice.

Assuming a hexagonal structure for the underlying lattice defines one possible distribution of pillaring cations. It is of interest to consider how different distributions of pillaring cations influence the reaction efficiency. Consider first a distribution of cations that leads to a layered lattice structure built up of triangular lattice arrays. The number of pathways available to the diffusing species in the one-layer system is $v = 6$; for the two-layer assembly, all sites are of valency $v = 7$; and, for the three-layer assembly, lattice sites in the upper and lower layers are of valency $v = 7$, whereas sites in the middle layer are of valency $v = 8$. For a distribution of cations

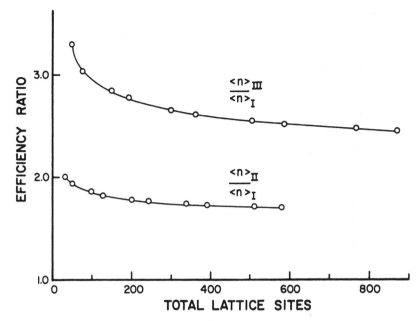

Figure 4.31. A plot of $\langle n \rangle_{II} / \langle n \rangle_I$ (lower curve) and $\langle n \rangle_{III} / \langle n \rangle_I$ (upper curve) versus the total number of lattice sites of the layered array.

built up from square-planar arrays, the valency of sites in the basal lattice is $v = 4$, whereas for the two layer assembly, all sites are of uniform valency $v = 5$.

The boundary conditions imposed and presentation of the data in Table III.22 for the triangular arrays follow the same conventions as that described earlier for the hexagonal arrays. In both cases the position of the stationary target molecule is assigned to be the centrosymmetric site in the basal plane of the array with its concentration $C(T) = 1/N$ bracketed between the values, $0.25 > C(T) > 0.00115$, for partially ordered arrays of hexagonal symmetry, and between the values $0.143 > C(T) > 0.00123$ for arrays of triangular symmetry.

Presented in Figures 4.32, 4.33, and 4.34 for the choice of initial conditions wherein reactant trajectories are initiated from all possible nontrapping sites of the system (the "entire lattice" case) are results calculated for $\langle n \rangle$ versus N for one-, two-, and three-layer lattices. In each figure, two profiles are displayed; the uppermost one corresponds to results calculated for an ensemble based on dimension $d = 2$ planar lattices of hexagonal symmetry $(v = 3)$ while the lower curve is based on $d = 2$ planar lattices of triangular symmetry $(v = 6)$. The ratios of reaction efficiency, $\langle n(II) \rangle / \langle n(I) \rangle$

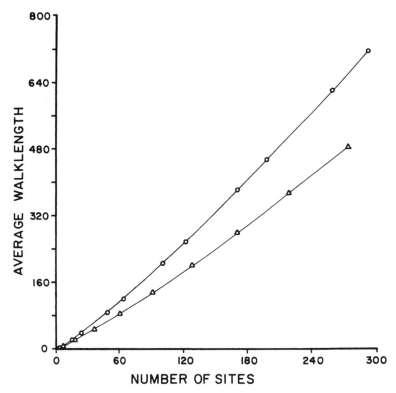

Figure 4.32. A plot of the average walklength $\langle n \rangle_{\mathrm{I}}$ versus the total number of lattice sites for a one-layer, $d = 2$ dimensional lattice subject to confining boundary conditions with the reaction center placed at the centrosymmetric site of the lattice. Displayed one results for lattices of coordination $v = 3$ (circles) and $v = 6$ (triangles).

versus N (Fig. 4.35), and $\langle n(\mathrm{III}) \rangle / \langle n(\mathrm{I}) \rangle$ versus N (Fig. 4.36), for these two lattice architectures can also be compared.

The first generalization that emerges from these calculations is that values of $\langle n \rangle$ for lattice assemblies built up from planar arrays of triangular symmetry are systematically lower than those calculated for layers of hexagonal symmetry. This result is consistent with the behavior noted previously for lattice systems of uniform valency, and is a consequence, again, of the fact that increasing the coordination number of the lattice allows one to cluster more satellite sites closer to the central target molecule. This conclusion seems not to depend on the choice of initial conditions for the lattice architectures considered. Assuming that the trajectories of the diffusing reactant are initialized from the upper, lower, or all nontrapping

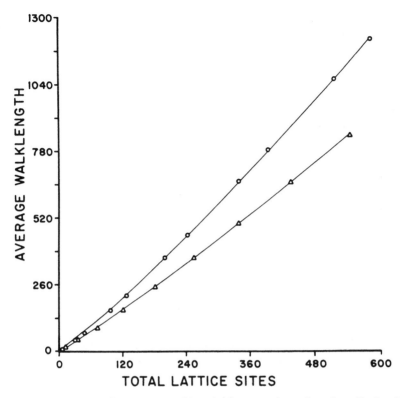

Figure 4.33. A plot of the average walklength $\langle n \rangle_{II}$ versus the total number of lattice sites for a two-layer lattice subject to confining boundary conditions with the reaction center placed at the centrosymmetric site of the lower (basal) layer. Displayed are results for lattices of in-plane coordination $v = 3$ (circles) and $v = 6$ (triangles).

sites of the array leads to only small differences in the values calculated for $\langle n \rangle$ (especially for larger lattices).

A second, rather striking result follows from an examination of Figures 4.35 and 4.36. Although the results for small systems are quite different (say, $N < 60$ for two-layer lattices and $N < 90$ for three-layer lattices), the results obtained in the limit of large $N (N = 600$ for two-layer lattices and $N = 900$ for three-layer ones) tend to coalesce. In particular, for two-layer systems, $\langle n(II) \rangle / \langle n(I) \rangle \sim 1.7$ and for three-layer systems, $\langle n(III) \rangle / \langle n(I) \rangle \sim 2.5$, for both distributions. To check further the apparent independence of the ratio $\langle n(II) \rangle / \langle n(I) \rangle$ on the valency of the defining lattice system, assemblies built up from square-planar lattices $(v = 4)$ can be constructed. As is evident from the data [61a] reported in Table III.23, the limiting ratio, $\langle n(II) \rangle / \langle n(I) \rangle \sim 1.7$ is also realized for these layered lattice systems.

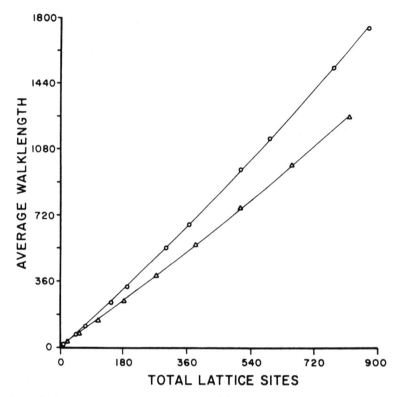

Figure 4.34. A plot of the average walklength $\langle n \rangle_{\mathrm{III}}$ versus the total number of lattice sites for a three-layer lattice subject to confining boundary conditions with the reaction center placed at the centrosymmetric site of the lowest (basal) layer. Displayed here are results for lattices of in-plane coordination $v = 3$ (circles) and $v = 6$ (triangles).

The above results show that, for large arrays, increasing the number of satellite sites from $N - 1$ (one layer) to $2N - 1$ (two layers) to $3N - 1$ (three layers) does not compromise the reaction efficiency by simple factors of 2 or 3 but by somewhat smaller factors that are lattice independent. This invariance is a consequence of the fact that flows on two- and three-layered lattices of given valency v remain effectively two dimensional, a point that was established quantitatively in Section II.C.2. However, a simple calculation also illustrates this point.

Focusing on the case of two hexagonal layers (i.e., a basal plane and one overlayer of sites), all sites in such a configuration are of valency $v = 4$, as noted previously. This valency is the same as for a one-layer, square-planar lattice. Accordingly, the asymptotic theory of Montroll, and den Hollendar and Kasteleyn can be used to compute values of $\langle n \rangle$ corresponding to the

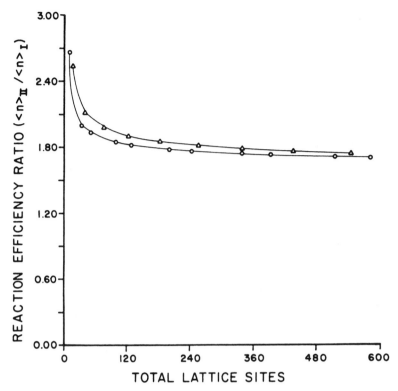

Figure 4.35. A plot of $\langle n \rangle_{II} / \langle n \rangle_{I}$ versus the total number of lattice sites for lattices of in-plane coordination $v = 3$ (circles) and $v = 6$ (triangles).

same number N of sites in two-layer arrays of hexagonal lattices of increasing (lateral) spatial extent. The results of this calculation are displayed in Table III.24. Given that the Montroll–den Hollander–Kasteleyn theory reproduces values of $\langle n \rangle$ to within 5% for large lattices, it is reasonable to conclude that the results reported in Table III.24 verify the effective $d = 2$ dimensionality of a two-layered hexagonal assembly (diffusion space). That is, the near constancy of the ratio $\langle n(II) \rangle / \langle n(I) \rangle$ with respect to lattice valency is a consequence of the fact that, despite the presence of the overlayer, the flow of the diffusing coreactant remains predominantly two dimensional.

These results are consistent with those obtained in Section II.C.2 where the number of planar lattices that must be "stacked" one above the other until the dimensionality of the flow converges to an effective dimension $d = 3$ was established. At a more qualitative level, note that in the time

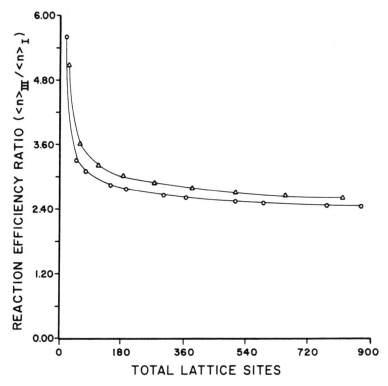

Figure 4.36. A plot of $\langle n \rangle_{\rm III}/\langle n \rangle_{\rm I}$ versus the total number of lattice sites for lattices of in-plane coordination $v = 3$ (circles) and $v = 6$ (triangles).

TABLE III.23
$\langle n \rangle_{\rm I}$ and $\langle n \rangle_{\rm II}$ values for $v = 4$ square-planar lattices

N (unit)	N (one-layer)	$\langle n \rangle_{\rm I}{}^a$	$\langle n \rangle_{\rm I}{}^b$	N (two-layer)	$\langle n \rangle_{\rm II}$	$\langle n \rangle_{\rm II}/\langle n \rangle_{\rm I}$
3×3	9	8.000000	9.000000	18	20.117647	2.235294
5×5	25	24.000000	31.666667	50	63.317966	1.999515
7×7	49	48.000000	71.615385	98	136.129999	1.900849
9×9	81	80.000000	130.604452	162	240.489894	1.841361
11×11	121	120.000000	209.937058	242	377.981585	1.800952
21×21	441	440.000000	942.792176	882	1601.018886	1.698167

^{a} $\langle n \rangle_{\rm I}$ is the average walklength before trapping for walks initiated at the nearest-neighbor site relative to the trap. Montroll [17] proved that $\langle n \rangle_{\rm I} = N - 1$ *exactly*, so the numerical values reported for $\langle n \rangle_{\rm I}$ are an important check on the accuracy of the calculations.

^{b} Values of $\langle n \rangle_{\rm I}$ here are in agreement with those reported in ref. 9 except for the 21×21 case; the value here is more accurate, since the $\langle n \rangle_{\rm I}$ value reported in ref. 9 was 439.8, whereas here 440.000000.

TABLE III.24

Comparison of $\langle n \rangle$ values for one-layer and two-layer $v = 4$ lattices

N	$\langle n \rangle_{\text{one-layer}}{}^{a}$	$\langle n \rangle_{\text{two-layer}}{}^{b}$	$\%^{c}$
8	8.000	7.701	-2.87
32	43.148	42.779	-0.86
50	72.507	73.375	1.20
98	159.95	163.70	2.34
128	217.70	224.30	3.03
200	364.42	378.09	3.62
242	453.19	471.86	3.96
338	663.79	694.36	4.40
392	785.47	822.53	4.62
512	1063.34	1118.63	4.94
578	1219.40	1284.91	5.10

$^{a}\langle n \rangle$ values calculated from the asymptotic expression, Eq. (4.21), and the coefficients in Table III.4.
$^{b}\langle n \rangle$ values reported in ref. 74.
$^{c}\% = [\langle n \rangle_{\text{two-layer}} - \langle n \rangle_{\text{one-layer}}] / \langle n \rangle_{\text{two-layer}} \times 100$.

regime before the asymptotic limit is reached, the dimension for a "stacked" n-layer system might well remain two-dimensional provided the third dimension has an effective length $\ell < (Dt)^{1/2}$, where D is the diffusion coefficient [61b]. Suppose that the relevant (physical/chemical) time scale for a dynamical process in a given microheterogeneous medium is of the order of 1 μsec. Then, for typical values of the diffusion coefficient of adsorbates migrating in a microheterogeneous media [see, for example, the experimental study of Caro et al. [75] on the diffusion of methane, ethane, and propane in ZMS-5], $10^{-8}\,\text{cm}^2/\text{sec} > D > 10^{-10}\,\text{cm}^2/\text{sec}$, the values of ℓ corresponding to this range of diffusion coefficients would be in the range $10\,\text{Å} > \ell > 1\,\text{Å}$. That is, for a 1-μsec timescale and a diffusion coefficient D of $10^{-10}\,\text{cm}^2/\text{sec}$, a "stacked" n-layer system will remain effectively two dimensional, provided the vertical lattice spacings are of the order of 1 Å. If the diffusivity of the adsorbate species were to increase to, say, $D = 10^{-8}\,\text{cm}^2/\text{sec}$, the vertical lattice spacing required to ensure that the process remains effectively two dimensional would be of the order of 10 Å. These values of ℓ are characteristic of the spatial dimensions of the microchannels in many structured media, specifically smectite clays and certain zeolites, and therefore suggest a physical rationale for determining the range of applicability of continuum theories which assume the diffusion space is of integer dimension (e.g., approaches based on a classical Fickian diffusion equation).

A third conclusion, which follows directly from the results presented here, is that if one purposefully wants to design intercalated systems for which different spatial distributions of pillaring cations influence in a significant way the overall reaction efficiency, the way to achieve this is to utilize small "crystallites" of smectite clays rather than extended, ordered arrays. Decreasing the size of the crystallite will also influence the relative efficiency of two- versus three-dimensional flows of the reactant to the target molecule. This is consistent with, and is another illustration of, the Adam-Delbruck concept of "reduction of dimensionality."

3. Zeolite Architectures

Several factors are known to influence the diffusivity [76–79] and reactivity [68–70,80–87] of guest molecules in zeolites. The free dimensions of the most restricted opening through which the molecules must pass, relative to the dimensions and shape of the diffusing species, are of critical importance. Also of significance for nonpolar molecules is that presorption of polar guest molecules (e.g., H_2O, NH_3, and alcohols) can change the diffusivities of nonpolar species, often by several orders of magnitude. A lattice approach can be mobilized to explore a further geometrical factor that can influence the efficiency of uptake of guest molecules in insulator host lattices, namely, the number and orientation of channels defining the internal structure of a zeolite. For example, one can examine quantitatively the efficiency of migration of guest molecules in zeolites having two topologically different channel structures: zeolite A with a simple cubic channel pattern and faujasite with a tetrahedral channel pattern. Some aspects of the consequences of presorption of polar molecules on the subsequent uptake, diffusion, and sorption of nonpolar guest species in these two zeolite structures can also be quantified.

The basic idea is that the channels and polyhedral cavities of a zeolite crystal can be placed in correspondence with the bonds and vertices of a $d = 3$ dimensional, finite Cartesian lattice. Hence, the problem of determining the trajectories of guest molecules through zeolite crystals of finite extent is modeled by studying the random walk problem on, here, finite lattices of two different topologies.

To place the study of topologically different structures on the same footing, it is useful to introduce the concept of a site-specific maximum sorption time τ_i, defined as the average time required for an internalized diffusing species, starting from cavity site i, to reach the cavity farthest removed from the surface of the zeolite (the one at the geometric center of the assembly). If a uniform jump time $\hat{\tau}$ is assigned for the diffusing guest to transit from a given polyhedral cavity to an adjacent one, then the characteristic times τ_i can be determined via calculation of the stochastic variable $\langle n \rangle_i$, the average walklength between site i and the centrosymmetric site of the system, that is, one specifies

$$\tau_i = \hat{\tau} \langle n \rangle_i \ . \tag{4.47}$$

When averaged over all cavity sites accessible initially to the guest, the consequent sorption time τ reflects the overall influence of different spatial distributions of cavities and attendant channel patterns on the efficiency of sorption.

Two sorts of ambient pressure/concentration conditions can be studied. First, consider the case where an internalized, diffusing species, upon encountering one of the exit ports, is passively confined to the zeolite, that is, it can remain for some time in a cavity near the surface before continuing its migration, but may not exit the system; in earlier discussion, this class of boundary conditions was referred to as "confining." In contrast, there is the possibility that the diffusing species, upon encountering a port, is actively internalized, that is, the external pressure/concentration conditions are such that the species suffers a displacement to a cavity site once removed from the zeolite surface; this amounts to imposing "reflecting" boundary conditions on the randomly diffusing species.

For the case of passive confinement, the site-specific (maximum) sorption times τ_i for zeolite A-type assemblies can be determined from the data presented in table III of ref. 9 and for faujasite assemblies in table 2 of ref. 14. Data for the case of active confinement are given in table IV of ref. 9 for the case $v = 6$, while data for $v = 4$ are given in table 4 of ref. 14. From τ_i the overall (average) maximum sorption time τ can be calculated and, for both passive and active confinement, a comparison made between zeolite A versus faujasite assemblies. Assuming that the cavity-cavity separation is approximately the same in both structures, results presented in ref. 14 show that for both types of confinement the maximum sorption time τ (a measure of the overall efficiency of sorption) is essentially the same for small crystals but becomes measurably different with increase in the number N of participating cavities, and document that sorption is a more efficient process in zeolite-A type structures, regardless of the ambient conditions. Moreover, the influence of "passive" versus "active" boundary conditions is propagated differently in the two structures. As is seen in Figure 4.37, where the quantity

$$\% = \frac{\tau(\text{passive}) - \tau(\text{active})}{\tau(\text{passive})} \times 100 \tag{4.48}$$

is plotted versus the number N of cavity sites for each type of structure, the persistence of surface (or boundary) effects is more noticeable in zeolite A- than in faujasite-type structures.

Bounds can be developed to bracket the increase in sorption time as a function of the number N of participating cavities when zeolite A is replaced by faujasite. Data on $\langle n \rangle$ (for N up to $N = 3375$) for finite $d = 3$ structures of coordination $v = 6$ can be represented by the formula, Eq. (4.23), whereas for coordination $v = 4$, the Eq. (4.24) pertains. Assuming a common jump time $\hat{\tau}$ for both zeolite structures, the following ratio can be constructed:

$$\frac{\tau(v = 4) - \tau(v = 6)}{\tau(v = 6)} = \frac{0.209318N + 0.646984N^{1/2}}{1.474131N - 2.812810N^{1/2}}. \quad (4.49)$$

The smallest zeolite A-type structure (the one that comprises six channels surrounding a central polyhedral cavity) is $N = 7$, and for this case the percent increase in maximal sorption time is 110%. Given the functional form displayed, a limiting value of this percentage when the number of polyhedral cavities increases indefinitely can also be computed; in the limit $N \to \infty$, the limiting percentage is 14.2%. Thus, the percent enhancement in efficiency of sorption when faujasite $(v = 4)$ is replaced by zeolite A $(v = 6)$ is given by the following bounds

$$14.2 < \% < 110. \quad (4.50)$$

For irreversible, encounter-controlled reactive processes of the type schematized by Eq. (4.2), the bound (4.50) leads to the prediction that, all other factors being equal, the reaction efficiency in zeolite A should be greater than in faujasite.

In obtaining the above estimates of the maximum sorption time, it was assumed that sorption occurred at one cavity only, viz., the one farthest removed from the surface of the zeolite crystal. The question arises whether and to what extent the above picture changes if sorption is permitted in any of the polyhedral cavities defining the assembly. Let s_i be the probability of sorption of the diffusing species in cavity i. In the case, all $s_i = s > 0$, the consequences of sorption at the satellite sites i are displayed in tables 3 and 5 of ref. 14 for faujasite-type structures for both ambient conditions noted earlier, and in table VI of ref. 9 for zeolite A-type structures. It is plain from these results that distinctions between different zeolite channel patterns or between different boundary conditions become inconsequential once the satellite sorption probability $s > 0.10$.

The observed, strong dependence of the walklength $\langle n \rangle$ on s can be confirmed analytically [9]. Suppose a walk is defined in which the diffusing particle is trapped at the site of the walk's origin as a length of one step. In this convention, the probability $p(n)$ that the walker is trapped at the nth site is

$$p(n) = (1 - s)^{n-1} s \qquad (4.51)$$

with

$$\langle n \rangle = \sum_{n=1}^{\infty} n \, p(n). \qquad (4.52)$$

Then,

$$\langle n \rangle = s \sum_{n=1}^{\infty} n(1 - s)^{n-1}$$
$$= s \sum n(1 - s)^{n-1}.$$

Because $(1 - s) < 1$ for $s \neq 0$ (the series is obviously not convergent for $s = 0$), then

$$\sum_{n=1}^{\infty} n(1 - s)^{n-1} = [1 - (1 - s)]^{-2} = \frac{1}{s^2}$$

or,

$$\langle n \rangle = 1/s. \qquad (4.53)$$

Thus, generally, in a system in which the background sites have been "switched on" or "activated" in such a way that the diffusing particle can be trapped (or can react) with finite probability $0 < s < 1$ at these sites, the expected walklength $\langle n \rangle$ is rigorously bounded by $1/s$. In contrast to the case $s = 0$ (the case where the satellite sites are neutral or passive) where the value of $\langle n \rangle$ increases as least as fast as the total number N of lattice sites (recall the analytic results, eqs. (4.22)–(4.24)), even a small absorption (or trapping) probability at the satellite sites causes a dramatic change in the overall efficiency of reaction. The presence of competing sorption sites (reaction centers) negates differences in the geometry of the host medium. It can be confirmed from the data reported in refs. 9 and 14 that the characteristic time τ for sorption does indeed approach the asymptotic value $1/s$ in the limit where the number N of accessible cavity sites becomes denumerably infinite.

Consider now the effect of presorbed polar molecules on the diffusivity of nonpolar molecules in host zeolites. Blockage of polyhedral cavities by strongly sorbed polar molecules is found experimentally to lead to a dramatic reduction in the observed diffusivities of nonpolar gases through these cavities. Suppose N_o represents the number of cavities accessible to a diffusing nonpolar molecule in the absence of presorbed polar species, and f the fraction of cavities that are blocked when presorption of polar molecules occurs. Then, writing $N = fN_o$, one infers from Figure 4.37 that differences between the two zeolite structures tend to be suppressed with increase in the fraction f. Moreover, suppose the migration of a nonpolar gas through a given polyhedral cavity is only partially obstructed owing to the presence of presorbed nonpolar species. Were a uniform fractional sorption of the diffusing nonpolar species possible in each of the cavities i (i.e., all $s_i = s > 0$), no more than a 10% sorption probability in each of the

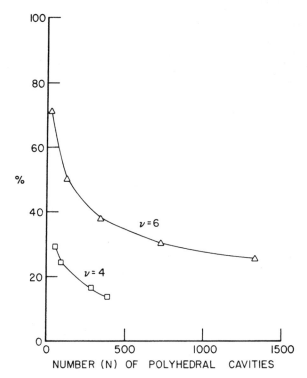

Figure 4.37. The effect of exterior boundary conditions on the sorption of guest molecules in a host zeolite system. Plotted is the percentage

$$\% = \frac{\tau \text{ (passive} - \tau \text{ active)}}{\tau \text{ (passive)}} \times 100$$

versus the total number N of polyhedral cavities for zeolites having the channel pattern of zeolite A ($v = 6$) and faujacite ($v = 4$).

polyhedral cavities would completely suppress differences arising either from system topology or the ambient conditions to which the zeolite is subjected.

The above results followed from a consideration of the overall number and orientation of channels in two representative zeolites. More detailed results can be obtained by considering the structure of particular channels in a given zeolite. As representative of the latter type of study, consider the situation where strictly channeled $d = 1$ dimensional flow is contrasted with a process in which the diffusing species has access to "side pockets," that is, a process in which, at periodic (spatial) intervals a migrating species has access to an expanded ($1 < d < 2$) diffusion space [22]. Mordenite is a zeolite that, in its hydrogen form, is characterized by an internal system of $d = 1$ dimensional (wide) channels. Each channel is lined by side pockets that can be entered through eight-ring windows of free dimension 2.9×5.7 Å. These side pockets provide a pathway for molecules to pass from one channel into (either of two) others. Given the size of these windows it is known that larger molecules (like benzene) cannot enter these side pockets; hence, larger molecules are confined to the wide channels. Smaller molecules, on the other hand, can occupy both the wide channels and the side pockets, and hence can migrate from one channel to another. In the sodium form of this aluminosilicate, the eight-ring windows are effectively blocked by Na^+ ions; in this case, all diffusion is effectively one dimensional, regardless of the size of the diffusing species.

In considering diffusion-controlled processes in mordenite, the extent to which the time scale for diffusion changes when the diffusing species can migrate along the channels (only) versus the case where the species has access to the side pockets can be estimated. Computer graphics displays of the wide channels of mordenite show that the side pockets are staggered along the channel; an idealized cross-sectional representation of this configuration is sketched in Figure 4.38. In terms of this representation, one possible trajectory of a diffusing species would be along the $d = 1$ channel

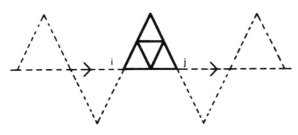

Figure 4.38.   Staggered configuration of side pockets along a $d = 1$ dimensional channel.

(horizontal as drawn) with random, left/right excursions characterizing the particle's motion. Diffusion into the side pockets would delay passage along the channel by providing an expanded diffusion space into which the migrating species could migrate. To estimate the dilation in time scale resulting from the latter possibility, imagine that a molecule is injected at point i, leaves eventually at point j, with the time scale for the restricted $(d = 1)$ motion to be compared with the time scale calculated assuming the diffusing species can migrate into the side pockets. These two processes can also be studied as a function of the spatial extent of the side pockets.

The data on $\langle n \rangle$ for a species migrating from the entry point i to the exit point j show that the time scale for diffusion can easily be extended by factors of 2.5–4.9, depending on the spatial extent of the reaction space defined by the side pocket [22]. Because excursions perpendicular to the plane of Figure 4.38 have been neglected, a degree of freedom that would dilate further the time scale, the sieving of small molecules or rare gases through the hydrogen versus sodium form of mordenite should be characterized by markedly different time scales.

A number of further insights can be drawn if the effect of a multipolar potential operative between coreactants in finite zeolite structures (crystallites) is considered. These studies will be reviewed in Section IV and Section V.B.

IV. POTENTIAL EFFECTS

Among the factors that can influence the efficiency of reaction between a diffusing coreactant A and a stationary target molecule B in a condensed phase, the strength and range of the correlations between coreactants, the temperature, and the dielectric constant of the medium play decisive roles. Although the interplay among these factors in modulating the reaction efficiency is complex, in certain simple cases predictions on the kinetic response of the system can be made. For example, as the temperature T of the medium increases, the down-range correlations between reaction partners will tend to get washed out (i.e., the motion of the diffusing coreactant will be essentially random), and the reactive event will be independent of the multipolarity of the species involved. Conversely, as the dielectric constant ε_r, of the medium is decreased, the influence of interparticle correlations between coreactants will persist over longer distances, and the efficiency of the underlying process will be correspondingly enhanced (for attractive potentials) or diminished (for repulsive interactions).

More generally, the onset of the regimes of behavior noted above is expected to be different for different reaction pairs, sensitive to the details of the governing potential function (viz., its strength and range), the temperature, and the dielectric constant of the medium as well as dependent on the dimensionality and spatial extent of the reaction space. Hence, it is not clear, a priori, how the separate influences of these factors are expressed when all are operating simultaneously. That is, whereas intuition may guide the understanding of how the reaction efficiency changes when one independent variable is changed, it is more difficult to unravel and interpret the complexities of diffusion-controlled reactive processes in condensed phases when several system parameters can influence the kinetic response of the system.

In this section, an approach to the resolution of the above dilemma is described using a lattice-statistical model [53]. For three topologically distinct geometries, it will be shown that the transition between two qualitatively different regimes of kinetic behavior (defined by certain asymptotic limits) can be quantified in terms of the interplay between two relevant "lengths" of the problem, a (generalized) Onsager (or thermalization) length S, and a length R defined by the separation between nearest-neighbor sites accessible to the diffusing coreactant in reaction space. The lattice-statistical results suggest that there is an underlying universality in descriptions of diffusion-reaction processes in condensed media, and that the kinetic behavior of systems subject to the combined influence of several variables can be predicted with some generality.

The three topologies for which results will be presented are now specified. Consider first a reaction pair compartmentalized within a $d = 3$ dimensional reaction space; for definiteness, the target molecule or reaction center is positioned at the center of the system, and the effect of system size on the efficiency of reaction between the diffusing coreactant and target is explored by considering sequentially systems of increasing spatial extent. Results are then presented for two topologically distinct geometries embedded in dimension $d = 2$. First, diffusion-reaction processes on $d = 2$ square-planar lattices, Euler characteristic $\chi = 0$ (where $\chi = F - E + V$, with F, E and V the number of faces (cells), edges, and vertices of the defining polyhedral complex [45]), are examined. Then, diffusion-reaction processes taking place on the surface of a colloidal catalyst or molecular organizate are studied by representing these media by polyhedral surfaces of dimension $d = 2$ and Euler characteristic $\chi = 2$ (a topology equivalent — i.e., homeomorphic — to the surface of a sphere), with N distinct sites on the surface organized into an array defined locally by the site valency of the lattice network. For processes in which potential interactions between reactants are restricted to surface correlations only, $d = 2$ and $\chi = 2$. However, when correlations can be propagated through the host medium, but the coreactant motion is strictly confined to the surface of the host, the topological dimension of the diffusion space remains $d = 2$, the Euler characteristic remains $\chi = 2$, but the interparticle correlations span the full $d = 3$ dimensionality of the host medium.

Central to the interpretation of the results reported later in this section is the specification of the variables calibrating the strength and range of the potential function $V(r)$ biasing the motion of the diffusing particle with respect to the target molecule. The probability of moving through a field of sites in the vicinity of the target molecule is defined by

$$p_{ij} = \exp[-\beta(V_j - V_i)]/q_i. \tag{4.54}$$

Here, $\beta = 1/kT$, p_{ij} is the probability of moving from site i to site j in the next step, and

$$q_i = \sum_{j=1}^{v} \exp\left[-\beta(V_j - V_i)\right] \tag{4.55}$$

is the finite temperature, local partition function; with v defined to be the number of nearest-neighbor sites surrounding site i, the normalization is

$$\sum_{j=1}^{v} p_{ij} = 1 . \tag{4.56}$$

The potential V_k sensed by the diffusing atom or molecule at the site k on a lattice characterized by N sites, dimensionality d and valency v can be expressed in reduced variables for each class of interactions considered. For the case of interacting ionic species, the Boltzmann factor appearing in Eq. (4.54) is

$$-\beta V(r) = -\frac{1}{(4\pi\varepsilon_o)\varepsilon_r} \cdot \frac{(Z_A e)(Z_B e)}{r} \cdot \frac{1}{kT}. \tag{4.57}$$

Here, k is Boltzmann's constant, e is the magnitude of the electronic charge, Z_i is the signed magnitude of the charge of species i, ε_o is the permittivity of free space, and ε_r is the dielectric constant of the medium at temperature T. In the lattice-statistical representation of the underlying diffusion-reaction process, one identifies a characteristic length R, usually taken to be the distance between adjacent locations in diffusion space accessible to the diffusing coreactant. If one then defines a reduced length ℓ,

$$\ell = \frac{r}{R} \tag{4.58}$$

which calibrates the spatial separation r between coreactants in terms of the lattice metric R, and introduces the Onsager length S,

$$S(C,C) = \frac{1}{(4\pi\varepsilon_o)\varepsilon_r} \cdot \frac{e^2}{kT} \tag{4.59}$$

which for Coulombic interactions gives the distance apart at which the mutual electrical potential energy of a pair of singly charged, point charges (C,C) has the magnitude of the thermal energy, then the Boltzmann factor, Eq. (4.57), can be expressed [53] in the dimensionless variables,

$$\exp[-V(r)/kT] = \exp\left[-\frac{W(C,C)}{\ell}\right] \tag{4.60}$$

where

$$W(C,C) = Z_A Z_B \left[\frac{S(C,C)}{R}\right]. \tag{4.61}$$

The advantage of the representation, Eqs. (4.60) and (4.61), is that it lends itself to a very useful, physical interpretation. For example, for singly charged species $(Z_A = Z_B = 1)$ in interaction, if $S < R (W < 1)$, the thermalization distance S is smaller than the lattice spacing R; in this regime, the down-range motion of the diffusing species would be essentially unperturbed by the biasing potential, that is, the motion would be essentially a random walk. If, on the other hand, $S > R (W > 1)$, the thermalization distance is greater than the lattice spacing, and the diffusing particle will be influenced significantly by the biasing potential, particularly in the near neighborhood of the trap. In this interpretation, the value $W = 1$ is of special interest since it signals the "crossover" between two, qualitatively different types of behavior.

Also considered below are angle-averaged ion-dipole and angle-averaged dipole-dipole potentials [88]. The Boltzmann factor for angle-averaged ion-dipole correlations can be expressed in terms of a (generalized) Onsager length

$$\exp[-V(r)/kT] = \exp\left[+\frac{W(C,\mu)}{\ell^4}\right] \tag{4.62}$$

where

$$W(C,\mu) = Z_A^2 \mu_B^2 \left[\frac{S(C,\mu)}{R}\right]^4. \tag{4.63}$$

and

$$S(C,\mu) = \left(\frac{1}{3}\right)^{1/4}\left\{\frac{1}{(4\pi\varepsilon_o)\varepsilon_r}\cdot\frac{ed}{kT}\right\}^{1/2} \tag{4.64}$$

where μ_i is the magnitude of the dipole moment of species i and d is the unit dipole (in Debyes). Similarly, for angle-averaged dipole-dipole interactions,

$$\exp\left[-V(r)/kT\right] = \exp\left[+\frac{W(\mu,\mu)}{\ell^6}\right] \tag{4.65}$$

where

$$W(\mu,\mu) + \mu_A^2\mu_B^2\left[\frac{S(\mu,\mu)}{R}\right]^6 \tag{4.66}$$

and

$$S(\mu,\mu) = \left(\frac{2}{3}\right)^{1/6}\left\{\frac{1}{(4\pi\varepsilon_o)\varepsilon_r}\cdot\frac{d^2}{kT}\right\}^{1/3}. \tag{4.67}$$

Representative values of $S(C,C)$, $S(C,\mu)$ and $S(\mu,\mu)$ are given in Table IV.1.

To study how the efficiency of a diffusion-reaction process is changed when a migrating particle is subjected to the combined influences of a nearest-neighbor short-ranged (chemical/cage) effect (recall the discussion in Section III.C.1 where the quantity $\langle n\rangle(C)$ was introduced) as well as a longer-range biasing potential, the quantity $\langle n\rangle(C/P)$ is calculated using site-to-site transition probabilities p_{ij} defined by Eq. (4.54) for each potential (P). Consider first the case where an attractive Coulombic potential is operative between a diffusing coreactant and a stationary target, that is, the case of ion pairs. Plots of the ratio $\langle n\rangle(C/P)/\langle n\rangle(C)$ versus $\log W$ for three representative lattices systems are displayed in Figure 4.39 for dimension $d = 3$ and in Figure 4.40 for $d = 2$, $\chi = 0$. The dramatic change of the profiles in these figures with respect to increase in the strength parameter W may be interpreted as a transition between two qualitatively different types of behavior in diffusion-reaction space, viz., a regime where the coreactant's motion is totally correlated with respect to the target ion, and a regime where the coreactant motion is totally uncorrelated. In this picture, the behavior in the vicinity of $W = 1$ is reminiscent of an order-disorder phase transition in phase space, an analogy that suggests a physical interpretation of the results

Table IV.1
Comparative values of $S(C,C)$, $S(C,\mu)$, and $S(\mu,\mu)$

Fix medium (H_2O), change temperature

T (°C)	$S(C,C)$ (Å)	$\langle S(C,\mu;\theta)\rangle \equiv S(C,\mu)$ (Å)	$\langle S(\mu,\mu;\theta,\phi)\rangle \equiv S(\mu,\mu)$ (Å)
0	6.960	0.915 $(Z_A = 1, \mu_B = 2.758)$	0.627 $(\mu_A = 1, \mu_B = 3.332)$ $(\mu_A = \mu_B = 1.825)$
20	7.110	0.924 $(Z_A = 1, \mu_B = 2.774)$	0.631 $(\mu_A = 1, \mu_B = 3.357)$ $(\mu_A = \mu_B = 1.832)$
100	8.058	0.984 $(Z_A = 1, \mu_B = 2.862)$	0.658 $(\mu_A = 1, \mu_B = 3.499)$ $(\mu_A = \mu_B = 1.870)$

Fix temperature ($T = 20°$), change medium

ε_r			
1	570.05	8.278 $(Z_A = 1, \mu_B = 8.298)$	2.722 $(\mu_A = 1, \mu_B = 14.472)$ $(\mu_A = \mu_B = 3.804)$
2	285.03	5.853 $(Z_A = 1, \mu_B = 6.978)$	2.161 $(\mu_A = 1, \mu_B = 11.485)$ $(\mu_A = \mu_B = 3.389)$
80	7.126	0.925 $(Z_A = 1, \mu_B = 2.776)$	0.632 $(\mu_A = 1, \mu_B = 3.357)$ $(\mu_A = \mu_B = 1.832)$

in terms of the relevant "lengths" of the problem. Specifically, for singly charged ion pairs, it is evident that when the Onsager length is in the range $S < 0.1\,R$, where R is the lattice spacing, the results are essentially indistinguishable from those calculated in the limit: $W \to 0$. Second, in the regime $S > R$, the limiting behavior $W \to \infty$ is essentially realized for values $S > 10\,R$. The "transition region," $0.1 < S/R < 10$ corresponds to the regime where the two relevant lengths of the problem, the Onsager length S and the lattice metric R, are of comparable magnitude. In this regime, small changes in the dielectric constant or in the temperature of the medium can produce dramatic changes in the reaction efficiency. Outside this (relatively) narrow range, such changes are expected to yield essentially no significant further changes in the efficiency of the irreversible process: $A^+ + B^- \to C$.

In the definition, Eq. (4.61), of the strength parameter W, specific values of the ratio S/R are scaled by the product $Z(A)Z(B)$; hence, the reaction efficiency can change quite dramatically, for a given (T, ε), when one (or both) of the reaction partners is multiply charged. For example, values of the

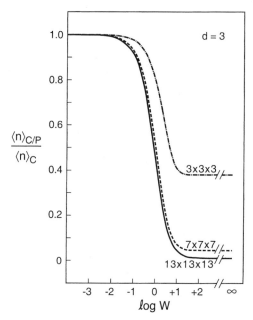

Figure 4.39. A plot of the ratio $\langle n \rangle (C/P)/\langle n \rangle (C)$ versus log W for reaction pairs interacting via the attractive ion-ion potential for the three simple cubic lattice systems: $3 \times 3 \times 3$ ($N = 27$), $7 \times 7 \times 7$ ($N = 343$), and $11 \times 11 \times 11$ ($N = 1331$).

ratio $\langle n \rangle (C/P)/\langle n \rangle (C)$ characteristic of the transition region $S/R = 1$ for singly charged ions collapse to the asymptotic value ($W \rightarrow \infty$) of the ratio for triply charged ion pairs. Conversely, values of $W = 1$ that characterize the transition region for multiply charged ion pairs, correspond to (essentially) uncorrelated motion of the diffusing coreactant with respect to the target molecule for singly charged ion pairs.

For all potentials, the following exact expressions can be derived [53] for $\langle n \rangle (C/P)$ in d dimensions in the limit $W \rightarrow \infty$, that is, the limit where the potential field experienced by the diffusing coreactant is so overwhelming that at each and every point in its trajectory the coreactant A moves toward the target molecule B (only):

$$\langle n \rangle (C/P) = (\ell + 1)(\ell - 1)(1/4)/(\ell - 1) \qquad d = 1 \qquad (4.68a)$$

$$\langle n \rangle (C/P) = (\ell + 1)(\ell - 1)\ell(1/2)/(\ell^2 - 1) \qquad d = 2 \qquad (4.68b)$$

$$\langle n \rangle (C/P) = (\ell + 1)(\ell - 1)\ell^2(3/4)/(\ell^3 - 1) \qquad d = 3 \qquad (4.68c)$$

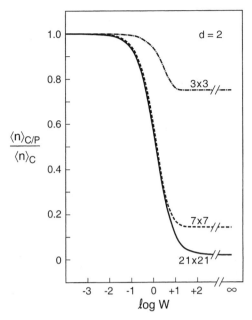

Figure 4.40. A plot of the ratio $\langle n \rangle (C/P)/\langle n \rangle (C)$ versus log W for reaction pairs interacting via the angle-averaged dipole-dipole potential for the three square-planar lattice systems: 3×3 ($N = 9$), 7×7 ($N = 49$), and 21×21 ($N = 441$).

When used in conjunction with the exact result, Eq. (4.38), the following asymptotic limits $(W \rightarrow \infty)$ are obtained (for all potentials) for the $d = 3$ and $d = 2$, $\chi = 0$ lattices:

$d = 3$:

$$\langle n \rangle (C/P)/\langle n \rangle (C) = 0.380282 \qquad N = 3 \times 3 \times 3 \qquad (4.69a)$$

$$\langle n \rangle (C/P)/\langle n \rangle (C) = 0.045093 \qquad N = 7 \times 7 \times 7 \qquad (4.69b)$$

$$\langle n \rangle (C/P)/\langle n \rangle (C) = 0.010664 \qquad N = 13 \times 13 \times 13 \qquad (4.69c)$$

$d = 2$:

$$\langle n \rangle (C/P)/\langle n \rangle (C) = 0.75 \qquad N = 3 \times 3 \qquad (4.70a)$$

$$\langle n \rangle (C/P)/\langle n \rangle (C) = 0.142188 \qquad N = 7 \times 7 \qquad (4.70b)$$

$$\langle n \rangle (C/P)/\langle n \rangle (C) = 0.020842 \qquad N = 21 \times 21. \qquad (4.70c)$$

Note that these limits are realized for interacting ion pairs (see Fig. 4.39 and 4.40).

354   JOHN J. KOZAK

Consider now the plots of the ratio $\langle n\rangle(C/P)/\langle n\rangle(C)$ versus log W for the shorter-range, angle-averaged dipole-dipole potential; see Figures 4.41 and 4.42 for $d = 3$ and $d = 2$, respectively. While, overall, the qualitative behavior uncovered for the longer-range ion-ion potential is also seen here, there are important differences. The calculations show that for lattices $\ell^d > 7^d$, the $W \to \infty$ asymptotic behavior is realized very slowly, with the emergence of a long-range "tail" in the curve $\langle n\rangle(C/P)/\langle n\rangle(C)$ versus log W clearly in evidence for the $13 \times 13 \times 13$ lattice in $d = 3$ and the 21×21 lattice in $d = 2$. The "tail" of the curves generated for short-range potentials for the largest lattices studied is so long that there is an apparent "inversion" in the order of the curves displayed for the lattices $7 \times 7 \times 7$ and $13 \times 13 \times 13$ in $d = 3$, Figure 4.41, and for the lattices 7×7 and 21×21 for $d = 2$, Figure 4.42. This "inversion" does not occur for curves generated for the long-range Coulombic potential, Figure 4.39 for $d = 3$ and Figure 4.40 for $d = 2$. The more extended transition region realized in the angle-averaged dipole-dipole case can be described in the language of phase transition theory by saying that the kinetic transition in diffusion-reaction

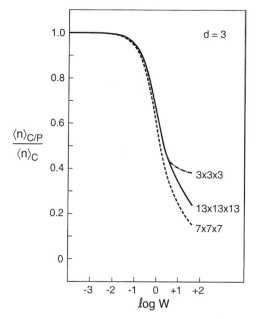

Figure 4.41. A plot of the ratio $\langle n\rangle(C/P)/\langle n\rangle(C)$ versus log W for reaction pairs interacting via the angle-averaged dipole-dipole potential for the three simple cubic lattice systems: $3 \times 3 \times 3\,(N = 27), 7 \times 7 \times 7\,(N = 343)$, and $11 \times 11 \times 11\,(N = 1331)$.

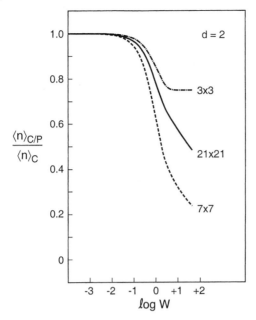

Figure 4.42. A plot of the ratio $\langle n\rangle(C/P)/\langle n\rangle(C)$ versus $\log W$ for reaction pairs interacting via the angle-averaged dipole-dipole potential for the three square-planar lattice systems: 3×3 $(N=9), 7\times 7$ $(N=49)$, and 21×21 $(N=441)$.

space is more "diffuse" than is the case for ion-ion correlations both in $d=2$ and $d=3$.

Despite the structural similarity in the profiles of $\langle n\rangle(C/P/\langle n\rangle(C)$ versus $\log W$ in dimensions $d=2,3$ for each potential, there is an underlying qualitative difference in the behavior in the regime $W\to 0$ in dimensions $d<3$ versus $d=3$. Using Montroll's results for $\langle n\rangle(RW)$ in dimensions $d=1,2,3$, viz., Eqs. (4.25), (4.21), and (4.22), respectively, and the exact result Eq. (4.38), in the limit $N\to\infty$

$$\frac{\langle n\rangle(C)}{\langle n\rangle(RW)}=1 \qquad\qquad d=1 \qquad\qquad (4.71a)$$

$$\frac{\langle n\rangle(C)}{\langle n\rangle(RW)}=1 \qquad\qquad d=2 \qquad\qquad (4.71b)$$

$$\frac{\langle n\rangle(C)}{\langle n\rangle(RW)}=0.340537330 \qquad d=3 \qquad\qquad (4.71c)$$

In 1921, Polya [89] proved that the probability of a return to the origin of a random walker is

$$p = 1 \qquad d = 1 \tag{4.72a}$$

$$p = 1 \qquad d = 2 \tag{4.72b}$$

$$p < 1 \qquad d = 3 \tag{4.72c}$$

In 1964, Montroll [17] showed that in the case $d = 3$, the Polya probability is

$$p = 0.340537330 \tag{4.73}$$

From these results, in $d = 2$ in the limit $W \to 0$ (i.e., in the limit where only a short-range "chemical" or "cage" effect pertains) the irreversible chemical event, $A + B \to C$, is guaranteed to occur (eventually). However, in $d = 3$ in the limit $W \to 0$ there is a finite probability that, despite the chemical reactivity of the species (A, B) involved, for systems sufficiently extended in space, the irreversible process, $A + B \to C$, will never occur. Hence, at the very least, the physical-chemical advantage gained in confining chemically reactive species to a surface is that the irreversible reaction, $A + B \to C$, will occur (eventually). In fact, later in this section numerical evidence will be presented that this result for dimension $d = 2$ is guaranteed even for the case where a repulsive potential biases the down-range motion of the diffusing coreactant.

The above transition behavior also emerges when the irreversible reaction, $A + B \to C$, takes place in a reaction space of distinctly different topology, namely one for which $d = 2$ and $\chi = 2$, as exemplified by the surface of a colloidal catalyst or molecular organizate [90]. One consequence of considering diffusion-reaction processes taking place on the surface of a host medium is that the (possible) differences in reaction efficiency can be studied when interparticle correlations between the diffusing coreactant and the stationary target B are restricted to the surface of the catalyst/organizate (only), versus the case where correlations can be propagated through the host medium. The former situation pertains if, for example, the catalyst simply provides an inert support on which the irreversible reaction takes place; the latter situation can arise in, for example, micelle kinetics, where molecular organizates can be designed with different dielectric interiors [91] (e.g., normal vs inverse micelles).

The polyhedral surface of Euler characteristic $\chi = 2$ used to specify the lattice-statistical problem is the surface of an $L \times L \times L$ cube (i.e., a Cartesian shell) with local valency $v = 3, 4$ (see below). The target molecule B is positioned at the centrosymmetric site of one face of the shell and the influence of system size on the efficiency of the underlying diffusion-reaction process is studied by calculating $\langle n \rangle (C/P)/\langle n \rangle (C)$ as a function of the number $N' = (N - 1)$ of surface sites available to the diffusing coreactant. Specifically, N' is set sequentially at $N' = 97$, 217, and 385, corresponding, respectively, to the surface of a cube with $L = 5$, 7, and 9. Later on in this section, by positioning the reaction center at sites of reduced symmetry (e.g., a vertex site), it will be shown how the reaction efficiency is modulated by the presence of surface imperfections or defects. Anchoring the target molecule at the centrosymmetric site on one face will be seen to represent the optimal condition for influencing reaction efficiency via interparticle correlations.

Physically, the image of a particle diffusing on a surface is not that of a rigidly constrained, two-dimensional motion but rather that of a quasi-three-dimensional motion with the particle "skipping" from site to site across the surface. Unless a diffusing coreactant A is physically bound to a membrane (say, via a hydrophobic tail) or unless physisorption at the surface of a catalyst restricts significantly the particle's motion, it is anticipated that for systems at finite temperature small excursions normal to the surface are not only possible but probable. Further, although one can envision a reactant on a micellar surface moving from "head group to head group" or a molecule on a platinum catalyst transiting from "atom to atom," given the geometrical volumes characterizing the molecules or atoms comprising the surface, it is reasonable to suppose that the channel structure which networks the surface is more appropriately described as a series of "hills and valleys" across which the diffusing coreactant must pass.

To acknowledge these realities, the lattice model described above can be augmented by permitting "up-down" excursions by the diffusing coreactant at each site of the lattice while, at the same time, assuming that surface forces are sufficiently important that the coreactant remains entrained in the vicinity of the surface. In a first approach, the "up-down" excursions at each site of the two-dimensional lattice can be regarded as "virtual" in the sense that such displacements simply reset the molecule at the original lattice site before the reactant undergoes an excursion to a neighboring site. In this case, the nonvertex sites in the expanded reaction space are denoted by the valency $v = 4 + 2$ to stress the fact that the particle's motion is spatially two-dimensional (with $v = 4$ pathways available to the diffusing coreactant at each accessible site) but is augmented by two additional degrees of freedom (which do not change the Cartesian position of the reactant on the lattice). In this notation the eight vertex (defect) sites are characterized by the valency $v = 3 + 3$.

A second generalization is to permit the diffusing coreactant, following an excursion away from the surface, to undergo lateral displacements above the surface before returning (eventually) to a (possibly) different point on the surface. In this case, the lattice model allows a quantification of the (physically intuitive) idea that at finite temperatures the diffusing particle can, in its random or biased motion across the surface, "skip over" the surface during parts of its trajectory while still being sufficiently entrained by surface forces to remain in the vicinity of the host colloidal catalyst particle or molecular organizate. The specification of the lattice model for this generalization will be deferred until later in this section, so that we can proceed to a discussion of differences seen in reaction efficiency when one changes from $(d, \chi) = (2, 0)$ to $(d, \chi) = (2, 2)$.

By examining the structure of the fundamental matrix $\mathbf{N}$ of Markov chain theory for the case of an unbiased, nearest-neighbor random walk, it can be shown [92] that $\langle n \rangle (RW, v = 6)/\langle n \rangle (RW, v = 4) = 3/2$ (exactly) and $\langle n \rangle (RW, v = 6)/\langle n \rangle (RW, v = 5) = 6/5$ (exactly). As is seen from the data presented in Table IV.2, for each of the geometries considered, numerical calculation of $\langle n \rangle (RW, v = 6)$, $\langle n \rangle (RW, v = 4)$, and $\langle n \rangle (RW, v = 5)$ yields ratios which are in exact agreement with the above results. Also in this table are reported values of $\langle n \rangle (RW)$ for the $N = 7 \times 14 = 98$ torus (all $v = 4$ and Euler characteristic $\chi = 0$) and for the corresponding square-planar lattice

TABLE IV.2

Comparison of $\langle n \rangle_{RW}$ versus $\langle n \rangle_C$ for diffusion-limited reactive processes. Host medium is modeled as a $L \times L \times L$ Cartesian shell of N surface sites, with v degrees of freedom available to the diffusing coreactant at each site (see text)

N	v	$\langle n \rangle_{RW}$	$\dfrac{\langle n \rangle_{RW} \, (v = 6)}{\langle n \rangle_{RW} \, (v)}$	$\langle n \rangle_C$	$\dfrac{\langle n \rangle_C \, (v = 6)}{\langle n \rangle_C \, (v)}$	$\dfrac{\langle n \rangle_C}{\langle n \rangle_{RW}}$
98						
5 cube	6	257.546212	1.000	113.046212	1.000	0.439
	5	214.621843	1.200	94.371843	1.198	0.440
	4	171.697475	1.500	75.697475	1.493	0.441
Torus	4	174.763441		78.715797		0.450
Plane	4	163.699044[a]		67.70034[b]		0.414
218						
7 cube	6	654.591734	1.000	330.091734	1.000	0.504
	4	436.394489	1.500	220.394489	1.498	0.505
Plane	4	417.963249[a]		201.9644[b]		0.483
386						
9 cube	6	1263.202673	1.000	686.702673	1.000	0.544
	4	842.135115	1.500	458.135115	1.499	0.544
Plane	4	809.051497[a]		425.0539[b]		0.525

[a] Refs. 13 and 20.
[b] Ref. 93.

(all $v = 4, \chi = 0$) subject to periodic boundary conditions. The presence of eight defect sites on each Cartesian shell ($\chi = 2$), and the absence of a centrosymmetric lattice site on the torus ($\chi = 0$) compromises somewhat the efficiency of trapping relative to a finite $d = 2$ square-planar lattice ($\chi = 0$). For the Cartesian shells considered here, the percentage of imperfections (defect sites) relative to the total number $N(= 98, 218, 386)$ of surface sites is 8.2%, 3.7%, and 2.1%. Correspondingly, the percent difference between results calculated for $\langle n \rangle (RW)$ for walks on a Cartesian shell versus those on the corresponding square-planar lattice should also decrease with increase in N, and they do, viz., from 4.9% to 4.4% to 4.1% as N increases from $N = 98$ to 218 to 386.

Turning next to calculations of the walklength $\langle n \rangle (C)$, note that values of the ratio $\langle n \rangle (C, v = 6)/\langle n \rangle (C, v)$ approximate the corresponding random walk ratios, with convergence to the exact $\langle n \rangle (RW, v = 6)/\langle n \rangle (RW, v)$ value realized as the number N of surface sites increases. By comparing results obtained for walks on the corresponding square planar lattice [93], it is seen once again that the symmetry breaking induced by the presence of the eight defect sites on the Cartesian shell leads to a somewhat less efficient diffusion-reaction process than on the corresponding finite $d = 2$ square planar lattice. The percent difference between results calculated for $\langle n \rangle (C)$ for walks on the Cartesian shell versus those on the corresponding square planar lattice decreases with increase in N, viz., from 11.8% to 9.1% to 7.9% as N increases from $N = 98$ to 218 to 386.

An immediate result follows from an examination of the ratio $\langle n \rangle (C)/\langle n \rangle (RW)$ for the three Cartesian shells considered here. For $N = 98$, $N = 218$, and $N = 386$, the ratios are $\langle n \rangle (C)/\langle n \rangle (RW) = 0.44$, 0.50, and 0.54, respectively. Since it was proved analytically (see earlier discussion) that in $d = 2$ this ratio is unity in the limit $N \to \infty$, it is evident that short-range focusing associated with a cage/chemical effect can have a dramatic influence on the efficiency of the underlying diffusion-reaction process in finite, compartmentalized systems. Different assumptions on the spatial character of a local cage or a specific realization of the short-range quantum-chemical effect can influence further the above ratios.

The most dramatic effects on the reaction efficiency here, as in the case ($d = 2, \chi = 0$), are uncovered in the regime where the Onsager (thermalization) length S is of magnitude comparable to the mean displacement R of the coreactant in diffusion space, that is, in the regime where $|W| = 1$. Listed in Table IV.3 are representative results for $\langle n \rangle (C/P)$ for attractive and repulsive Coulombic interactions, angle-averaged ion-dipole and angle-averaged dipole-dipole interactions, for values of the dimensionless Onsager length, $|W| = 1.0$. Reported as a function of system size, both "through space" and "surface only" results, designated, respectively, as $\langle n \rangle (T)$ and $\langle n \rangle (D)$, are

TABLE IV.3

Values of $\langle n \rangle$ for diffusion-reaction processes guided by multipolar correlations propagated "through space" versus "surface only" for the setting $|W| = 1$. Host medium is modeled as a $L \times L \times L$ Cartesian shell of N surface sites, with $v = 4 + 2$ degrees of freedom available to the diffusing coreactant at each site (see text)

N	Potential	$\langle n \rangle(T)$ Through Space	$\langle n \rangle(D)$ Surface Only	$\dfrac{\langle n \rangle(T)}{\langle n \rangle(D)}$
98	Coulombic (attractive)	69.661504	65.624232	1.062
	Coulombic (repulsive)	211.465785	234.078895	0.903
	Ion-Dipole (angle averaged)	76.332488	76.402771	0.999
	Dipole-Dipole (angle averaged)	79.942356	79.308960	1.008
218	Coulombic (attractive)	200.678286	191.158819	1.050
	Coulombic (repulsive)	657.501110	708.706366	0.928
386	Coulombic (attractive)	418.845749	402.350040	1.041
	Coulombic (repulsive)	1401.261695	1484.879888	0.944

given for three Cartesian shells. Although differences between "through space" versus "surface only" results are essentially negligible for angle-averaged ion-dipole and angle-average dipole-dipole correlations, for attractive Coulombic interactions, potential correlations propagated across the surface (only) yield walklengths that are slightly smaller (and hence the attendant diffusion-reaction process somewhat more efficient) than when correlations are propagated through the host medium. The opposite ordering pertains when repulsive Coulombic interactions are considered. These results can be understood by noting that the "through space" Cartesian distances between sites on, for example, the face opposite the one which the target molecule B is anchored are numerically smaller than the separation of such sites as measured along the surface of the shell. Consequently, when the diffusing reactant A is on the face of the shell opposite the one on which the target molecule is localized, "through space" attractive correlations between the coreactants are somewhat more enhanced than in the "surface only" case. In the former case the diffusing coreactant will tend to spend relatively more time in the antipodal region of the host media before finding its way, eventually, to the target molecule. Conversely, since the coreactant A will experience somewhat greater repulsive correlations on the antipodal face when correlations are propagated through the host medium, there will be an enhanced tendency for the coreactant A to migrate to a different point in the diffusion space, that is, a site somewhat closer to the trap. In any case, as may be seen from the ratios $\langle n \rangle(T)/\langle n \rangle(D)$ displayed in Table IV.3, even for Coulombic interactions, the differences between these two calculations are only a matter of a few percent ($<10\%$), with convergence to $\langle n \rangle(T)/\langle n \rangle(D) \rightarrow 1$ as the system size increases.

To compare the effect on the reaction efficiency upon changing the multipolarity of the attractive potential, the ratios $\langle n \rangle(C/P)(C, \mu)/\langle n \rangle(C/P)(C, C)$ and $\langle n \rangle(C/P)(\mu, \mu)/\langle n \rangle(C/P)(C, C)$ can be calculated for the attractive Coulombic $\langle n \rangle(C/P)(C, C)$, the angle-averaged ion-dipole $\langle n \rangle(C/P)(C, \mu)$, and the angle-averaged dipole-dipole $\langle n \rangle(C/P)(\mu, \mu)$ potentials. For the case $N = 98$, the shorter walklengths calculated for the attractive Coulombic case yield values of these ratios greater than unity. For "through space" correlations, the ratios are 1.10 and 1.15 for the ion-dipole and dipole-dipole cases, respectively; for "surface only" correlations, the corresponding ratios are 1.16 and 1.21. Thus, differences in reaction efficiency induced by changing the multipolarity of the potential interaction between the coreactants A and B are somewhat more pronounced when the host medium provides simply an inert support on which the diffusion-reaction process takes place.

Finally, as noted earlier, ratio $\langle n \rangle(v = 6)/\langle n \rangle(v = 4)$ equals 3/2 exactly for the case of the unbiased, nearest-neighbor random walk. As is seen in Table IV.4, for attractive Coulombic interactions there is a systematic convergence to the exact limiting value of 3/2 with an increase in system size. Interestingly, the value 1.500 is realized for repulsive Coulombic interactions, even for the smallest system size reported.

Consider now the specific case where attractive Coulombic forces govern the down-range motion of a diffusing ion with respect to a stationary target ion [90]. As for the topology ($d = 2$, $\chi = 0$), the results obtained are most conveniently displayed as graphs of the ratio $\langle n \rangle(C/P)/\langle n \rangle(C)$ versus the dimensionless Onsager length W (see Figures 4.43 through 4.45 below). In all cases, a limiting behavior emerges in the regimes $W \rightarrow 0$ and $W \rightarrow \infty$, representing two qualitatively different types of behavior in the diffusion-reaction process. In the limit $W \rightarrow 0$, the coreactant's motion is totally uncorrelated with respect to the target molecule, and the ratio $\langle n \rangle(C/P)/\langle n \rangle(C) \rightarrow 1$. In the limit $W \rightarrow \infty$, the coreactant's motion is totally correlated with respect to the target ion, that is, in each displacement the diffusing ion moves always toward the trap; the value of the limiting ratio $\langle n \rangle(C/P)/\langle n \rangle(C)$ depends on the spatial extent of the reaction space and can

TABLE IV.4

Values of $\langle n \rangle$ for diffusion-reaction processes guided by multipolar "surface only" correlations for the setting, $|W| = 1$. Host medium is modeled as a $L \times L \times L$ Cartesian shell of N surface sites, with $v = 4$ degrees of freedom available to the diffusing coreactant at each site (see text)

N	Potential	$\langle n \rangle (D)$ Surface Only	$\dfrac{\langle n \rangle (v = 6)}{\langle n \rangle (v = 4)}$
98	Coulombic (attractive)	44.263734	1.483
	Coulombic (repulsive)	156.043516	1.500
	Ion-Dipole (angle averaged)	51.487416	1.484
	Dipole-Dipole (angle averaged)	53.874771	1.472
218	Coulombic (attractive)	127.978474	1.494
	Coulombic (repulsive)	472.411898	1.500
	Ion-Dipole (angle averaged)	160.819735	
	Dipole-Dipole (angle averaged)	167.781214	
386	Coulombic (attractive)	268.786702	1.497
	Coulombic (repulsive)	989.835897	1.500
	Ion-Dipole (angle averaged)	348.643023	
	Dipole-Dipole (angle averaged)	362.259618	

be calculated analytically for each of the three Cartesian shells, viz.

$$\langle n \rangle (C/P)/\langle n \rangle (C) = 0.043777 \qquad N = 98 \qquad (4.74a)$$

$$\langle n \rangle (C/P)/\langle n \rangle (C) = 0.033455 \qquad N = 218 \qquad (4.74b)$$

$$\langle n \rangle (C/P)/\langle n \rangle (C) = 0.021408 \qquad N = 386 \qquad (4.74c)$$

Displayed in Figure 4.43 is the ratio $\langle n \rangle (C/P)/\langle n \rangle (C)$ versus log W for the case $N = 98$ (the 5 cube with the target ion at a centrosymmetric site on one face) for the case where attractive correlations between the target ion and the diffusing coreactant are transmitted through the host medium. The

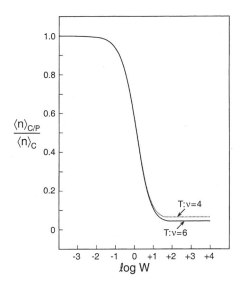

Figure 4.43. A plot of the ratio $\langle n \rangle (C/P)/\langle n \rangle (C)$ versus log W for reaction pairs interacting via the attractive ion-ion potential. The curves displayed describe the differences in reaction efficiency when the number of degrees of freedom available to the diffusing ion at each site on the surface is restricted to $v = 4$, or augmented by (two) virtual displacements so that $v = 4 + 2 = 6$.

profile labeled $v = 4$ represents the case where the diffusing ion is confined in its motion to the surface of the catalyst (only), while the profile labeled $v = 4 + 2 = 6$ refers to surface trajectories augmented by possible displacements of the coreactant away from the surface. As the (dimensionless) thermalization length W increases, there occurs a transition between the behavior exhibited in the two limiting regimes $W \rightarrow 0$ and $W \rightarrow \infty$; differences in the profiles corresponding to $v = 4$ and $v = 4 + 2$ appear only in the regime log $W > 1$. This transition is similar to that uncovered above in the case $(d = 2, \chi = 0)$, as well as in the study presented in ref. 94. As seen from Tables IV.3 and IV.4, in the vicinity of $|W| = 1$, the extra degrees of freedom associated with the setting $v = 6$ lead to a slightly less efficient process, relative to the case $v = 4$.

The results displayed in Figure 4.43 can be compared with those displayed in Figure 4.44; in the latter figure, $v = 6$ and "through space" versus "surface only" correlations between the diffusing ion and the target ion are compared, again for the case $N = 98$. A slight differentiation in the profiles begins to appear in the regime log $W > -1$ and persists until log $W = 2$, where the correct limiting behavior of the ratio $\langle n \rangle (C/P)/\langle n \rangle (C) = 0.44$ is realized. As seen from the data in Table IV.4, allowing the possibility

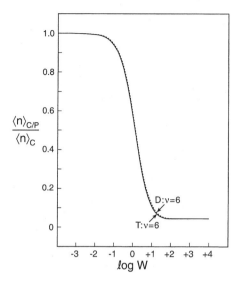

Figure 4.44. A plot of $\langle n \rangle (C/P)/\langle n \rangle (C)$ versus log W for reaction pairs interacting via ion-ion potential. The total number of surface sites is $N = 98$; the lattice connectivity is 4 and the total number of degrees of freedom available to the diffusing ion at each lattice site is $v = 4 + 2$. The curve labeled D refers to the case of surface-mediated correlations (only); the curve labeled T describes the behavior when correlations can be propagated through the host medium.

of "through space" correlations between coreactants compromises some-what the efficiency of the diffusion-reaction process, relative to the case of "surface only" correlations; again, the rather restricted range of W over which the differences are numerically significant should be noted.

Finally, to explore the effect of system size, displayed in Figure 4.45 are results obtained assuming surface-mediated correlations only; the local motion of the diffusing ion is restricted to the surface of the catalyst/ organizate (only), that is, $v = 4$. Apart from the different limiting behavior anticipated in the regime $W \to \infty$, the profiles of the three curves display minimal differences until the Onsager length S becomes of the order of the lattice metric R, that is, in the vicinity of log $W = 0$. Once again, significant differences in the profiles in Figure 4.45 appear over a rather restricted range of W.

A natural way of addressing how the reaction efficiency is changed by the presence of surface imperfections is to design a surface on which imperfections are interspersed in a more-or-less regular way and then to examine how the probability distribution function governing the diffusion-reaction process changes as a function of the (relative) number of defect sites. This can be implemented by considering a catalyst particle for which

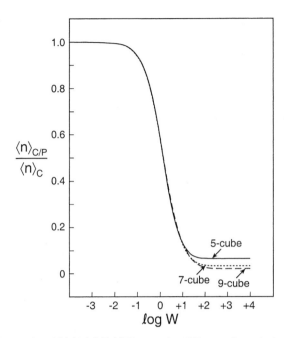

Figure 4.45. A plot of $\langle n \rangle (C/P)/\langle n \rangle (C)$ versus log W for reaction pairs interacting via the attractive ion-ion potential. Displayed is the effect of system size on the reaction efficiency; the total number of surface sites for the 5-cube, the 7-cube, and the 9-cube is $N = 98$, $N = 218$, and $N = 386$, respectively. In each case, $v = 4$ and surface-only correlations are considered.

surface imperfections (defects) characterized locally by the valency $v = 3$ are distributed symmetrically on a surface characterized globally by the coordination $v = 4$, that is, by positioning sites of valency $v = 3$ at the vertices of a Cartesian shell, the great majority of whose sites are characterized by a valency $v = 4$. Given this specification, the percentage of imperfections relative to the total number N (here, $N = 26, 98, 216, 386$) of surface sites is 30.8%, 8.2%, 3,7%, and 2.1%. The influence of the $v = 3$ defect sites in modulating the surface-mediated reactive process can then be determined in a straightforward way by studying the reaction efficiency as a function of the total number N of surface sites.

Three limiting cases can be studied. In the first, the target molecule is anchored at a site on the surface farthest removed from any site of valency $v = 3$, that is, at the centrosymmetric site of a face. This is the case examined extensively in the preceding discussion; this location will be referred to as site 1. A second case specifies the location of the target molecule at a site ($=2$) somewhat closer to a (pair of) defect sites, viz., at the midpoint of an edge. And, in the third case, the target molecule is anchored at a site $(=3)$ of valency $v = 3$, that is, at one of the surface imperfections. For a Cartesian shell of edge length L, the distance between the target molecule and the nearest surface defect in each case would be $0.71 L$, $0.5 L$ and L for cases (sites) 1, 2, and 3, respectively. Again, to account more realistically for the motion of the diffusing coreactant on the surface, calculations can be performed for the augmented displacements defined by the valencies, $v = 3 + 3$ for the 8 defect sites and $v = 4 + 2$ for the remaining $N - 8$ surface sites.

Displayed in Figure 4.46 is the first moment $\langle n \rangle (C/P)$ of the probability distribution function versus system size for a coreactant A diffusing on a Cartesian shell where the biasing potential is an attractive ion-ion potential of strength $W = -1$; this value of the (normalized) Onsager length is at the midpoint of the transition region displayed in Figure 4.43. The observed monotonicity with respect to N is anticipated from Eq. (4.21). For each system size the longest walklength (or equivalently, the longest reaction time) is that for which the target molecule is anchored at site 3, that is, at one of the vertex (defect) sites of the Cartesian shell. Positioning the reaction center away from a defect site has about the same effect on the overall reaction efficiency whether the target molecule is positioned at site 1 or site 2, with the influence of the lattice imperfection felt a bit more at the (closer) site 2 than at site 1. The same trends are observed [92] when angle-averaged

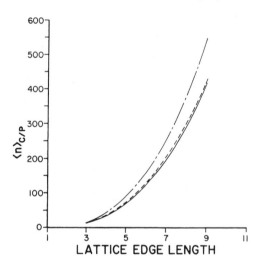

Figure 4.46. The mean walklength $\langle n \rangle (C/P)$ versus the lattice edge length ℓ for a target molecule anchored at the center of a face (solid line), the midpoint of an edge (dashed line), or a vertex position (hyphenated line). The governing potential is an attractive ion-ion potential with $W = -1$.

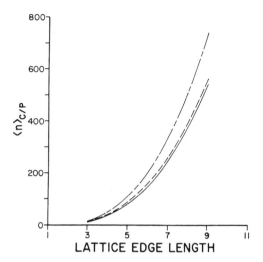

Figure 4.47.   The same conventions as in Figure 4.46 except that the governing potential is an attractive dipole-dipole potential with $W = -1$.

ion-dipole potentials, or angle-averaged dipole-dipole potentials (see Figure 4.47) are considered.

Whereas attractive correlations between a diffusing coreactant and a target molecule result in profiles of $\langle n \rangle (C/P)$ versus N that are concave upward, the curves describing the change in reaction efficiency with respect to system size for repulsive potentials are concave downward (see Fig. 4.48). The behavior displayed by curves of $\langle n \rangle (C/P)$ versus N for repulsive potentials suggests that repulsions can play an important role in influencing the reaction efficiency in small organizates. In fact, for all system sizes the influence of repulsive correlations on the reaction efficiency is much more pronounced than is the case for attractive interactions. Notice that the ordinate in Figure 4.48 is the logarithm of the quantity $\langle n \rangle (C/P)$. Finally, consistent with the results for attractive potentials, the most significant effect on the reaction efficiency is realized when the target molecule is situated at a defect site; the relative ordering of the curves for a target molecule anchored at sites 1, 2 is also the same.

For attractive ion-ion potentials, the consequence of increasing the potential strength parameter W is that the ratio $\langle n \rangle (v, W; 4+2, 0)/\langle n \rangle$ $(v, W; 4, 0) = 3/2$ shifts away from the $W = 0$ limiting value of 3/2 to (larger) values in the neighborhood of 2.00. The value of this ratio calculated for the defect site 3 in the limit $W \rightarrow 0$ is also 2, suggesting that spatial imperfections associated with sites of valency lower than the global valency (and in the absence of reactant correlations) play the same role as attractive

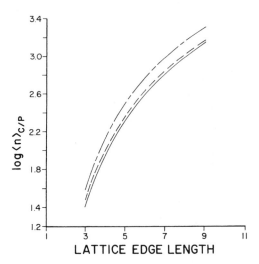

Figure 4.48. The same conventions as in Figure 4.46 except that the governing potential is a repulsive ion-ion potential with $W = +1$.

ion-ion correlations (with the target molecule positioned away from a defect site) in mediating the efficiency of encounter-controlled reactive processes on $(d = 2, \chi = 2)$ surfaces [92].

Finally, it is of interest to assess the consequences as regards reaction efficiency of permitting the diffusing coreactant, following an excursion away from the surface, to undergo lateral displacements above the surface before returning (eventually) to a (possibly) different point on the surface [95]. Although there is a continuum of sites above the surface accessible to a diffusing coreactant, the reaction space exterior to the surface of the host particle can be sampled by designing an extended lattice structure. Consider the case of a coreactant migrating on an $N = 26$ Cartesian shell, with the target molecule anchored either at a defect site $(v = 3 + 3)$ or at a site $(v = 4 + 2)$ farthest removed from the (eight) defect sites; consistent with the notation used above, these sites will be referred to as sites 3 and 1, respectively.

Calculations for $N = 124$, 342, and 728, to be reviewed, refer to the number of sites accessible to a coreactant diffusing on or above the Cartesian shell, $N = 26$. The case $N = 124$ comprises the $N = 26$ basal sites and the $N' = 98$ sites defining the first overlayer. The case $N = 342$ includes the $N = 26$ basal sites and the first two overlayers, $N' = 98$ and 218; the case $N = 728$ encompasses the $N = 26$ basal sites and the first three overlayers, $N' = 28$, 218, and 386. Thus, the motion of the diffusing coreactant, strictly confined in the previous discussion to the $d = 2$ dimensional surface of a series of Cartesian shells of ever increasing spatial extent (surface area), here takes on more and more three-dimensional character $(2 < d < 3)$ as one considers sequentially the augmented reaction spaces, $N = 124, 342$, and 728.

Two scenarios can be considered. In the first, it is assumed that in its migration through reaction space, the diffusing coreactant can move on or off the surface of the host catalyst particle until, ultimately, an irreversible reaction occurs with the localized target molecule. Then, in a second series of calculations, the diffusing coreactant in its migration through the available reaction space, upon encountering the surface of the host catalyst particle for the first time, is constrained to move on the surface of the host (only) until reaction with the surface-bound target molecule occurs. The consequences of these two scenarios can be quantified in terms of the ratio of (diffusion controlled) rate constants

$$\frac{k_D[V = V(r)]}{k_D[V = 0]} \sim \frac{\langle n \rangle [V = 0]}{\langle n \rangle [V = V(r)]} \tag{4.75}$$

or, more explicitly, in terms of the ratios

$$\frac{k(C; W)}{k(W = 0)} \sim \frac{\langle n \rangle (W = 0)}{\langle n \rangle (C; W)}$$

and

$$\frac{k(C, Tr; W)}{k(C; W)} \sim \frac{\langle n \rangle (C; W)}{\langle n \rangle (C, Tr; W)}$$

respectively, with Tr denoting the "tracking" boundary condition.

Considering first the data reported in Table IV.5, in the presence of a short-range chemical or cage effect (C) but in the absence of down-range correlations $(W = 0)$, the enhancement in the above ratio of rate constants ranges from a factor of 3.3 when the diffusing coreactant is strictly confined to the surface of the host catalyst particle (the case $N = 26$), through a maximum of 3.9 as the coreactant begins to take advantage of an expanded reaction space in the near environment of the host particle, to a value of 3.7 when the coreactant drifts further and further away from the host. The falloff in the ratio, Eq. (4.75), with expansion in the available reaction space follows from the fact that short-range chemical effects are operative only over distances $\sim R$.

As the down-range biasing potential is "turned on," the enhancement in the ratio of rate constants increases dramatically for attractive potentials, indeed by an order of magnitude (or more) as W increases. That the chemical affinity between reactant pairs can dominate the kinetics even in the presence of ion-ion repulsive interactions is seen from the data reported for $W > 0$. For like-charged ion pairs with correlations scaled by the parameter $W = 1$, one still finds an enhancement in the rate of reaction, thereby reflecting the importance of short-range chemical interactions.

The enhancement in the ratio of rate constants for shorter-range attractive potentials (the angle-averaged ion-dipole and angle-averaged dipole-dipole potentials) as a function of the extent of the reaction space follows the same trends as that noted above for $W = 0$, viz., an increase in the ratio through a maximum followed by a falloff as the reaction space accessible to the diffusing coreactant is expanded. This turnover in the ratio of rate constants is not yet realized for ion-ion attractive potentials for the extended reaction space considered in these calculations (a consequence of the persisting spatial influence of long-range attractive correlations), whereas for repulsive ion-ion interactions the ratio simply falls off with expansion of the reaction space.

The above trends describe the consequences of anchoring the target molecule at a surface site at a location maximally separated from the (eight) defect sites on the $N = 26$ surface. On repositioning the target molecule at one of the defect sites and carrying through the calculation of Eq. (4.75) as a function of the strength/range of the governing potential function and the spatial extent of the available reaction space, the data set at the bottom of Table IV.5 is generated. Comparing values of the ratio of rate constants for corresponding values of (W, N) one finds that the enhancement in the ratio of rate constants is systematically smaller (in 64 of the 68 cases considered) when the target molecule is positioned at the defect site.

The results reported in Table IV.5 can be contrasted with those reported in Table IV.6, where the effect on the kinetics of restricting the motion of the diffusing coreactant to the surface of the host catalyst particle upon first encounter is considered. This "entrainment" or "tracking" constraint on the motion of the diffusing coreactant is a realization of the effect analyzed by Adam and Delbruck [51] within the framework of a continuum theory based on Eq. (4.1) and referred to as "reduction of dimensionality." The ratio $k(C, Tr; W)/k(C; W)$ can be calculated, again as a function of the down-range potential operative between coreactants, and as a function of the spatial extent of the reaction space. Corresponding to a fixed setting of the control parameters (N, W), entrainment of the diffusing coreactant on the surface upon first encounter plays a relatively more important role in influencing the kinetics the shorter the range of the biasing attractive potential. Whereas the long-range ion-ion attractive potential influences dramatically the efficiency of the diffusion-reaction process in the absence

TABLE IV.5

The ratio $k(C; W)/k(W = 0)$ for a target molecule anchored at sites 1 and 3

Potential	N	$\langle n \rangle (W = 0)/\langle n \rangle (C; W) \sim k(C; W)/k(W = 0)$				
		(a) Site 1				
		$W = -9$	$W = -4$	$W = -2$	$W = -1$	$W = 0$
Ion-ion	26	16.68	10.30	6.58	4.80	3.28
(attractive)	124	36.08	19.35	10.48	6.71	3.88
	342	54.23	25.92	12.88	7.59	3.87
	728	65.42	29.21	14.00	7.89	3.74
Ion-dipole	26	16.59	12.35	8.36	5.68	3.28
(angle-averaged)	124	23.29	15.50	10.35	6.99	3.88
	342	22.17	14.87	10.26	7.04	3.87
	728	19.30	13.36	9.52	6.69	3.74
Dipole-dipole	26	13.86	10.96	7.94	5.56	3.28
(angle-averaged)	124	16.23	12.44	9.23	6.60	3.88
	342	14.84	11.71	8.96	6.55	3.87
	728	13.02	10.54	8.29	6.20	3.74
		$W = +9$	$W = +4$	$W = +2$	$W = +1$	$W = 0$
Ion-ion	26	0.007	0.347	1.22	2.08	3.28
(repulsive)	124	0.001	0.168	0.949	2.01	3.88
	342	0.000	0.086	0.678	1.71	3.87
	728	0.000	0.055	0.535	1.51	3.74
		(b) Site 3				
		$W = -9$	$W = -4$	$W = -2$	$W = -1$	$W = 0$
Ion-ion	26	16.77	10.81	6.98	4.98	3.21
(attractive)	124	26.49	14.38	8.23	5.42	3.16
	342	39.69	19.32	10.32	6.42	3.45
	728	49.49	22.65	11.60	6.90	3.45
Ion-dipole	26	12.92	9.67	7.28	5.36	3.21
(angle-averaged)	124	13.59	9.64	7.06	5.17	3.16
	342	14.32	10.24	7.64	5.68	3.45
	728	14.01	10.18	7.68	5.72	3.45
Dipole-dipole	26	10.08	8.37	6.76	5.16	3.21
(angle-averaged)	124	9.66	7.94	6.33	4.87	3.16
	342	10.00	8.33	6.79	5.32	3.45
	728	9.84	8.27	6.79	5.34	3.45
		$W = +9$	$W = +4$	$W = +2$	$W = +1$	$W = 0$
Ion-ion	26	0.004	0.220	0.974	1.85	3.21
(repulsive)	124	0.001	0.113	0.729	1.62	3.16
	342	0.000	0.077	0.634	1.58	3.45
	728	0.000	0.053	0.519	1.44	3.45

TABLE IV.6

The ratio $k(C, Tr; W)/k(C; W)$ for a target molecule anchored at sites 1 and 3

Potential	N	$\langle n\rangle(C;W)/\langle n\rangle(C, Tr; W) \sim k(C, Tr; W)/k(C; W)$					$\langle n\rangle(W=0/$ $\langle n\rangle_T(W=0)$
		$W=-9$	$W=-4$	$W=-2$	$W=-1$	$W=0$	
				(a) Site 1			
Ion-ion	26	1.00	1.00	1.00	1.00	1.00	1.00
(attractive)	124	0.996	1.14	1.41	1.66	1.92	3.22
	342	1.00	1.13	1.41	1.75	2.03	5.58
	728	1.00	1.10	1.35	1.67	1.94	6.69
Ion-Dipole	26	1.00	1.00	1.00	1.00	1.00	1.00
(angle-averaged)	124	1.13	1.31	1.50	1.70	1.92	3.22
	342	1.10	1.25	1.45	1.73	2.03	5.58
	728	1.07	1.19	1.36	1.64	1.94	6.69
Dipole-dipole	26	1.00	1.00	1.00	1.00	1.00	1.00
(angle-averaged)	124	1.30	1.43	1.57	1.75	1.92	3.22
	342	1.23	1.34	1.51	1.77	2.03	5.58
	728	1.17	1.27	1.42	1.68	1.94	6.69
		$W=+9$	$W=+4$	$W=+2$	$W=+1$	$W=0$	
Ion-ion	26	1.00	1.00	1.00	1.00	1.00	1.00
(repulsive)	124	18.24	6.06	3.41	2.60	1.92	3.22
	342	175.8	17.86	5.76	3.52	2.03	5.58
	728	702.2	26.11	6.57	3.69	1.94	6.69
				(b) Site 3			
		$W=-9$	$W=-4$	$W=-2$	$W=-1$	$W=0$	
Ion-ion	26	1.00	1.00	1.00	1.00	1.00	1.00
(attractive)	124	1.04	1.23	1.47	1.68	1.97	2.35
	342	1.04	1.20	1.48	1.77	2.12	4.17
	728	1.02	1.15	1.39	1.69	2.01	5.27
Ion-dipole	26	1.00	1.00	1.00	1.00	1.00	1.00
(angle-averaged)	124	1.32	1.47	1.63	1.79	1.97	2.35
	342	1.27	1.20	1.62	1.86	2.12	4.17
	728	1.20	1.33	1.50	1.74	2.01	5.27
Dipole-dipole	26	1.00	1.00	1.00	1.00	1.00	1.00
(angle-averaged)	124	1.48	1.57	1.70	1.84	1.97	2.35
	342	1.43	1.54	1.70	1.91	2.12	4.17
	728	1.33	1.42	1.57	1.79	2.01	5.27
		$W=+9$	$W=+4$	$W=+2$	$W=+1$	$W=0$	
Ion-ion	26	1.00	1.00	1.00	1.00	1.00	1.00
(repulsive)	124	15.10	4.57	2.93	2.38	1.97	2.35
	342	82.62	13.13	5.12	3.30	2.12	4.17
	728	354.5	25.43	6.53	3.61	2.01	5.27

of "surface entrainment," reduction of dimensionality is of less consequence for ion pairs than for reactants correlated by the shorter-range ion-dipole and dipole-dipole potentials. In fact, the effect of "tracking" is most pronounced for repulsive ion-ion interactions, where entrainment of the diffusing coreactant on the surface of the host catalyst particle can increase significantly the possibility of reaction between like-charged ion pairs simply by "concentrating" the reactants on the surface.

Finally, for attractive potentials, the "concentration effect" induced by imposing the tracking constraint ("reduction of dimensionality") on the motion of the diffusing coreactant plays a relatively more important role in influencing the kinetics if the target molecule is anchored at a defect site on the surface. For repulsive ion-ion interactions, the opposite situation pertains, that is, somewhat greater enhancements in the ratio $k(C, Tr; W)/k(C; W)$ are found when the target molecule is localized on a site maximally removed from the surface defect sites.

V. MORE GENERAL SITUATIONS

The studies described in the preceding sections focused on disentangling geometrical factors characterizing the system topology (the Euler number, the size, dimensionality, and connectivity of the reaction space), from correlation effects (strength, range, and multipolarity of the potential), in influencing the efficiency of encounter-controlled reactive processes in compartmentalized systems. Since in laboratory experiments several (and perhaps all) of these variables may be at play simultaneously, it is of interest to examine diffusion-reaction processes in more complex systems, and to use the insights gained in the above studies to understand trends in reaction efficiency. Accordingly, we consider first the case of multiple reaction centers (in the athermal regime) and then take up the problem of reactive processes mediated by interparticle correlations in/on porous catalyst particles.

A. Multiple Reaction Centers

Theoretical approaches to the study of spatial effects in chemical kinetics tend to focus either on the detailed description of a few molecules in interaction or, at the other extreme, to account for the net effect of many-body interactions by constructing a statistical average over all possible configurations of reactants. The intermediate situation wherein a diffusing coreactant encounters a nonrandom distribution of reaction centers (clusters or multiplets) is a problem of considerable importance in physics, chemistry and biology [96–109]. The difficulty of analyzing theoretically the role of multiplets of reactions centers in influencing the progress of a diffusion-controlled reactive process is that the number of distinct ways in which clusters (of given size and orientation) can be distributed on a reaction space of finite spatial extent escalates dramatically with increase in system size. Indeed, it is this very complexity that is responsible for the subtlety of experimental effects uncovered in studies on chemical and biological systems in which "sequestering" of active sites occurs.

To make some progress in understanding this problem, one can consider systematically small constellations of active sites (singlets, doublets, triplets, quartets, hexamers, and combinations thereof) distributed on a reaction space (a lattice) of integral dimensionality [110]. By identifying representative configurations and analyzing the efficiency of reaction in each case (as gauged, once again, by calculating the mean walklength $\langle n \rangle$ before an irreversible (or quasi-reversible) athermal reaction takes place at one of the active sites defining an individual cluster), the first outlines of a unifying description emerge.

For example, it was demonstrated [110] that the most dramatic changes in reaction efficiency induced by nonrandom arrangements of multiplets on a $d = 2$ lattice of hexagonal symmetry occur in the concentration regime:

$$N_T/N = C_T < 0.1$$

where N_T is the number of reaction centers distributed on a lattice of N sites overall. Above this concentration, there appears to be little discrimination in the reaction efficiency with respect to the geometry and relative orientation of active sites. However, below the concentration $C_T \sim 0.1$, the diffusion-reaction event was found to be exquisitely sensitive both to the cluster geometry and orientation, and to the degree of reversibility of reaction, with the most dramatic changes in reaction efficiency occurring when there is less than a 50% probability that an irreversible reaction takes place on first encounter.

Geometrical and/or statistical considerations can be used to interpret trends in the reaction efficiency in the concentration regime $C_T < 0.1$. Following a strategy introduced by Guggenheim [111] in his work on "mixtures," the simplest correlation is to enumerate the number of clusters of a given configuration and then to differentiate the spatial distribution of reaction centers by identifying a discretized length κ that specifies the overall distribution of traps with respect to a fixed (reference) point in reaction space. This was the program carried out in ref. 110.

A somewhat more sensitive measure can be determined by constructing first a "reference" lattice [112]; suppose that the N_T trapping sites of a given array are redistributed on an N-site lattice in such a way that each trap is surrounded by the maximum number of nontrapping sites. Such a distribution defines one for which the reaction centers are maximally separated; their overall spatial relationship is the one most ordered (symmetrized) with respect to a given setting of $[N_T, N]$. Corresponding to this distribution, one can identify an effective (reduced) lattice of $N' = N/N_T$ sites having one centrally disposed deep trap. The average walklength $\langle n \rangle_M$ for a coreactant diffusing on such a N'-site lattice (of given symmetry and subject to periodic boundary conditions) can be calculated using Eq. (4.21). Then, for each multiplet configuration, the following ratio can be defined:

$$\delta = \frac{\langle n \rangle_M}{\langle n \rangle} \tag{4.76}$$

If the reaction centers are maximally separated, $\delta = 1$, whereas values of $\delta < 1$ reflect clustering of reaction centers. The parameter δ is therefore a measure of the efficiency of a diffusion-reaction process for a nonrandom configured system, relative to the efficiency for a companion, fully symmetrized distribution of reaction centers.

In equilibrium and nonequilibrium thermodynamics, the spatial organization of a system is reflected in the configurational entropy. As has been discussed by Denbigh [113] and Jaynes [114], there are several statistical analogues of the entropy, each of which has the desired thermodynamic properties. For example, the function S_i, defined by

$$S_i = -k \sum_n P_n \ln P_n, \tag{4.77}$$

where P_n is the probability of a system i being in the state n, is clearly a function of state. Using the representation (4.77) it can be demonstrated that for two (uncoupled) independent systems characterized by the entropies S_1 and S_2, the entropy S of the composite system is additive:

$$S = S_1 + S_2, \tag{4.78}$$

even if the two subsystems are not at equilibrium [115]; if statistical independence is not satisfied

$$S - [S_1 + S_2] < 0. \tag{4.79}$$

The existence of statistical correlations between the subsystems leads to an overall situation less random than when the subsystems are statistically independent of (or very weakly coupled to) each other.

Consider the probability that a coreactant diffusing on a discretized space (a lattice) undergoes an irreversible reaction at a given target site. For the discrete (geometric) distribution [116], the (normalized) probability of being trapped in n steps is

$$P_n = \frac{1}{\langle n \rangle} \left(1 - \frac{1}{\langle n \rangle} \right)^{n-1} \tag{4.80}$$

for which the first four moments are, respectively, the mean walklength $\langle n \rangle$, the relative width $(1-p)^{1/2}$, the skewness $(2-p)/(1-p)^{1/2}$, and the kurtosis $[6 + p^2/(1-p)]$, where $p = 1/\langle n \rangle$. Given this representation for P_n, the entropy S_i reduces, eventually, to the simple form

$$S_i/k = \ell n \langle n \rangle_i + 1 \tag{4.81}$$

and the general expression for the change in the entropy for a process involving r distinct subsystems (or states) is

$$\Delta S = S - \sum_{i=1}^{r} S_i = k \, \ell n \frac{\langle n \rangle e^{-(r-1)}}{\langle n \rangle_1 \langle n \rangle_2 \cdots \langle n \rangle_r} \tag{4.82}$$

Extensive calculations were reported in refs. 110 and 112 for a wide variety of configurations of active sites (monomers, dimers, triplets, quartets, hexamers, and various — binary and ternary — combinations) distributed on a finite lattice of hexagonal symmetry (overall 48 different configurations), and the efficiency of the underlying diffusion-reaction process studied as a function of the concentration C_T of reaction centers, the discretized length κ (ref. 110), and the statistical parameter δ (ref. 112). Also reported in ref. 112 were values of S_i for each configuration.

Qualitatively, the dependence of the system entropy on the concentration C_T of active sites and on the parameter δ is straightforward. Consider first the fate of a coreactant migrating through a field of sites, a certain fraction of which are reaction centers. The greater the number of trapping sites, the sooner the particle is trapped (on average); fewer paths (trajectories through reaction space) translate into a smaller statistical entropy. Similarly, the more symmetrically distributed the reaction centers ($\delta \rightarrow 1$), the sooner the particle is likely to encounter one of the active sites; interrupting the flow of the coreactant through reaction (phase) space necessarily decreases the entropy (the randomness) of the process.

Going beyond these qualitative statements, it is of interest to determine whether changes in concentration C_T are relatively more (or less) important than changes in the parameter δ in influencing the magnitude of the statistical entropy. Plotted in Figure 4.49 is the entropy S (per unit volume, in units of Boltzmann's constant) as a function of δ for two representative

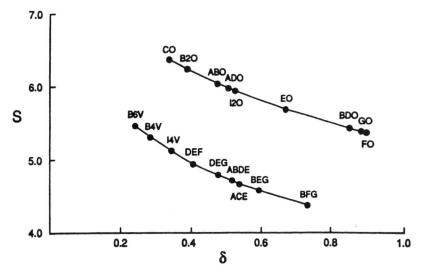

Figure 4.49. Plot of the statistical entropy S (per unit volume, in units of Boltzmann's constant) as a function of the statistical order parameter δ for two representative concentrations in the range $C_T < 0.1$. The upper curve displays the results for nine configurations characterized by $C_T = 6/256$, while for the different set of nine configurations noted on the lower curve, $C_T = 16/256$. Several configurations noted on the upper curve are diagrammed in Figure 4.50a and several configurations noted on the lower curve are shown in Figure 4.50b; the remaining configurations are specified in ref. 113.

concentrations in the range, $C_T < 0.1$. The upper curve displays results for nine multiplet configurations characterized by $C_T = 6/256$ while the lower curve shows the behavior for nine different configurations corresponding to $C_T = 16/256$. The multiplet distributions referenced in the figure are either displayed explicitly in Figure 4.50, or can be constructed from information specified in ref. 112.

Calculations show that the multiplets I2O and ABDE are characterized by (essentially) the same value of δ, viz., 0.513 and 0.515, respectively. However, the concentration C_T for the configuration ABDE is a factor of 2.7 greater than for I2O, and the increased number of trapping sites leads to an attendant reduction of 21% in the entropy on going from I2O to ABDE (or a factor of 3.4 in the value of $\langle n \rangle$). Alternatively, one can fix the concentration C_T and consider changes in the entropy resulting from a similar (quantitative) changes in δ. For the common setting $C_T = 0.625$ (lower curve), δ changes by a factor of 2.6 on going from the multiplet configuration B4V to BFG; this is accompanied by a decrease of 18% in the entropy (or a factor of 2.6 in the value of $\langle n \rangle$). The percentage change computed in these two cases is similar enough (21% vs 18%) that (to a fair approximation) one may

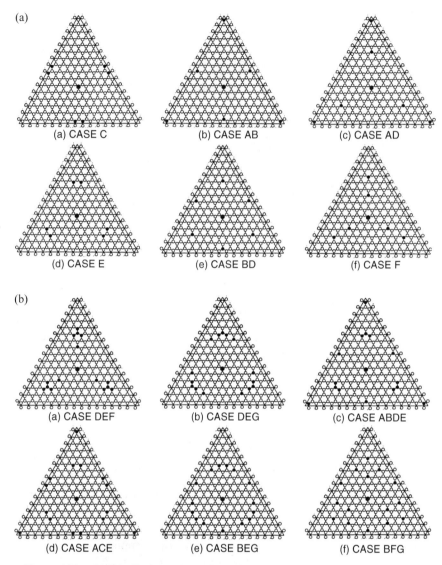

Figure 4.50. (a) Distributions of competing reaction centers consistent with the fixed concentration, $C_T = 6/256 = 0.0273$, of traps; the configurations illustrated here, but *excluding* the centrosymmetric trap, are among those noted on the upper curve of Figure 4.49. (b) Distributions of competing reaction centers consistent with the fixed concentration, $C_T = 16/256 = 0.0625$, of traps; the configurations illustrated here, but *including* the centrosymmetric trap, are among those noted on the lower curve of Figure 4.49.

conclude that comparable, quantitative changes in the number and/or distribution of reaction centers (as parametrized by C_T and δ, respectively) lead to similar changes in the entropy of the underlying diffusion-controlled process. Thus, given an initial nonrandom distribution of active sites, changes in catalyst efficiency may be a consequence of "aging," that is, a change in the concentration C_T of active sites, or may be the result of a redistribution of reaction centers (leading to a concomitant change in the parameter δ), and that both effects are of comparable importance. In either case, one will move along the curve of S versus δ, with different S values reflecting a process governed by different system geometries (and hence different turnover numbers and effective rate constants).

The above quantitative "tradeoff" between the concentration of active sites on the one hand, and their distribution on the other, in influencing the reaction efficiency is relevant to the interpretation of experiments on the catalytic properties of bimetallic systems (e.g., the studies of Sinfelt et al. [103] on Ni/Cu alloys) and to studies on catalyst deactivation [106]. It is also known that surface recognition events in cellular systems may be regulated by changes in the sequestering of receptors (e.g., processes at synapses [117], the studies of Edelman et al. [118] on cell-adhesion molecules [CAMs]). Although in these experiments energetic considerations are obviously of great importance, the above calculations show, as in studies of protein structure where hydrophobic effects play a fundamental role [119] that organizational changes resulting from the sequestering (or not) of reaction centers, and the associated entropy differences, can be significant in driving the irreversible process.

There is another aspect of "sequestering" that can be illuminated by the calculation of the entropy differences. Highly configured distributions of reaction centers can often be decomposed into a number of disjoint (geometrically uncoupled) subsystems. The question is: If, for each of the subsystems comprising the overall assembly, one calculates the statistical entropy associated with an irreversible (or quasi-reversible) process, is the entropy S for the composite system a simple additive function of the entropies $[S_i, S_j, \ldots]$ calculated for the individual subsystems? That is, is

$$\Delta S = S - [S_i + S_j + \cdots] = 0 \,, \qquad (4.83)$$

thereby implying that diffusion-controlled reactive processes taking place on highly configured systems can interpreted in terms of statistically independent events?

The unequivocal conclusion drawn from the data reported in ref. 112 is that this is not the case. ΔS is always negative and, in fact, satisfies an invariance relation that has interesting physical consequences. Given a final configuration built up from different initial and intermediate configurations, although the overall change ΔS is invariant, entropy changes characterizing the individual stages realized in achieving the final configuration (state) are different. The practical implications of this result are immediate. Since different subsystems may be thought of as different stages in the activation (or deactivation) of a heterogeneous catalyst, laying down (activating) monomers first, and then adding (activating) hexamers has different entropic consequences than laying down hexamers first and then activating monomeric reaction centers individually or in sets, even though the overall change in the statistical entropy of the process will be the same irrespective of the sequencing. In general, one finds that in an n-step process, the slowest (rate determining) step is the one associated with reactive events involving highly sequestered, localized clusters.

All of the above leads to the conclusion that an analysis of catalytic processes in systems for which the reactive sites are distributed in a nonrandom array needs to consider explicitly the entropic features of the diffusion-reaction process. However accurately the energetics of the system are determined in quantum chemical calculations, neglect of these entropic considerations will lead to an incomplete description of the diffusion-reaction event, particularly in those situations where "aging" of the catalyst is important.

Once a diffusing atom/molecule has been immobilized at a particular site, there are a variety of subsequent events that can occur. For example, in the early stages of thin-film growth [120], it is possible that the deposition event activates the surrounding sites, so that these sites become "traps" for species subsequently deposited on or in the system. It is also possible that the immobilization of a diffusing adatom at a specific site triggers the deactivation of nontrapping sites in the neighborhood of that site. Here, in effect, all pathways, ingoing and outgoing, to the set of deactivated lattices sites are cut or "blocked," thus changing the average distance a newly deposited adatom travels in this restricted reaction space before encountering the initially immobilized species. Thirdly, a newly deposited adatom, rather than joining an existing island, may nucleate a new island; previously active, nucleation sites may be blocked and an adatom will have to negotiate these blocked sites before being immobilized at some alternative nucleation site.

The nucleation and growth of islands during submonolayer deposition is, of course, a problem that has been studied for decades [121,122]. Analyses based on mean-field rate equations have led to an understanding of the dependence of mean island density, N_{av}, on deposition conditions. More recently, kinetic Monte Carlo simulation studies have been used to test various nucleation models [123,124]. In each of the three cases mentioned above, there occur changes in the organization or structure of the reaction space induced by a precursor species. On creating a new trap or blocking a neutral site on a lattice of N sites, only a relatively few (say, v) sites in the neighborhood of the affected site are sensitive to the activation or deactivation induced by the precursor species. Hence, as shown in Section II, changes in the efficiency of trapping/reaction/nucleation of a newly deposited atom or molecule, determined by calculating the matrix $\mathbf{\Pi} = (\mathbf{I} - \mathbf{P})^{-1}$, can be computed by inverting a $v \times v$ matrix instead of an $N \times N$ one. Realization of this program for the three cases cited above has been described in ref. 11. Here we focus on the consequences of assuming different sequences for generating a final nucleation pattern or island morphology on a finite, planar array.

Consider a randomly diffusing particle that is trapped irreversibly at a site within the boundary of a finite domain. The diffusion of a second species in the same domain is now studied, but subject to the constraint that once the first atom/molecule has been immobilized at a particular site, that site is inaccessible to a second diffusing species (i.e., the site occupied by the first species is "blocked"). The $\langle n \rangle_j$ ($j = 2$), corresponding to this second event in a j-state nucleation process can be calculated, as can the $\langle n \rangle_j$ for each subsequent stage in the process, all subject to the constraint that, as the island morphology begins to unfold, all sites occupied at that stage of the pattern formation are inaccessible to subsequently deposited diffusing species.

Two features of the above description deserve emphasis. First, as sites become "blocked," the diffusional space accessible to subsequent diffusing adatoms contracts; in fact, on finite planar arrays, not only do subsequent adatoms have available a smaller space within which to diffuse, but they have to negotiate more "obstacles" (i.e., the "blocked" sites). Second, in a j-step deposition process, depending on the symmetry, there are as many as $j!$ different ways in which a given final morphology can be generated (i.e., there are many developing configurations of blocked sites (defining the island morphology) that can lead (eventually) to a given final pattern in the deposition process).

To develop a theory for sequential traps [12], consider first a lattice with a single trap. A walker moves over the finite lattice until being trapped. Next, another trap opens, often on an adjoining site, and the (filled) original trap is then blocked. Another walker moves over the modified lattice until it, too, is trapped. The second trap is then blocked and a third opened, and so on, for some predetermined number of traps.

We number lattice states in such a way that the trap to be blocked will be indexed by 0, while the trap to be opened will have index 1. The original $\mathbf{P}$

matrix can be written in partitioned form as follows:

$$\mathbf{P} = \begin{bmatrix} 0 & 0 & \mathbf{0}_{(1,N-2)} \\ p_{10} & p_{11} & \mathbf{P}_1 \\ \mathbf{p}_0 & \mathbf{p}_1 & \mathbf{Q} \end{bmatrix} \qquad (4.84)$$

Here $\mathbf{P}_1$ is a $1 \times (N-2)$ row, containing the probabilities of going from state 1 to the states other than 0 or 1, $\mathbf{p}_0$ and $\mathbf{p}_1$ are $(N-2) \times 1$ columns containing the probabilities for going from the states other than 0 or 1 to 1 or 0, respectively, and $\mathbf{Q}$ is a $(N-2) \times (N-2)$ matrix. The whole first row of $\mathbf{P}$ is zero precisely because state 0 is a trap. Similarly, the original $\mathbf{\Pi}$ matrix can be written as

$$\mathbf{\Pi} = \begin{bmatrix} 1 & 0 & \mathbf{0}_{(1,N-2)} \\ 1 & \pi_{11} & \boldsymbol{\rho}_1 \\ \boldsymbol{\iota}_{(N-2,1)} & \boldsymbol{\pi}_1 & \hat{\mathbf{\Pi}} \end{bmatrix} \qquad (4.85)$$

where $\boldsymbol{\iota}$ is a column vector with all elements equal to 1. All of the elements of column 0 of $\mathbf{\Pi}$ are 1 because state 0 is the only trap on the lattice: regardless of where a walk starts, it always pays exactly one visit to state 0 and is then trapped. All the elements of row 0, except the 0 element itself, are 0 because starting from the trap, there is nowhere else to go.

When the new trap opens at state 1, the change to $\mathbf{P}$ can be written as

$$\Delta\mathbf{P} = -\mathbf{e}_1 [p_{10} \quad p_{11} \quad \mathbf{P}_1]$$

where $\mathbf{e}_1$ is an $N \times 1$ vector with all elements zero except that indexed by 1, which equals 1. It is easy to see that $\mathbf{P}' \equiv \mathbf{P} + \Delta\mathbf{P}$ has a row indexed by 1 which is zero: state 1 is a trap. $\Delta\mathbf{\Pi}$, the matrix of changes to $\mathbf{\Pi}$ caused by the changes $\Delta\mathbf{P}$, can be written as

$$\Delta\mathbf{\Pi} = \frac{1}{\pi_{11}} \begin{bmatrix} 0 \\ \pi_{11} \\ \boldsymbol{\pi}_1 \end{bmatrix} ([1 \quad \pi_{11} \quad \boldsymbol{\rho}_1] - \mathbf{e}_1^{\tau}) \qquad (4.86)$$

At the end of this calculation, only the rows and columns of $\Delta\mathbf{\Pi}$ after those indexed by 0 and 1 are retained, because those rows and columns of the changed $\mathbf{\Pi}$ are trivial. In the meantime, though, we retain everything.

The next step is to block the trapping state 0. This involves changes to $\mathbf{P}'$ as follows:

$$\Delta\mathbf{P}' = \sum_{i \in \mathbf{N}_0} p_{i0} \mathbf{e}_i (\mathbf{e}_i^{\tau} - \mathbf{e}_0^{\tau})$$

Here $\mathbf{N}_0$ denotes the set of states that, according to $\mathbf{\Pi}' = \mathbf{\Pi} + \Delta\mathbf{\Pi}$, can communicate directly with state 0 before it is blocked. Note that state 1 cannot belong to $\mathbf{N}_0$, because it is a trap in $\mathbf{P}'$. Note also that the operation of blocking is defined in such a way that any step which, before blocking, would have gone to state 0, now simply stops where it is for one step. If $\mathbf{N}_0$ is empty, this means that, with the opening of state 1 as a trap, state 0 is inaccessible from any other lattice site, and so no further changes to $\mathbf{\Pi}'$ are needed. Otherwise, let v_0 be the number of states contained in $\mathbf{N}_0$.

Changes in $\Delta\mathbf{\Pi}'$ that are to be added to $\mathbf{\Pi}'$ for the blocking of 0 are given by

$$\Delta\mathbf{\Pi}' = \begin{bmatrix} 0 & \mathbf{0}_{(1,N-1)} \\ -\tilde{\pi}'_0 & \tilde{\mathbf{\Pi}}'_0 \mathbf{J}^{-1} \mathbf{K} \tilde{\mathbf{R}}'_0 \end{bmatrix} \tag{4.87}$$

where $\tilde{\mathbf{\Pi}}'$ is the $(N-1) \times v_0$ block of $\mathbf{\Pi}'$ with all rows except row 0 and only the columns indexed by indices in $\mathbf{N}_0$; symmetrically, $\tilde{\mathbf{R}}'_0$ is a $v_0 \times (N-1)$ block, $\mathbf{K}$ is a $v_0 \times v_0$ diagonal matrix, with ith diagonal element $p'_{i0} = p_{i0}$, since $\mathbf{P}$ and $\mathbf{P}'$ differ only in row 1; and, finally, the $v_0 \times v_0$ matrix $\mathbf{J}$ is given by

$$\mathbf{J} = \mathbf{I}_{v_0} - \mathbf{K}\hat{\mathbf{R}}'_0 \tag{4.88}$$

where $\hat{\mathbf{R}}'_0$ is the $v_0 \times v_0$ block of $\mathbf{\Pi}'$ with both rows and columns indexed by indices in $\mathbf{N}_0$.

After the double operation of opening the new trap and blocking the old, the resulting matrix of expected numbers of visits can be denoted as $\mathbf{\Pi}''$, and it takes the form

$$\mathbf{\Pi}'' = \begin{bmatrix} 1 & 0 & \mathbf{0}_{(1,N-2)} \\ 0 & 1 & \mathbf{0}_{(1,N-2)} \\ \mathbf{0}_{(N-2,1)} & \mathbf{1}_{(N-2,1)} & \tilde{\mathbf{\Pi}}'' \end{bmatrix}$$

where the forms of rows and columns 0 and 1 are dictated by the facts that 0 is a blocked site and that 1 is the sole trap open on the lattice. The $(N-2) \times (N-2)$ block $\tilde{\mathbf{\Pi}}''$ can be constructed by starting from the corresponding block of $\mathbf{\Pi}$ and adding the corresponding blocks of the two matrices of changes, $\Delta\mathbf{\Pi}$ and $\Delta\mathbf{\Pi}'$.

From Eq. (4.86), the appropriate block of $\Delta\mathbf{\Pi}$ is

$$-\frac{1}{\pi_{11}}\pi_1\rho_1 \tag{4.89}$$

and, from Eq. (4.87), that of $\Delta\mathbf{\Pi}'$ is

$$\mathbf{L}\mathbf{J}^{-1}\mathbf{K}\mathbf{R} \tag{4.90}$$

where $\mathbf{L}$ and $\mathbf{R}$ are respectively $(N-2) \times v_0$ and $v_0 \times (N-2)$ blocks of $\mathbf{\Pi}'$. What we really want is an expression involving only blocks and elements of $\mathbf{\Pi}$, without the need to calculate $\mathbf{\Pi}'$ in an intermediate step. Given Eq. (4.86), $\mathbf{L}$ and $\mathbf{R}$ can be constructed as follows. Let $\boldsymbol{\pi}_i$ and $\boldsymbol{\rho}_i$ be respectively the ith column and row of $\mathbf{\Pi}$. Then for $i \in \mathbf{N}_0$, the ith column of $\mathbf{L}$ is

$$\boldsymbol{\pi}_i - \frac{\pi_{1i}}{\pi_{11}} \boldsymbol{\pi}_1 \qquad (4.91)$$

but without rows 0 and 1, and the ith row of $\mathbf{R}$ is

$$\boldsymbol{\rho}_i - \frac{\pi_{i1}}{\pi_{11}} \boldsymbol{\rho}_1 \qquad (4.92)$$

without columns 0 and 1. The $\hat{\mathbf{R}}$ used in computing $\mathbf{J}$ in Eq. (4.88) can then be obtained by selecting the v_0 columns of $\mathbf{R}$, and $\mathbf{K}$ is defined directly in terms of the original $\mathbf{P}$ matrix. The changes to $\mathbf{\Pi}$ brought about by the double operation are then given by the sum of Eqs. (4.89) and (4.90) and can be computed directly from the original $\mathbf{P}$ and $\mathbf{\Pi}$ matrices.

The stochastic consequences of assuming different sequences in generating a given nucleation pattern or morphology on a finite planar array can be quantified. To visualize the processes we wish to study, consider first the finite array sketched in Figure 4.51. The shaded circles in this figure define atoms of the underlying substrate or host lattice; arrays having both square planar and triangular symmetry will be considered, with the total number N of atoms defining the host lattice fixed at $N = 48$. Suppose now that a first atom, after deposition, diffuses randomly on this lattice and becomes immobilized in a certain location; the open circle labeled "one" in the display at the top of Figure 4.51 denotes such an adatom. The average number $\langle n \rangle_1$ of random displacements before immobilization (or "trapping") on a $N = 48$ lattice with square planar symmetry is $\langle n \rangle_1 = 94.295\,750$. Once this first adatom is locked in place, the average number $\langle n \rangle_2$ of steps required for a second diffusing atom to be immobilized, but subject to the constraint that now one site is "blocked" or inaccessible to the second random walker, can be calculated; the number N' of sites available to the second atom is $N' = N - 1 = 47$ and, using the method just described, $\langle n \rangle_2 = 95.033\,096$ before trapping on the same square planar lattice.

Recall that, in nucleation theory [125–134], the size at which for the first time an island becomes more stable with the addition of just one more atom is defined as the critical island size i. In describing the statistical consequences of continuing this process of sequential trapping, in effect $i = 1$ in this calculation.

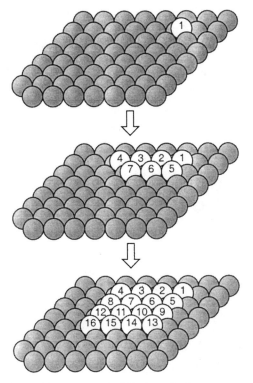

Figure 4.51. The "row filling" nucleation sequence.

Referring again to Figure 4.51, the nucleation pattern after the deposition of 7 adatoms on the underlying lattice is depicted in the center illustration, and the pattern after deposition of 16 adatoms on the substrate is depicted at the bottom of the figure. Noting the order $(1 \to 2 \to 3 \to \cdots \to 16)$ in which the 16 atoms are deposited sequentially, this case is referred to as the "row filling" case. Two other nucleation sequences are displayed in Figures 4.52 and 4.53. That specified in Figure 4.52 is designated as the "dendrite" case and that in Figure 4.53 as the "compact" case. Values of $\langle n \rangle_j$ for these three cases for deposition in a square planar array are given in Table IV.7; in Table IV.8, are listed the $\langle n \rangle_j$ values for deposition in a triangular lattice array. The calculations reported in Tables IV.7 and IV.8 were carried out assuming that when a diffusing atom encounters a site on the boundary of the $N = 48$ host lattice, the diffusing atom moves away from that site in its next displacement. For comparison, one can also consider the possibility that the diffusing atom can remain at the same boundary site before moving

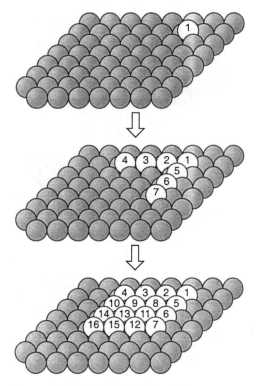

Figure 4.52. The "dendrite" nucleation sequence.

away, and the results obtained imposing this alternative boundary condition are given in ref. 12 for deposition in a square planar array.

The three growth patterns illustrated in Figures 4.51, 4.52, and 4.53 can be characterized in terms of the average branch thickness b of a fractal island. In the "hit and stick," diffusion-limited-aggregation (DLA) model [135,136], (regime I fractal growth), $b = 1$. Experimentally [137–141], deposition is characterized by wider branch thicknesses ($b \approx 4$); it has been suggested by Zhang, Chen, and Lagally [142] that this regime (II) of extended fractal growth can be understood if every adatom, upon reaching the edge of a developing island, can relax to the extent that it has found at least two nearest neighbors with atoms belonging to the island. They note that this regime can be defined on a triangular lattice, but is absent on a square lattice. The center parts of Figures 4.51 and 4.52 are characterized by $b = 2$ and $b = 1$, respectively, and the bottom panel of all three figures is characterized by $b = 4$. These different growth patterns, as well as developing branch thicknesses, can be discussed from an entropic point of view.

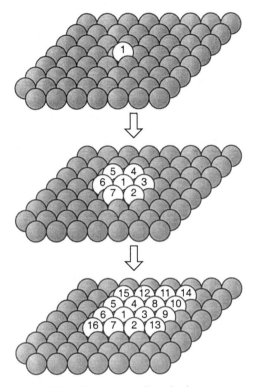

Figure 4.53. The "compact" nucleation sequence.

To begin, as diagrammed, there are $16!/2!$ different ways in which the final morphology can be realized, so the cases considered here are only representative of the kind of efficiencies that can be gained (or not) when different assumptions are made on the cohesive energy among adatom species on a given lattice. As a first step in understanding the results, $\langle n \rangle_j$ is plotted versus the number B of sites blocked in the nucleation process. Displayed in Figure 4.54 are the results obtained for deposition in a square-planar array for the "row filling" case and for the "compact" case. A similar plot is displayed in Figure 4.55 for deposition in a triangular array. For both symmetries, a pattern in the data emerges in the row filling case, namely, a systematic decrease in $\langle n \rangle_j$, once a given row has been initiated, followed by a "jump" in $\langle n \rangle_j$, once a given row has been filled and the next value of B is considered. More generally, one finds that for $B \geq 5$, values of $\langle n \rangle_j$ for the "compact" case are systematically larger than for the "row filling" case and, concomitantly, the overall time scale associated with the

TABLE IV.7

Values of $\langle n \rangle_j$ for three nucleation sequences forming a 16-atom square-planar array

Trap site	Number of blocked sites	Effective lattice size	Row	Dendrite	Compact
1	0	48	94.295750	94.295750	58.480519
2	1	47	95.033906	95.033906	70.427292
3	2	46	93.240683	93.240683	130.58908
4	3	45	100.89593	100.89593	85.475202
5	4	44	91.157266	91.157266	89.158769
6	5	43	113.08878	83.323382	130.57460
7	6	42	107.65152	86.545244	139.15966
8	7	41	104.27111	124.32039	127.14026
9	8	40	83.280712	106.03066	107.09015
10	9	39	102.51249	96.205496	151.45274
11	10	38	95.862836	111.34925	123.67753
12	11	37	90.260874	94.035788	148.57080
13	12	36	83.427533	101.43189	135.04622
14	13	35	99.766017	86.556127	193.74668
15	14	34	91.970378	91.970378	124.17620
16	15	33	84.268109	84.268109	84.268109

TABLE IV.8

Values of $\langle n \rangle_j$ for three nucleation sequences forming a 16-atom triangular array

Trap site	Number of blocked sites	Effective lattice size	Row	Dendrite	Compact
1	0	48	103.73724	103.73724	51.983646
2	1	47	85.414248	85.414248	63.378585
3	2	46	75.196521	75.196521	86.994866
4	3	45	74.105288	74.105288	82.049082
5	4	44	102.03347	102.03347	83.539571
6	5	43	99.084704	74.404799	88.852943
7	6	42	85.265801	66.234068	113.03597
8	7	41	65.838360	148.36967	89.331439
9	8	40	88.960599	91.108973	100.72069
10	9	39	87.543000	61.397650	117.45498
11	10	38	74.451960	121.71118	123.28048
12	11	37	55.684651	64.218968	139.16097
13	12	36	84.016225	102.41832	128.57889
14	13	35	83.194300	53.960768	182.34293
15	14	34	70.396682	70.396681	117.57398
16	15	33	52.677996	52.677996	52.677996

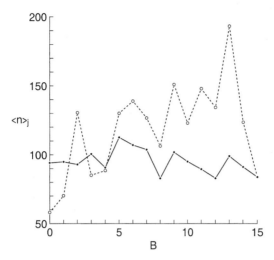

Figure 4.54. A plot of $\langle n \rangle_j$ versus the number B of blocked sites for a square-planar array. The "row filling" case is denoted by the solid line and the "compact" case by the dashed line.

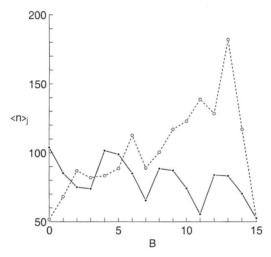

Figure 4.55. A plot of $\langle n \rangle_j$ versus the number B of blocked sites for a triangular array. The "row filling" case is denoted by the solid line and the "compact" case by the dashed line.

development of the pattern in the "compact" case is longer than for the other two cases studied here.

The overall entropy change for a sequence involving r distinct precursor states is given by Eq. (4.82). Values of S_j/k for the three nucleation sequences studied here for the case of deposition in a square-planar array show that for $j > 5$, the entropies calculated for the "compact" case are systematically larger than for the "row filling" and "dendrite" cases. A further insight can be derived by recalling that for two uncoupled (independent) subsystems characterized by entropies S_1 and S_2, the entropy of the composite system is additive (see Eq. (4.78)) whereas if statistical independence is not satisfied Eq. (4.79) pertains. Hence, the presence of statistical correlations between two subsystems leads to an overall situation less random than when the systems are statistically independent of (or weakly coupled to) each other. For which nucleation sequence are the statistical correlations most pronounced? The final pattern is the same in each case and the last site filled in each nucleation sequence is the same; the state entropy S_{16}/k associated with that site ($j = 16$), calculated using the result, Eq. (4.81), is determined to be 5.709 057. Adopting $j = 16$ as the final state of the system, the quantity

$$\Delta(S/k) = S_{16}/k - \sum_{j=1}^{15} (S_j/k)$$

can be calculated for each of the nucleation sequences considered here. The results are

$$\Delta(S/k)(\text{"row filling"}) = -80.729\ 551$$

$$\Delta(S/k)(\text{"dendrite"}) = -80.880\ 874$$

$$\Delta(S/k)(\text{"compact"}) = -83.356\ 057$$

Thus, the unfolding of the nucleation sequence defined by the "compact" case is characterized by a specificity which demands more significant statistical correlations between successive stages in the j-step process than the other two cases studied here.

Finally, we comment on the relevance of the calculations to the discussion in the review of Zhang and Lagally [120] on the manipulation of growth kinetics. A number of factors were identified that can influence the quality of $d = 2$ films grown by vapor-phase expitaxy: reduction in the island-edge barrier, hindering diffusion along island edges by impurities, increasing the island density, and enhancing atom mobility on top of an island with respect to that on the lower layers. From Figures 4.54 and 4.55, and keeping in mind the relationship between the magnitude of $\langle n \rangle$ and the time scale of the process, the values calculated for deposition in a square-planar array are systematically larger than the values calculated for deposition in a triangular array, regardless of the mechanism assumed (i.e., "row filling," "dendrite," or "compact"). Of the two mechanisms, the

$\langle n \rangle$-values for the "compact" case are larger than the "row filling" case, regardless of the lattice assumed. From these two observations, it may be concluded that deposition in an array having square-planar symmetry via the "compact" mechanism requires the longest time to nucleate. Deposition in an array having triangular symmetry via the "row filling" mechanism requires the shortest time. Thus, the most entropically favorable process for growth is one in which the morphology is generated by a "row filling" mechanism on lattices with triangular geometry (e.g., on face-centered-cubic (fcc) (111) or hexagonal close-packed (hcp) (0001) substrates).

The generality of the above conclusion is limited for several reasons. First a particular domain size was assumed. Second, although the role of temperature was not considered, it is probable that the analysis is portraying nucleation in the very low temperature regime. Zhang and Lagally [120] analyzed the regime of fractal islands with "thin arms," but in practice there is a continuous increase in arm thickness, and a transition from fractal to compact islands. A model for this transition has been presented by Bartelt and Evans [143]. At higher temperature, there is also a significant restructuring of the island during growth, its shape being determined by a competition between the DLA instability and equilibrating processes. And, very recently, it has been demonstrated [144] that postdeposition coarsening of the adlayer is dominated typically by diffusion and subsequent coalescence of large $d = 2$ clusters rather than Ostwald ripening involving a diffusion-mediated transfer of adatoms from smaller to larger islands. There are, however, no matters of principle standing in the way of applying the "shorthand" method described above to the study of aspects of these experimental findings.

B. Processes In/On Porous Catalysts

"Reaction rates and selectivity in porous catalyst pellets are strongly influenced by rate limitations associated with diffusion of reactants between the surface of the pellet and active sites deep within the structure." Thus, began the introduction to the treatise on "Transport in Porous Catalysts," by Jackson [145] a monograph which gave a summary of the theoretical progress made in this area prior to 1977. Also unfolding in the 1970s was an area of experimental investigation, the study of kinetic processes in micro-heterogeneous systems, an area that evolved so rapidly that a series of review papers and full-length monographs could hardly keep pace with developments in the field. As representative of the latter, there appeared in succession the monographs of Fendler [91], Thomas [68], and Kalyanasundaram [69], each stressing the remarkable kinetic advantages gained in studying photochemical/photophysical processes in compartmentalized systems (e.g., micelles, zeolites, clays, vesicles, colloidal catalyst particles, silica, silica gel, porous vycor glass, cellulose, starch). A glance at the current literature [68] reveals that activity in this area of experimental study has not waned and, if anything, has accelerated in recent years, leading to fundamental insights in understanding reactive processes in structured media and to the possibility of exploiting this understanding in the design of materials of commercial interest.

In dealing with the problem of transport and reactive processes in porous media, the theoretical strategies described in Jackson [145], in the monographs of Scheidegger [146], Dullien [147], and Collins [148], and in the seminal treatise of Aris [62], centered on two main approaches, the "dusty gas" model (developed by Mason and his colleagues [149]) and the "capillary network" model. Generalizations of the capillary network model are more relevant to the study of the structured inorganic systems cited in refs. 68,69, and 91. The chief aim of this section will be to describe how the methods and results described earlier in this review cast light on photoinduced or radiation-induced reactive process taking place within or on porous catalyst materials. Specifically, the synergetic interplay of both geometrical and potential effects in influencing these processes will be described and documented.

As a first example, one can quantify the interplay among system topology, multipolar correlations and "reduction of dimensionality" by generalizing the lattice model of encounter-controlled reactive processes in porous zeolite catalyst particles introduced in Section III.D.3. Focusing on the cavity/capillary structure of zeolite A, processes in which a stationary target molecule is localized either within an interior cavity of the catalyst particle or housed within a port on the external surface can be considered. For the case where a mobile coreactant is confined in its motion to the interior and/or surface of the particle, it is found that reactive processes where the target molecule is anchored to the exterior surface are characterized by longer time scales than those for which the particle is situated within a cavity site in the interior of the porous catalyst; this result is valid either in the presence or in the absence of multipolar correlations [93]. It is also found that the ratio of diffusion-controlled rate constants, Eq. (4.75), is systematically smaller the closer the target molecule is positioned to the exterior surface of the catalyst. Decreases in the temperature and/or the (local) dielectric constant of the host medium tend to accentuate the importance of interparticle correlations, and calculations show that the influence of these two variables is more pronounced if the target molecule is embedded on an interior surface of the porous catalyst.

To assess the relevance of these results to actual experimental situations, the assumptions made in carrying out these model calculations can be reviewed. First, the cubic lattice adopted in the above study applies only to zeolites having the regular channel pattern of zeolite A. This is not a fundamental limitation, however, inasmuch as it was shown in Section III.D.3 that other lattice geometries can be studied straightforwardly as well, for example, the capillary network in faujasite which was modeled as a $d = 3$ tetrahedral lattice. Moreover, processes in/on fractal lattices of dimension $2 < d < 3$ can also be studied (see Section III.B.2). The latter possibility is of interest in light of the study of Yang, El-Sayed, and Suib [80] who used the Klafter-Blumen theory [46] for one-step energy transfer processes on fractal sets to suggest a corresponding theoretical relationship for energy-transfer on zeolite systems. In particular, using an equation similar to the KB equation to interpret the observed decay of the uranyl ion (donor) emission in the presence of europium ion as acceptor, these authors showed that the "apparent fractional" dimensions of the zeolites A, Y, mordenite and ZMS-5 assuming a dipolar exchange mechanism were 2.43, 2.47, 2.66, and 1.30, respectively. The first three values are close to the (exact) fractal dimension of the Menger sponge (viz. $\ell n\, 20/\ell n\, 3 = 2.73$) suggesting that this fractal lattice might serve as a model for reactive processes in certain zeolite systems.

Second, the diffusing coreactant in the above study is assumed to be a structureless particle or, at least, one for which the particle's motion is safely described by the motion of its center of mass (as would be the case if the molecular diameter of the coreactant is smaller than the diameter of the capillary "windows"). Some progress has been made in analyzing the random walk characteristics of a dimer diffusing through a structured medium [150] (where steric effects cannot be neglected) so that, in principle, this limitation of the above study can also be removed.

Given these caveats, it is of interest to use the insights generated in the calculations described above as a qualitative guide in understanding the experimental effects uncovered by Turro and his colleagues [81] in their classic study of the size/shape selectivity of o-methyldibenzyl ketone (o-ACOB) absorbed on ZMS zeolites. At room temperature, on photolysis of o-ACOB, fragments o-$\dot{A}$ and $\dot{B}$ are formed and can then recombine; owing to steric effects, the products o-A-o-A and o-A-B are formed on the external zeolite surface (only) while the product B-B is found on the internal surface. Specifically, the B-B product can result by recombination of $\dot{B}$ fragments within the zeolite on an interior surface, or on the external surface followed by sieving of B-B into the zeolite. The B-B reaction is found to exhibit a pronounced temperature effect with the percentage of B-B formed on the external surface increasing as the temperature decreases.

On the basis of the calculations reported in ref. 151, exactly the opposite behavior would have been predicted, viz. a decrease in temperature should result in a relative enhancement of the recombination event in the interior of the catalyst relative to an exterior (only) surfacial process. That is, if the effect of changes in temperature is reflected only through the Boltzmann factor in the local partition function, Eq. (4.55), then (both for attractive and repulsive correlations between reactants), the model predicts an enhanced accumulation of B-B products in the interior of the catalyst with decrease in temperature. It is clear, therefore, that the origin of the temperature effect in the Turro experiment must be sought elsewhere.

One possibility is the presence of an activation barrier, as Turro noted [81]. He argues that the experimental effect can be understood if there is "a relatively high activation energy for sieving of $\dot{B}$ radicals from the external to the internal surface." The motion of highly structured $\dot{B}$ fragments is expected to be constrained by the effective dimension of the ports (and the sinusoidal character) of the ZMS-5 capillary network, a steric effect reminiscent of the one found in studies of dimer diffusion in constrained geometries [150] (where significant differences in transit times were found when monomer vs dimer mobilities were compared). The barrier to entry (or re-entry) into the internal network of the catalyst would then be overcome by an increase in temperature; conversely, entry would be retarded by a decrease in temperature. Note that the presence of an activation barrier would also result in an effective "reduction of dimensionality" in the flow of the diffusing $\dot{B}$ fragment. From the discussion presented in Section II.C.1, although the efficiency of reaction is higher the more centrally located the target (here, a site on an interior surface of the catalyst), confinement of the diffusing species to the exterior surface, once reached, would result in the lower-dimensional, surficial process being more efficient. Finally, recalling the earlier discussion of diffusional flows within/on the Menger sponge (Section III.B.3), if indeed a $\dot{B}$ fragment formed on the surface is able to overcome an activation barrier and enter the internal pore structure of the zeolite catalyst, the fact that $\langle n \rangle_S$ and $\langle n \rangle_D$ are both smaller than $\langle n \rangle_M$ confirms that even "short cuts" through the catalyst particle to active sites anchored on the surface do not lead to recombination events more efficient than for surface-only processes.

The selective separation of products discovered in the Turro experiment represents one of a wide range of properties that make crystalline zeolites industrially important. In another study, Thomas and co-workers [84] have examined imperfectly ordered (quasi-crystalline or semiamorphous) zeolites, noting that these too possess attractive properties. As pointed out by Thomas and Bursill, "it is rather puzzling that a material which lacks long-range order can nevertheless exhibit marked cation-exchange activity and catalytic activity, which in the case of the crystalline varieties, arise because of the unit-cell porosity of the aluminosilicate."

In fig. 1 of ref. 84a, Thomas and Bursill show a high-resolution electron micrograph of a thin film (<40 Å thick) of quasi-crystalline zeolite A. From this evidence they extract the following noteworthy features: (i) a "raft" (ca. 10^4 Å^2 in projected area) of crystalline material consisting of an ordered array of supercages (about 100 in all) surrounded by essentially amorphous aluminosilicate; (ii) a smaller "raft" (ca. 3×10^3 Å^2 in area), less well ordered than the other and not in registry with it, consisting of some 30 supercages, also surrounded by amorphous material; and (iii) several isolated supercages in the amorphous background.

The lattice model of zeolite A described previously can be extended to study the role of multipolar correlations in influencing the efficiency of reaction between a fixed target molecule and a diffusing coreactant in zeolites that exhibit both crystalline and semiamorphous behavior. The basic idea is to couple the lattice-based stochastic theory with a version of the Debye-Smoluchowski theory of encounter-controlled reactions, and to quantify the differences in the diffusion-controlled rate constant k_D for reactions taking place in crystalline regions of finite extent ("rafts") versus those occurring in crystalline regions surrounded by an amorphous aluminosilicate [93]. A natural parameter in this approach is R_c, the distance at which the discretized (lattice-based) cell model is replaced by the continuum representation, and to study the reaction efficiency as a function of the reduced length $r_i = R_c/R$, where R is the distance between adjacent cavities.

Calculations [93] demonstrate that for a crystallite of finite extent, the most dramatic changes in the ratio of steady-state, diffusion-limited rate constants, $k_D[V = V(r)]/k_D[V = 0]$, occur for values of r_i in the range, $1 < r_i < 8$, for reactants interacting via ion-dipole and dipole-dipole forces, with a longer range characterizing ion-ion interactions. For the choice $r_i = 8$, calculations show that for "rafts" characterized by this range parameter, ion-dipole and dipole-dipole assisted reactions are enhanced by factors of 2–3 when carried out in finite crystallites. When $r_i = 8$, the corresponding enhancement in the above ratio of rate constants for ion pairs is about the same as for a homogeneous solution characterized by the same value of W (Eq. (4.61)). The calculations also show that the focusing effect of interparticle interactions as mediated by the "raft" structure of a semiamorphous zeolite fairly approximates an extended crystalline zeolite structure precisely because the influence of the potential is of greatest consequence over distances less than the spatial extent of the "raft" structure observed experimentally. Thus, calculations based on the cell/continuum model suggest that much of the marked catalytic activity characterizing fully crystalline zeolites is already captured by semiamor-

phous zeolites containing raft structures of the spatial extent identified by
Thomas and Bursill [84].

VI. CONCLUSIONS

The general field with which this review is concerned is currently one of the
most exciting in chemical physics, the study of kinetic processes in systems
of finite size and/or of restricted dimensionality. Problems ranging from the
study of organized molecular assemblies (micelles, vesicles, microemul-
sions), biological systems (cells, microtubules, chloroplasts, mitochondria),
structured media such as clays and zeolites, and nucleation phenomena in
finite domains are among those under active investigation.

This review has explored the influence of system geometry on the
efficiency of an encounter-controlled, irreversible reaction of the form, Eq.
(4.2), where A is a diffusing coreactant and B is a target molecule anchored
at a site within or on the boundary of a compartmentalized system. The
approach taken is to represent the geometry of the system by a lattice, and
then to determine the dynamics using the stochastic master equation, Eq.
(4.3). The advantage of this approach is that geometries of quite arbitrary
shape and dimensionality (integral or fractal) can be studied, a variety of
conditions can be imposed on the motion of the diffusing coreactant and,
from a technical point of view, the necessity in continuum theories, Eq.
(4.1), of specifying the Laplacian in spaces of intermediate or fractal
dimension is bypassed.

In this section, we shall distill from the examples presented a number of
generalizations which, hopefully, may be of use to the experimentalist in
understanding irreversible processes in confined systems. Considered
sequentially will be the role of system size, dimensionality, shape, and
internal connectivity, all facets of the underlying topology of the system.
Then, the role of intermolecular potentials and the effects of changes in the
multipolarity of the potential, the temperature and the dielectric constant of
the ambient medium will be reviewed.

From the classical literature on continuum theories of diffusion-reaction
processes based on Eq. (4.1), it is anticipated that the larger the system size,
the longer the time scale required for the reactive event, Eq. (4.2), to occur.
The corresponding dependence for lattice systems was first proved
analytically by Montroll and Weiss [17–19] who studied nearest-neighbor
random walks on finite lattices of integral dimension subject to periodic
boundary conditions. In a lattice-based approach to diffusion-controlled
processes, one can also examine the influence of the number of pathways (or
reaction channels) available to the diffusing coreactant at each point in the

reaction space, and these authors showed that the more pathways available and/or the higher the dimensionality of the lattice, the more efficient the process, Eq. (4.2). Whereas, in the main, these conclusions also pertain to reactions in compartmentalized systems, the results presented here reveal that the restricted spatial extent of finite assemblies can modulate significantly the interplay among these variables.

For example, in considering finite, $d = 2$ dimensional arrays, although the system size N remains the dominant variable in influencing the reaction efficiency, it is shown in Section III.B.1 that for fixed N, the shape of the domain is also important. Specifically, the smaller the number N_b of vertices defining the boundary of the planar array, the smaller the value of the overall average walk length $\langle n \rangle$ of the random walker before trapping. Second, for fixed N and fixed N_b, the smaller the value of the root-mean-square distance $\langle r^2 \rangle^{1/2}$ of the N lattice sites relative to the center of the array, the smaller the value of $\langle n \rangle$. It is only after the size and shape variables have been set that one finds for fixed $\{N, N_b, \text{ and } \langle r^2 \rangle^{1/2}\}$, that $\langle n \rangle$ decreases with increase in the overall average valency $\langle v \rangle$ (or coordination) of the array.

The seminal importance of system size vis a vis the local connectivity of the reaction space can be compromised in surficial processes if the number of triangular domains relative to domains of higher coordination is increased; in fact, an inversion in the relative importance of N and $\langle v \rangle$ can be found. Moreover, if the flow of the diffusing coreactant is biased (owing to the presence of a uniform field or a concentration gradient), simple examples show that there are situations where reactions can occur more efficiently when the target molecule is anchored at a site of lower valency relative to the overall valency of the lattice.

In studying size/shape effects in $d = 3$ compartmentalized systems, the calculations presented in Section III.C.1 demonstrate that the topology of the space, the location of the target molecule and the fate of the diffusing coreactant as it confronts the boundary of the system are of critical importance. The general conclusion is that the internal, centrosymmetric site in the reaction space is always optimal as regards the reaction efficiency, even when multipolar interactions between the coreactants are at play. If, however, the trajectory of the diffusing coreactant is restricted to $d < 3$ upon encountering the boundary of the system for the first time, that is, if one invokes the Adam–Delbruck concept of "reduction of dimensionality," one finds that, with increase in system size, there is a size beyond which the timing and efficiency of the process, Eq. (4.2), is enhanced if the target molecule is moved from an internal centrosymmetric site to a centrosymmetric site on the boundary of the system. In considering processes taking place in asymmetrical compartmentalized systems, there is a crossover in reaction efficiency when the space is elongated along one dimension to yield a tubular structure or, conversely, when one extends the face perpendicular to that dimension to form a platelet. For small tubules or platelets, reaction at an internal target is optimal, whereas for larger systems, placing the target on the surface and invoking "reduction of dimensionality" is the more efficient process. Even here, however, if the target molecule is positioned at a relatively inaccessible site on the boundary (namely a site of restricted valency), the advantages gained upon reducing the dimensionality of the flow are lost.

Regarding the dependence of the reaction efficiency on the dimensionality of the compartmentalized system, the studies reported in Sections III.B.3 and III.B.4 on processes taking place on sets of fractal dimension showed that, consistent with the results found for spaces of integer dimension, the higher the dimensionality of the lattice, the more efficient the trapping process, ceteris paribus. Processes within layered diffusion spaces, which can be characterized using an approach based on the stochastic master equation (4.3), show a gradual transition in reaction efficiency from the behavior expected in $d = 2$ to that in $d = 3$ as the number k of layers increases from $k = 1$ to $k = 11$.

Turning next to the role of multipolar potentials in influencing the reaction efficiency between coreactants confined to a compartmentalized system, interesting new effects arise that are a consequence of the interplay among different length scales in the problem. The first is the spatial extent of the system; the second is the average site-to-site displacement of the diffusing molecule in the reaction space (as calibrated by the lattice metric R); and, the third is the Onsager or thermalization length S. The (generalized) Onsager length is a composite variable defined in terms of the specific multipolar potential, and the medium parameters of temperature and dielectric constant (Section IV).

Physically, S is the distance apart at which the mutual potential energy between reactants has the magnitude of the thermal energy. For lengths smaller than S, electrostatic interactions dominate, whereas for lengths greater than S, thermal motion tends to wash out correlation effects. In terms of the dimensionless parameter, $W = S/R$, the value $W = 1$ is that for which the balance between electrostatic and thermal energies is realized in each and every displacement of the diffusing coreactant. The value $W \to 0$, which defines the regime of an unbiased random walk, can be approached in the regime of very high temperature, very large dielectric constant, and/or in the case where site-to-site excursions of the diffusing coreactant are much larger than the thermalization length. Conversely, the behavior realized in the limit of $W \to \infty$ can be approached in the regime of very low temperature, in a medium characterized by small values of the dielectric constant, and/or the case where coreactant displacements are much smaller than the thermalization length.

It is noteworthy that in calculations of the normalized walklength, $\langle n \rangle (C/P)/\langle n \rangle (C)$, as a function of log W, the profiles generated (Section IV) under a variety of assumptions on host system topology (viz., dimensionality $d(=2,3)$, Euler characteristic $\chi(=0,2)$), on its spatial extent, on the nature of the possible short-range displacements of the diffusing coreactant (viz., allowing or not out-of-plane excursions of the diffusing particle), and on the nature of the spatial correlations between reaction partners (in organizates, surface-mediated only or propagated through the host medium) are surprisingly similar. Significant numerical differences between the profiles occur only over a rather restricted range of S/R values. The universality of this behavior leads to the conclusion that in the regime of (very) small or large W, changes in the temperature, dielectric character of the medium, and charge type are relatively inconsequential in influencing the efficiency of reaction in compartmentalized systems. Conversely, small changes in T, ε, and Z_i can have a dramatic effect on the reaction efficiency in the range of intermediate W, viz. for values in the range $0.1 < W < 10$.

With respect to the multipolarity of the potential, the longer the range of the potential, the greater the influence on the reaction efficiency in compartmentalized systems, as expected. However, in small compartmentalized systems (i.e., systems of restricted spatial extent), calculations show that the relative importance of short- versus long-range potentials in influencing the reaction efficiency can be inverted [53], a consequence of the fact that the gradient of a short-ranged potential (e.g., an angle-averaged dipole-dipole potential) will be greater than that of a long-range potential (the couloumbic potential) in the vicinity of the reaction center. The most dramatic changes in the reaction efficiency in compartmentalized systems occur when the three length scales of the problem are of comparable magnitude.

In conclusion, presented in this review has been a summary of an extensive body of work focused on the question posed in the Introduction, viz. how does a system's morphology influence the efficiency of a diffusion-controlled chemical reaction taking place within or on a compartmentalized system. An effort was made to establish unequivocally the interplay among system variables such as its size, shape, dimensionality, and local connectivity, and then to quantify how these geometrical effects conspire to modulate the importance of multipolar correlations between coreactants in enhancing (or not) the efficiency of the process. It should be emphasized that the theoretical approach described in Section II allows the calculation of numerically exact results for the Markovian models studied, so the conclusions that follow from the analysis of these results may be regarded as "rigorous." *Y a–t–il un cheval meilleur?*

In discussing the results obtained in the study of the experimental problems highlighted in the previous section, an effort was made not only to extract insights from the calculations, but also to detail the limitations of the models studied. As noted, there are no questions of principle standing in the way of constructing more realistic models. Factors such as the shape of the diffusing species and target molecule, the possibility that the coreactant can undertake non-nearest neighbor "hops" in its diffusion through reaction space, and a more precise characterization of the short-range quantum-chemical interactions of the two reactants all can be incorporated and explored within the framework of the Markovian theory presented here. It is anticipated, therefore, that the theoretical approach described herein will continue to be fruitful in describing and understanding the factors that influence reaction efficiency in compartmentalized systems.

REFERENCES

1. D. W. Thompson, *On Growth and Form*, ed. J. T. Bonner (Cambridge University Press, Cambridge, 1992).

2. (a) G. Nicolis and I. Prigogine, *Self-Organization in Nonequilibrium Systems. From Dissipative Structures to Order through Fluctuations* (John Wiley and Sons, New York, 1977); (b) G. Nicolis, *Introduction to Nonlinear Science* (Cambridge University Press, Cambridge, 1995).

3. (a) H. Haken, *Synergetics, An Introduction*, 3rd ed. (Springer, Berlin, Heidelberg, New York, 1983); (b) H. Haken, *Advanced Synergetics* (Springer, Berlin, Heidelberg, New York, 1993).

4. For a comprehensive review of the literature on random walks see: G. H. Weiss, *Aspects and Applications of the Random Walk* (North Holland, Amsterdam, 1994); for an earlier review, see: G. H. Weiss and R. J. Rubin, Adv. Chem. Phys. **52**, 363 (1983).

5. (a) J. G. Kemeny and J. L. Snell, *Finite Markov Chains* (Van Nostrand, Princeton, 1960). (b) D. T. Gillespe, *Markov Processes, An Introduction for Physical Scientists* (Academic Press, New York, 1992).

6. Texts and monographs on stochastic processes include: (a) R. Bellman, *Introduction to Matrix Analysis* (McGraw-Hill, New York, 1970); (b) I. I. Gihmann and A. V. Skorohod, *Stochastic Differential Equations* (Springer, Berlin, Heidelberg, New York, 1973); (c) L. Arnold, *Stochastic Differential Equations* (Oldenbourg, Munich, 1973); (d) N. G. van Kampen, *Stochastic Processes in Physics and Chemistry* (North-Holland, Amsterdam, 1981); (e) C. W. Gardiner, *Handbook of Stochastic Methods* Springer Ser. Synergetics, Vol. 13 (Springer, Berlin, Heidelberg, New York, 1983).

7. R. S. Knox, *J. Theo. Biol.* **21**, 244 (1968).

8. C. A. Walsh and J. J. Kozak, *Phys. Rev. Lett.* **47**, 1500 (1981).

9. C. A. Walsh and J. J. Kozak, *Phys. Rev. B* **26**, 4166 (1982).

10. P. A. Politowicz and J. J. Kozak, *Phys. Rev. B* **28**, 5549 (1983).

11. R. Davidson and J. J. Kozak, *J. Phys. Chem. B* **102**, 7393 (1998).

12. R. Davidson and J. J. Kozak, *J. Phys. Chem. B* **102**, 7400 (1998).

13. P. A. Politowicz and J. J. Kozak, *Langmuir* **4**, 305 (1988).

14. P. A. Politowicz and J. J. Kozak, *Mol. Phys.* **62**, 939 (1987).

15. P. A. Politowicz, J. J. Kozak and G. H. Weiss, *Chem. Phys. Lett.* **120**, 388 (1985).

16. J. J. Kozak, *Phys. Rev. A* **44**, 3519 (1991).

17. E. W. Montroll, *Proc. Symp. Appl. Math. Am. Math. Soc.* **16**, 193 (1964).

18. E. W. Montroll and G. H. Weiss, *J. Math. Phys.* **6**, 167 (1965).

19. E. W. Montroll, *J. Math. Phys.* **10**, 753 (1969).

20. W. Th. F. den Hollander and P. W. Kasteleyn, *Physica A* **112A**, 523 (1982).

21. R. A. Garza-López and J. J. Kozak, *Phy. Rev. E* **49**, 1049 (1994).

22. R. A. Garza-López, J. K. Rudra, R. Davidson and J. J. Kozak, *J. Phys. Chem.* **94**, 8315 (1990).

23. B. B. Mandelbrot, *The Fractal Geometry of Nature* (Freeman, San Francisco, 1983).

24. M. Schroeder, *Fractals, Chaos, Power Laws* (Freeman, New York, 1991).

25. H.-O. Peitgen, H. Jürgens and D. Saupe, *Chaos and Fractals* (Springer, New York, 1992).

26. Y. Gefen, A. Aharony and S. Alexander, *Phys. Rev. Lett.* **50**, 77 (1983). There is now a well-developed literature on this class of problems; see, for example: (a) A. Blumen, J. Klafter and G. Zumofen in *Optical Spectroscopy of Glasses,* ed. I. Zschokke (Reidel, Dordrecht, 1986) pp 199–265; (b) S. Havlin and D. Ben-Avraham, Adv. Phys. **36**, 695 (1987); (c) J. M. Drake and J. Klafter, Phys. Today **43**, 46 (1990).

27. I. Prigogine, C. George, F. Henin and L. Rosenfield, *Chem. Scr.* **4**, 5 (1973).

28. G. J. Staten, M. K. Musho and J. J. Kozak, *Langmuir* **1**, 443 (1985).

29. T. Loughran, M. D. Hatlee, L. K. Patterson and J. J. Kozak, *J. Chem. Phys.* **72**, 5791 (1980).

30. See: M. Kac in *Mathematics in the Modern World* (W. H. Freeman, San Francisco, 1968) pp 165–174.

31. S. Wolfram, *Rev. Mod. Phys.* **55**, 601 (1983).

32. G. D. Abowd, R. A. Garza-López and J. J. Kozak, *Phys. Lett. A* **127**, 155 (1988).

33. R. A. Garza-López and J. J. Kozak, *Phys. Rev. A* **40**, 7325 (1989).

34. H. Takayasu, *Fractals in the Physical Science* (Manchester University Press, Manchester, 1990).

35. T. Viscek, *Fractal Growth Phenomena* (World Scientific, London, 1989).

36. J. Feder, *Fractals* (Plenum, New York, 1988).

37. (a) S. Cornell, M. Droz and b. Chopard, *Physica A* **188**, 322 (1992); (b) B. Chopard and M. Droz, *J. Stat. Phys.* **64**, 859 (1991).

38. (a) K. Lindenberg, B. J. West and R. Kopelman, in *Noise and Chaos in Nonlinear Dynamical Systems,* eds. I. Moss, L. A. Lugiato and W. Schleich (Cambridge University Press, Cambridge, 1990); (b) K. Lindenberg, W.-S. Shen and R. Kopelman, *J. Stat. Phys.* **65**, 1269 (1991).

39. C. R. Doering, M. Burschka and W. Horsthemke, *J. Stat. Phys.* **65**, 953 (1991).

40. A. Tretyakov, A. Provata and G. Nicolis, *J. Phys. Chem.* **99**, 2770 (1995).

41. (a) K. Menger, *Dimensiontheorie* (Leipzig, 1928); (b) L. Blumenthal and K. Menger, *Studies in Geometry* (W. H. Freeman, San Francisco, 1970); (c) K. Menger, *J. Graph Theory* **5**, 341 (1981).

42. J. J. Kozak, *Chem. Phys. Lett.* **275**, 199 (1997).

43. R. A. Garza-López, M. Ngo, E. Delgado and J. J. Kozak, *Chem. Phys. Lett.* **306**, 411 (1999).

44. (a) R. A. Garza-López and J. J. Kozak, *J. Phys. Chem.* **103**, 9200 (1999); (b) R A. Garza-López, L. Naya and J. J. Kozak, *J. Chem. Phys.* **112**, 9956 (2000).

45. M. Henle, "A Combinatorial Introduction to Topology" (Freeman, San Francisco, 1979).

46. (a) J. Klafter, J. M. Deake and A. Blumen, J. Lumin. **31**, 642 (1984); (b) J. Klafter and A. Blumen, *J. Chem. Phys.* **80**, 874 (1985).

47. *Molecular Dynamics in Restricted Geometries,* eds. J. Klafter and J. M. Drake (John Wiley and Sons, New York, 1989).

48. *The Fractal Approach to Heterogeneous Catalysis,* ed. D. Avnir (John Wiley and Sons, Chichester, 1989).

49. *Fractal and Disordered Systems,* eds. A. Bunde and S. Havlin (Springer-Verlag, Heidelberg, 1991).

50. D. Avnir, J. J. Carberry, O. Citri, D. Farin, M. Grätzel and A. V. McEvoy, Chaos **1**, 397 (1991).

51. G. Adam and M. Delbrück, "Reduction of dimensionality in biological diffusion processes in *Structural Chemistry and Molecular Biology,* eds. A. Rich and N. Davidson (Freeman, New York, 1968). For a review of the experimental work relevant to this concept, see: M. A. McCloskey and M. J. Poo, *Cell Biol.* **102**, 88 (1986).

52. M. D. Hatlee and J. J. Kozak, *Proc. Natl. Acad. Sci. U.S.A.* **78**, 972 (1981).

53. J. J. Kozak and R. Davidson, *J. Chem. Phys.* **101**, 6101 (1994).

54. P. H. Lee and J. J. Kozak, *J. Chem. Phys.* **80**, 705 (1984).

55. B. A. O'Shaughnessy and I. Procaccia, *Phys. Rev. A* **32**, 3073 (1985).

56. R. Orbach, *Science* **231**, 814 (1986).

57. S. Alexander and R. Orbach, *J. Phys. Lett.* **43**, L625 (1982).

58. R. Rammal and G. Toulase, *J. Phys. Lett.* **44**, L13 (1983).

59. (a) P. Argyrakis, *Phys. Rev. Lett.* **59**, 1729 (1987); (b) G. Pitsianis, L. Bleris, P. Argyrakis, *Phys. Rev. B* **39**, 7097 (1989); (c) P. Argyrakis, A. Coniglio and G. Paladin, *Phys. Rev. Lett.* **61**, 2156 (1988); (d) P. Argyrakis, *Phys. Rev. Lett.* **61**, 2157 (1988).

60. (a) J. K. Rudra and J. J. Kozak, *Bull. Am. Phys. Soc.* **35**, 826 (1990); (b) J. K. Rudra and J. J. Kozak, *Phys. Lett. A* **151**, 429 (1990).

61. (a) R. A. Garza-López and J. J. Kozak, *J. Phys. Chem.* **95**, 3278 (1991); (b) R. A. Garza-López and J. J. Kozak, *J. Phys. Chem.* **96**, 6027 (1992).

62. R. Aris, *The Mathematical Theory of Diffusion and Reaction in Permeable Catalysts* (Clarendon Press, Oxford), 1975); Vols. I and II.

63. (a) V. A. Bloomfield and S. Prager, *Biophys. J.* **27**, 447 (1979); (b) M. D. Hatlee, J. J. Kozak, G. Rothenburger, P. P. Infelta and M. Grätzel, *J. Phys. Chem.* **84**, 1508 (1980).

64. P. A. Politowicz and J. J. Kozak, *Proc. Natl. Acad. Sci.* (USA) **84**, 8175 (1987).

65. P. A. Politowicz, R. A. Garza-López, D. E. Hurtubise and J. J. Kozak, *J. Phys. Chem.* **93**, 3728 (1989).

66. T. J. Pinnavaia, *Science* **220**, 365 (1983).

67. P. Lazlo, *Science* **235**, 1473 (1987).

68. (a) J. K. Thomas, *The Chemistry of Excitation at Interfaces* (American Chemical Society, Washington, DC, 1984); (b) J. K. Thomas, *Chem. Rev.* **93**, 301 (1993) : (c) see also the

studies presented in the volume of *J. Phy. Chem.* dedicated to Professor J. K. Thomas, viz. *J. Phys. Chem.* **103**, 9055–9381 (1999).

69. (a) K. Kalyanasundarm, *Photochemistry in Microheterogeneous Systems* (Academic Press, New York, 1987); (b) M. Gratzel and K. Kalyanasundaram, *Kinetics and Catalysis in Microheterogeneous Systems* (Decker, New York, 1991).

70. Representative papers from this group are: (a) P. K. Ghosh and A. J. Bard, *J. Phys. Chem.* **88**, 5519 (1984); (b) P. K. Ghosh and A. J. Bard, *J. Am. Chem. Soc.* **105**, 5691 (1983); (c) P. K. Ghosh, A. W.-H. Mau and A. J. Bard, *J. Electroanal. Chem. Interfacial Electrochem.* **169**, 315 (1984); (d) A. J. Bard, *Integrated Chemical Systems: A Chemical Approach to Nanotechnology* (John Wiley and Sons, New York, 1994).

71. Representative papers from this group are: (a) R. A. Schoonheydt, P. DePauw, D. Vliers and F. C. DeSchrijver, *J. Phys. Chem.* **88**, 5113 (1984); (b) R. A. Schoonheydt, J. Cenens and F. C. DeSchrijver, *Chem. Soc. Faraday Trans.* 1 **82**, 281 (1986); (c) K. Viane, J. Caigui, R. A. Schoonheydt and F. C. DeSchrijver, *Langmuir* **3**, 107 (1987).

72. Representative papers from this group are: (a) D. Krenske, S. Abdo, H. Van Damme, M. Cruz and J. J. Fripat, *J. Phys. Chem.* **84**, 2447 (1980); (b) A. Habti, D. Keravis, P. Levitz and H. Van Damme, *J. Chem. Soc. Faraday Trans.* 2 **80**, 67 (1984); (c) H. Nijs, J. J. Fripat and H. Van Damme, *J. Phys. Chem.* **87**, 1279 (1983).

73. P. A. Politowicz and J. J. Kozak, *J. Phys. Chem.* **92**, 6078 (1988).

74. P. A. Politowicz, L. B. S. Leung and J. J. Kozak, **93**, 923 (1989).

75. J. Caro, M. Bülow, W. Schirmer, J. Kärger,W. Heink, H. Pfeifer and S. P. Zdanov, *J. Chem. Soc. Faraday Trans.* **81**, 2541 (1985).

76. (a) R. M. Barrer and L. V. C. Rees, *Trans. Faraday Soc.* **50**, 852 (1954); *ibid.* **50**, 989 (1954); (b) R. M. Barrer in *Inclusion Compounds,* eds. J. L. Atwood, J. E. D. Davies and D. D. MacNicol (Academic Press, New York, 1984); Vol. I.

77. R. Schöllhorn in *Inclusion Compounds,* eds. J. L. Atwood, J. E. D. Davies and D. D. MacNicol (Academic Press, New York, 1984), Vol. I, pp. 248–350.

78. J. M. Newsam, *Science* **231**, 1093 (1986).

79. J. Kärger and D. M. Ruthven, *Diffusion in Zeolites and Other Microporous Solids,* (J. Wiley and Sons, New York, 1992).

80. (a) C.-I. Yang, M. A. El-Sayed and S. L. Suib, *J. Phys. Chem.* **91**, 4440 (1987); (b) C.-I. Yang, P. Evesque and M. A. El-Sayed, "Effects of Finite Volumes on Electronic Energy Transfer" in *Molecular Dynamics in Restricted Geometries,* eds. J. Klafter and J. M. Drake (John Wiley and Sons, New York, 1989).

81. (a) N. J. Turro and P. Wan, *J. Am. Chem. Soc.* **107** , 678 (1985); (b) B. H. Barretz and N. J. Turro, *J. Photochem.* **24**, 201 (1984); (c) V. Ramamurthy, D. R. Corbin, C. V. Kumar and N. J. Turro, *Tetrahedron Lett.* **31**, 47 (1990); (d) N. J. Turro, *Pure Appl. Chem.* **58**, 1219 (1986).

82. F. Wilkinson, H. L. Casal, L. J. Johnson, J. C. Scaiano, *Can. J. Chem.* **64**, 539 (1986).

83. X. Liu, K.-K. Iu and J. K. Thomas, *J. Phys. Chem.* **93**, 4120 (1989).

84. (a) J. M. Thomas and L. A. Bursill, *Angew. Chem. Int. Ed. Engl.* **19**, 745 (1980); (b) L. A. Bursill, J. M. Thomas and K. J. Rao, *Nature* (London) **289**, 157 (1981); (c) L. A. Bursill and J. M. Thomas, *J. Phys. Chem.* **85**, 3007 (1981).

85. (a) N. J. Turro, C.-C. Cheng, X.-G. Lei, *J. Am. Chem. Soc.* **107**, 3739 (1985); (b) N. J. Turro, C.-C. Cheng, L. Abrahms and D. Corbin, *J. Am. Chem. Soc.* **109**, 2449 (1987).

86. (a) V. Ramamurthy, J. V. Caspar, D. R. Corbin and D. F. Eaton, *J. Photochem. Photobiol. A: Chem.* **50**, 157 (1989); (b) V. Ramamurthy, J. V. Caspar and D. R. Corbin, *J. Am. Chem. Soc.* **113**, 594 (1991); (c) J. V. Caspar, V. Ramamurthy and D. R. Corbin, *J. Am. Chem. Soc.* **113**, 600 (1991).

87. *Photochemistry in Organized and Constrained Media,* ed. V. Ramamurthy (VCH, New York, 1991).

88. J. O. Hirschfelder, C. F. Curtiss and R. B. Bird, *Molecular Theory of Gases and Liquids* (John Wiley and Sons, New York, 1954).

89. G. Polya, *Math. Ann.* **84**, 149 (1921).

90. J. J. Kozak, *J. Chem. Phys.* **110**, 3056 (1999).

91. J. H. Fendler, *Membrane Mimetic Chemistry* (John Wiley and Sons, New York, 1982).

92. J. B. Mandeville, D. E. Hurtubise, R. Flint and J. J. Kozak, *J. Phys. Chem.* **93**, 7876 (1989).

93. J. B. Mandeville, J. Golub and J. J. Kozak, *J. Phys. Chem.* **92**, 1575 (1988).

94. R. A. Garza-Lopez, D. Byun, K. Orellana, A. Partikian, D. Siew, A. Yu and J. J. Kozak, *J. Chem. Phys.* **103**, 9413 (1995).

95. J. B. Mandeville and J. J. Kozak, *J. Am. Chem. Soc.* **114**, 6139 (1992).

96. G. A. Somorjai, *Chemistry in Two Dimensions: Surfaces* (Cornell University Press, Ithaca, 1981).

97. (a) W. M. H. Sachtler, Faraday Discuss. *Chem. Soc.* **72**, 7 (1982); (b) V. Ponec and W. M. H. Sachtler, *J. Catal.* **24**, 250 (1972).

98. *Adsorption on Metal Surfaces,* ed. J. Bénard (Elsevier, Amsterdam, 1983).

99. A. J. Forty, *Contemp. Phys.* **24**, 271 (1983).

100. A. A. Balandin, *Adv. Catal.* **10**, 96 (1958).

101. D. A. Dowden, *Catal., Proc. Int. Congr.,* 5th **1972**, 621 (1973).

102. (a) M. Boudart, *Adv. Catal.* **20**, 153 (1969); (b) M. Boudart, *Am. Sci.* **57**, 1 (1969); (c) M. Boudart, *Top. Appl. Phys.* **4**, 275 (1975); (d) M. Boudart and M. A. McDonald, *J. Phys. Chem.* **88**, 2185 (1984).

103. (a) J. H. Sinfelt, J. L. Carter and D. J. C. Yates, *J. Catal.* **24**, 283 (1972); (b) J. H. Sinfelt, *Acc. Chem. Res.* **10**, 15 (1977); (c) J. H. Sinfelt in *Many-Body Phenomena at Surfaces,* eds. D. Langreth and H. Suhl (Academic Press, New York, 1984).

104. M. J. Cardillo, *Springer Ser. Chem. Phys.* **20**, 149 (1982).

105. (a) F. Rosenberger, *Fundamentals of Crystal Growth I* (Springer-Verlag, Berlin, 1979); (b) *Crystal Growth: A Tutorial Approach,* eds. W. Bardsley, D. T. J. Hurle and J. B. Mullin (North Holland, Amsterdam, 1979); (c) V. V. Voronkov in *Modern Theory of Crystal Growth I,* eds. A. A. Chernov and H. Müller-Krumbhaar (Springer-Verlag, Berlin, 1983); (d) J. Van der Eerden in *Modern theory of Crystal Growth I,* eds. A. A. Chernov and H. Müller-Krumbhaar (Springer-Verlag, Berlin, 1983).

106. See, for example: *Catalytst Deactivation,* eds. B. Delmon, G. F. Froment (Elsevier, Amsterdam, 1980).

107. E. Shustorovich, R. C. Baetzold and E. L. Muetterties, *J. Phys. Chem.* **87**, 1100 (1983).

108. *Many-Body Phenomena at Surfaces,* eds. D. langreth and H. Suhl (Academic Press, New York, 1984).

109. E. Sackmann in *Biophysics,* eds. W. Hoppe, W. Lohmann, H. Markl and H. Zuegler (Springer-Verlag, Berlin, 1983) pp 425–457.

110. P. A. Politowicz and J. J. Kozak, *Langmuir* **1**, 429 (1985).

111. E. A. Guggenheim, *Mixtures* (Clarendon Press, Oxford, 1952).

112. P. A. Politowicz and J. J. Kozak, *J. Phys. Chem.* **94**, 7272 (1990).

113. K. Denbigh, *The Principles of Chemical Equilibrium,* 3rd ed. (Cambridge University Press, Cambridge, 1971) Chapter 11.

114. (a) E. T. Jaynes, *Phys. Rev.* **106**, 620 (1957); (b) E. T. Jaynes, *Phys. Rev.* **108**, 171 (1957).

115. G. Wannier, *Statistical Physics* (John Wiley and Sons, New York, 1966).

116. (a) E. U. Condon and M. Odishaw, *Handbook of Physics,* (McGraw-Hill, New York, 1958); (b) M. Abramowitz and I. A. Stegun, *Handbook of Mathematical Functions* (Dover Publications, New York, 1971), Chap. 26; (c) M. K. Musho and J. J. Kozak, *J. Chem. Phys.* **81**, 3229 (1984).

117. (a) E. R. Kandel and J. H. Schwarz, *Principles of Neural Science* (Elsevier, North-Holland, New York, 1981); (b) S. W. Kuffler, J. G. Nicholls and A. R. Martin, *From Neuron to Brain,* 2nd ed. (Sinauer Associates, Sunderland, MA, 1984).

118. (a) G. M. Edelman and J. A. Cpally in *The Neurosciences,* ed. F. Schmitt (Rockefeller University Press, New York, 1970); (b) G. M. Edelman, *Immunol. Rev.* **100**, 11 (1987); (c) S.-S. Jan, K. L. Crossin, S. Hoffman and G. M. Edelman, *Proc. Natl. Acad. Sci. U.S.A.* **84**, 7977 (1987); (d) G. M. Edelman, *Topobiology: An Introduction to Molecular Embryology* (Basic Books, New York, 1988).

119. (a) W. Kauzmann, *Adv. Protein Chem.* **14**, 1 (1959); (b) C. Tanford, *The Hydrophobic Effect* (Wiley, New York, 1973); (c) J. H. Fendler and E. J. Fendler, *Catalysis in Micellar and Macromolecular Systems* (Academic Press, New York, 1975).

120. Z. Zhang and M. G. Lagally, *Science* **276**, 377 (1997).

121. J. Venables, *Philos. Mag.* **27**, 697 (1973).

122. S. Stoyanov and S. Kashchiev, *Curr. Top. Mater. Sci.* **7**, 69 (1981).

123. (a) M. C. Bartelt and J. W. Evans, *Surf. Sci.* **298**, 421 (1993); (b) M. C. Abrtelt and J. W. Evans, *Mater. Res. Soc.Symp.* **312**, 255 (1993).

124. (a) J. W. Evans and M. C. Bartelt, *J. Vac. Sci. Technol. A* **12**, 1800 (1994); (b) C. Ratsch, A. Zangwill, P. Smilauer and D. D. Vvedensky, *Phys. Rev. Lett.* **72**, 3194 (1994); (c) J. Amar and F. Family, *Phys. Rev. Lett.* **74**, 2066 (1995).

125. Y.-W. Mo, J. Kleiner, M. B. Webb, M. G. Lagally, *Phys. Rev. Lett.* **66**, 1998 (1991); Surf. Sci. **268**, 275 (1992).

126. A. Pimpinelle, J. Villain and D. E. Wolf, *Phys. Rev. Lett.* **69**, 985 (1992).

127. J. E. Vasek, Z. Y. Zhang, C. T. Salling, M. G. Lagally, *Phys. Rev. B* **51**, 17207 (1995).

128. P. E. Quesenberry and P. N. First, *Phys. Rev. B* **54**, 8218 (1996).

129. L. Andersohn, Th. Berke, U. Köhler and B. J. Voigtländer, *Vac. Sci. Technol. A* **14**, 312 (1996).

130. M. Fehrenbacher, J. Spitzmüller, U. Memmert, H. Rauscher and R. J. Behm, *Vac. Sci. Technol. A* **14**, 1499 (1996).

131. (a) J. A. Stroscio, D. T. Pierce and R. A Dragoset, *Phys. Rev. Lett.* **70**, 3615 (1993); (b) J. A. Stroscio and D. T. Pierce, *Phys. Rev. B* **49**, 8522 (1994).

132. (a) J.-K. Zuo, J. F. Wendelken, H. Dürr and C.-L. Liu, *Phys. Rev. Lett.* **72**, 3064 (1994); (b) H. Dürr, J. F. Wendelken and J.-K. Zuo, *Surf. Sci.* **328**, L527 (1995).

133. H. Brune, H. Röder, C. Boragno and K. Kern, *Phys. Rev Lett.* **73**, 1955 (1994).

134. M. Bott, M. Hohage, M. Morgenstern, Th. Michely and G. Cosma, *Phys. Rev. Lett.* **76**, 1304 (1996).

135. T. A Witten and L. M. Sandler, *Phys. Rev. Lett.* **47**, 1400 (1981).

136. P. Meakin, *Phys. Rev. A* **27**, 1495 (1983).

137. R. Q. Hwang, J. Schröder, C. Günther and R. J. Behm, *Phys. Rev. Lett.* **67**, 3279 (1991).

138. Th. Michely, M. Hohage, M. Bott and G. Cosma, *Phys. Rev. Lett.* **70**, 3943 (1993).

139. (a) H. Röder, E. Hahn, H. Brune, J.-P. Bucher and K. Kern, *Nature* **366**, 141 (1993); (b) H. Röder, K. Bromann, H. Brune, C. Boragno and K. Kern, *Phys. Rev. Lett.* **74**, 3217 (1995).

140. M. Hohage, *Phys. Rev. Lett.* **76**, 2366 (1996).

141. H. Brune, *Surf. Sci.* **349**, L115 (1996).

142. Z. Y. Zhang, X. Chen and M. G. Lagally, *Phys. Rev. Lett.* **73**, 1829 (1994).

143. M. C. Bartelt and J. W. Evans, *Surf. Sci.* **L829**, 314 (1994).

144. See: L. Bardotti, M. C. Bartelt, C. J. Jenks, C. R. Stoldt, J.-W. Wen, C.-M. Zhang, P. A. Thiel and J. W. Evans, *Langmuir* **14**, 1487 (1998), and references cited therein.

145. R. Jackson, *Transport in Porous Catalysts* (Elsevier, Amsterdam, 1977).

146. A. E. Scheidegger, *The Physics of Flow Through Porous Media*, 3rd ed. (University of Toronto Press, Toronto, 1974).

147. F. A. L. Dullien, *Porous Media, Fluid Transport and Pore Structure* (Academic Press, New York, 1979).

148. R. E. Collins, *Flow of Fluids Through Porous Media* (Reinhold, New York, 1961).

149. (a) R. B. Evans III, G. M. Watson and E. A. Mason, *J. Chem. Phys.* **35**, 2076 (1961); (b) R. B. Evans III, G. M. Watson and E. A. Mason, *J. Chem. Phys.* **36**, 1894 (1962); (c) E. A. Mason, R. B. Evans III, G. M. Watson, *J. Chem. Phys.* **38**, 1808 (1963); (d) E. A. Mason and A. D. Malinauskas, *J. Chem. Phys.* **41**, 3815 (1964); (e) E. A. Mason, A. D. Malinauskas and R. B. Evans III, *J. Chem. Phys.* **46**, 3199 (1967).

150. (a) R. A. Garza-Lopez, J. K. Rudra and J. J. Kozak, *Chem. Phys. Lett.* **174**, 278 (1990); (b) R. A. Garza-Lopez and J. J. Kozak, *J. Phys. Chem.* **96**, 9457 (1992).

151. J. B. Mandeville and J. J. Kozak, *J. Phys. Chem.* **96**, 796 (1992).

AUTHOR INDEX

Numbers in parentheses are reference numbers and indicate that the author's work is referred to although his name is not mentioned in the text. Numbers in *italic* show the pages on which the complete references are listed.

SUBJECT INDEX